Agricultural Development in the
Middle East

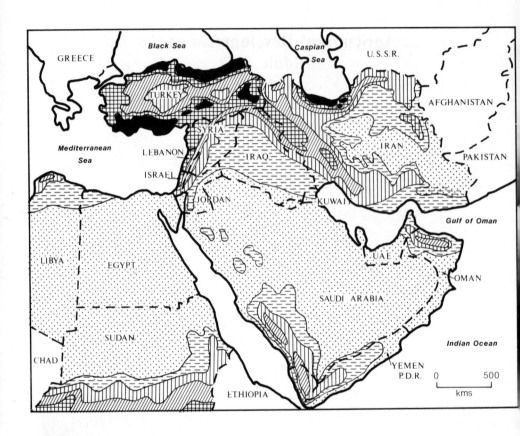

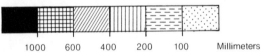

1000   600   400   200   100   Millimeters

Agricultural Development in the Middle East

Edited by

Peter Beaumont

St. David's University College, Lampeter

and

Keith McLachlan

School of Oriental & African Studies, London

JOHN WILEY & SONS

Chichester · New York · Brisbane · Toronto · Singapore

Copyright © 1985 by John Wiley & Sons Ltd

Library of Congress Cataloging in Publication Data
Main entry under title:

Agricultural development in the Middle East.

  Includes index.
  1. Agriculture—Economic aspects—Near East.
2. Agriculture—Near East. I. Beaumont, Peter.
II. McLachlan, K. S. (Keith Stanley)
HD2056.5.A64   1985   338.1'0956   85–6414

ISBN 0 471 90762 6

British Library Cataloguing in Publication Data
Agricultural development in the Middle East.
  1. Agriculture—Economic aspects—Near East
  I. Beaumont, Peter II. McLachlan, K. S.
  338.1'0956   HD2056.5

ISBN 0 471 90762 6

Phototypeset by Dobbie Typesetting Service, Plymouth, Devon.
Printed and bound in Great Britain.

Contents

List of Contributors

MARTIN E. ADAMS,	*Hunting Technical Services Ltd, Agricultural Consultants, UK*
MUHAMMAD RASHID AL-FEEL,	*Professor, Department of Geography, University of Kuwait, Kuwait*
J. A. ALLAN,	*Reader, Department of Geography, School of Oriental and African Studies, University of London, UK*
PETER BEAUMONT,	*Professor, Department of Geography, Saint David's University College, University of Wales, Lampeter, UK*
BRIAN W. BEELEY,	*Senior Lecturer, Open University, UK*
SHEILA CARAPICO,	*Professor, Department of Political Science, University of Richmond, Virginia, USA*
RODERIC DUTTON,	*Director of the Centre for Overseas Research and Development (CORD), University of Durham, UK*
ECKART EHLERS,	*Professor, Department of Geography, Philipps-Universität, Marburg, West Germany*
CHARLES GURDON,	*Researcher and broadcaster on Middle East affairs, currently at the School of Oriental and African Studies, University of London, UK*
KINGSLEY E. HAYNES,	*Professor, School of Public and Environmental Affairs, and Director, Center for Urban and Regional Analysis, Indiana University, Bloomington, USA*
JAMES M. HOLT,	*Binney and Partners, Consulting Engineers, UK*
E. G. H. JOFFE,	*Consultant Editor for Middle East and North Africa with the Economist Publications.*
RICHARD I. LAWLESS,	*Senior Lecturer and Assistant Director, Centre for Middle Eastern and Islamic Studies, University of Durham, UK*
KEITH MCLACHLAN,	*Senior Lecturer, Department of Geography, School of Oriental and African Studies, University of London, UK*

IAN R. MANNERS,   *Associate Professor, Department of Geography, University of Texas, Austin, USA*

HERBERT W. OCKERMAN,   *Professor, Department of Animal Science, Ohio State University, Columbus, Ohio, USA*

SHOKO OKAZAKI,   *Professor, Osaka University of Foreign Studies, Japan*

TAGI SAGAFI-NEJAD,   *Associate Professor of International Business, Sellinger School of Business and Management, Loyola College in Maryland, USA*

SHIMOON G. SAMANO,   *formerly of Department of Horticulture—Food Technology, Ohio State University, Columbus, Ohio, USA, currently in Iraq*

DALE WHITTINGTON,   *Assistant Professor, Department of City and Regional Planning, University of North Carolina at Chapel Hill, USA*

Acknowledgements

This volume is the product of considerable efforts by the several authors who have so enthusiastically taken part in its preparation. Their hard work and in some cases patience is acknowledged by the editors. The encouragement of Michael Coombs of John Wiley & Sons deserves fulsome thanks. We are grateful, too, for the considerable assistance of Sue Harrop of SOAS in preparing the maps and diagrams. While the editors have endeavoured to ensure that spelling, transliteration of place names, and punctuation are standard throughout the text, it was felt important that the character of each contribution should not be diminished by editorial amendments. Wherever possible, the original text has been used despite obvious variations in style and idiosyncrasies of expression deployed by individual authors from seven different countries. However, any errors must be accounted as the responsibility of the editors alone.

Peter Beaumont
St. David's University College, Lampeter

Keith McLachlan
School of Oriental & African Studies, London

Preface

The Middle East has become a principal centre of the international oil industry during recent decades. Many economies of the region appear on the surface to be dominated by oil production and exports, with the main sources of foreign exchange income, treasury receipts and value-added in national income deriving from petroleum activities. Certainly, the Middle East has experienced rapid expansion as measured by foreign trade or growth of Gross Domestic Product. Other changes in population geography and social structure have taken place in the wake of the arrival of oil wealth.

However, petroleum is a very recent event. Oil exports began from Iran in the early years of the twentieth century and it was not until the 1950s and 1960s that Arab states became deeply affected by the oil business. Traditional society and economy in the non-oil states of the region, while not entirely untouched by growth of the oil industry in neighbouring countries, persisted. Rural communities and their agricultural production remained extremely important elements in all the non-oil states, providing employment, basic foodstuffs and earnings through commodity exports. The 'large population' countries illustrate the point well, with high proportions of their labour forces in agriculture—70 per cent in Sudan, 54 per cent in Turkey and 50 per cent in Egypt. Even in the oil-based economies themselves, agriculture retained considerable strength as an employer. Some 42 per cent of the workforce in Iraq and 39 per cent in Iran remained on the land, while in Saudi Arabia, the archetypal oil economy, no less than 60 per cent of employed adults were to be found in cultivation and herding in 1983.

The last decade has witnessed a powerful movement in the Middle East region both to diversify the oil-based economies and to create a greater degree of food self-sufficiency. Governments have been forced to look with greater realism at the scope for food security and the potential within agriculture for job creation as oil revenues have fallen. Ideological factors, too, have been important motivations for turning to traditional economic activities in so far as they are viewed as less damaging to religious and social values than modern industry imported from the West. The reappraisal of agriculture has been prompted by scientific considerations that reinforce cultural prejudices. Oil is an example *par excellence* of a wasting asset with a short life cycle. Agriculture, on the other hand, makes use of renewable resources, in many cases the only permanent economic assets available, and appears to offer a ready and cheap vehicle for long-term development.

The editors of this volume find that the radical changes in Middle Eastern perceptions of their own agricultural resources and the nature of those resources have been inadequately evaluated in the literature. There have been only limited attempts to examine the recent history of land and water use in agriculture in the region in a way that would help create a deeper understanding of the history of, modern background to and difficulties affecting so vital a sector. This book is offered as a first step towards redressing the gap in the literature. An international group of contributors has been assembled, each with specific expertise either by topic or country. All are well known in their various fields. In an attempt to escape from the straitjacket of standard presentations for each country review, authors were encouraged to approach their study through their own speciality. There are authors drawn from the Middle East, Japan, the United Kingdom, the United States of America and West Germany, each representing differing points of view, which it is hoped, broadens and enhances the analyses contained in the book.

The volume has evolved around a four-tiered structure. As an introduction, the editors present overviews of resources and institutional constraints on agriculture, respectively, which act as a general framework for the subsequent debate. A second section brings together authors interested in systematic topics of agricultural development, ranging from livestock to planning. Each chapter in this section looks at the region as a whole. In the third part of the volume a variety of authors take up the question of agricultural change within individual countries, including Egypt, Sudan, Iran, Iraq, Saudi Arabia, Turkey, Syria, Oman and the sheikhdoms of the Gulf. The final point is a conclusion which draws together the threads of argument emerging from the core of the volume in Sections Two and Three.

Part I
The Middle East Region

Agricultural Development in the Middle East
Edited by P. Beaumont and K. McLachlan
© 1985 John Wiley & Sons Ltd

Chapter 1

The Agricultural Environment: An Overview

PETER BEAUMONT

INTRODUCTION

Over the last forty years there has been a revolution in farming activities throughout the Middle East as many rural societies have changed from subsistence agriculture based on traditional methods to modern farming techniques relying on fossil fuel subsidies. This transition has occurred at different times and at different rates, but today there are few parts of the region which have not been affected by Western technology. As with all changes it is difficult to ascribe exact dates as to when particular events occurred. Mechanization in certain areas began in the inter-war period, but it was really after 1950 that the pace of change began to accelerate markedly as contact with the industrialized West increased. In some countries the beginning of this transition is later still. For example, in Oman the impact of Western technology on agriculture did not commence until the early 1970s, following the accession of Sultan Qaboos. In general it has been in the oil-rich countries of the Gulf where the pace of change has been greatest. Here extremely expensive high technology farming systems have been introduced on a quite remarkable scale, including controlled environment crop production and zero-grazing animal rearing systems.

Another major impact on agriculture in the Middle East has been rapid population growth especially since the Second World War (Clarke & Fisher, 1972). In the 1920s the population of the region was around 7.5 million. By 1950 it is estimated at 130 million, and by the early 1980s this had grown to 270 million. In many countries annual rates of population increase remain in excess of 2.5 per cent and a figure of 420 million is likely to be reached by the end of the century (Population Reference Bureau Inc., 1983). Indeed in Egypt, Turkey, and Iran the population increase is close to 1 million each year. Given this massive increase in population, agricultural land has been subjected to ever-growing pressures to produce more food.

Prior to the mid-part of the twentieth century, agriculture in the Middle East had probably not changed greatly in terms of the tools and cultivation techniques employed over the previous 6,000 years since seed agriculture became established in the region. Two types of arable farming can be recognized. The first, and probably the most ancient, is 'dry-farming', whereby a crop is sown and raised using water from precipitation and no further water is supplied. This kind of farming is located in the more northerly and higher parts of the region where precipitation totals are greatest and evapotranspiration rates lowest. It covers a wide range of environments from intermontane basins in Turkey and Iran to undulating plateaux and plains in Syria and Iraq. In the wetter parts of this region agriculture has prospered with few natural checks, other than drought periods. In contrast, on the drier margin of this belt precipitation fluctuations from year to year have imposed considerable stresses on the inhabitants, causing famine and even abandonment of the land. This is, in fact, the true desert fringe, which changes its position periodically in response to environmental conditions. With increasing population pressures over the last 50 years, it is this marginal region which has been settled in many countries, often with disastrous results as studies of desertification have shown (Mabbutt & Floret, 1980; Tomaselli, 1977; Le Houerou, 1977; Floret & Hadjej, 1977).

Irrigated agriculture probably originated in the Middle East, with the town of Jericho having one of the strongest claims to possess the oldest-known irrigation system (Kenyon, 1969). This type of farming is located in those areas where annual precipitation totals are less than 240 mm and inter-annual variability is greater than 37 per cent (Brichambaut & Wallen, 1963). Here long-term cultivation is only possible when precipitation amounts are supplemented by extra water supplies from either groundwater or surface water sources. The use of irrigation, whether small- or large-scale, requires greater capital investment, in terms of water control and transmission works, and greater labour input, than is needed for rain-fed agriculture. The physical constraints on irrigated agriculture, particularly land slope, water availability and soil moisture-holding capacity, mean that irrigation can only be carried out in limited areas where favourable conditions prevail. In effect this confines it to flood plains and adjacent alluvial plains in lowland areas and to alluvial fans on the upland margins.

On whatever scale it operates, irrigation requires organization of society to perform specific tasks associated with water distribution and control to ensure an equitable distribution of scarce resources at different times of year. The smallest scale of operation is usually at an individual holding level with cultivation made possible with water from, perhaps, a small well. In a traditional society water would be raised by human or animal energy sources, though nowadays diesel or electric power is more common. Throughout much of the region though it is the village unit which provides the normal-sized operating unit for irrigation, with water being obtained from either a ground water source or from rivers and streams. Along the major flood plains much larger irrigation systems predominate with canal systems covering many hundreds of villages. These can be centred around major cities to form classic oasis settlements, such as Isfahan, Damascus or Kairouan, or else dispersed along the flood plains as on the lower courses of the Nile, the Euphrates, and the Tigris. The successful operation of these systems requires careful planning and continuous maintenance of the water intakes and the irrigation canals themselves.

A feature of the last thirty years has been the extension of irrigation into areas not formerly cultivated. This has met with varying degrees of success. In Iran, the projects bringing irrigation to the Khuzestan lowlands have been conspicuous failures, owing to complex physical and social conditions. In contrast, the development of irrigation along the Ghor in the Jordan valley, seems to be more successful, though this too is not without its problems.

Pastoralism plays a key role in the agriculture of the Middle East and the nomadic tribes of Arabia and southern Iran are well known throughout the world (Barth, 1962; Johnson, 1969; Wilkinson, 1977). With the low precipitation which prevails over much of the southern part of the region biological productivity is low. This means that animals cannot be supported on natural vegetation at any one point for long and as a consequence these animals have to be moved from one area to another to crop available biomass. Many writers distinguish two types of nomadic activity dominated by horizontal or vertical movement. This distinction is more apparent than real as both types involve altitudinal movements as well as spatial changes. The great importance of nomadism is that it is often the only way in which the existing low density vegetation can be harvested in an efficient manner. The animals eat the plants and convert them to meat, wool, and hides which can be moved 'on the hoof' and utilized at a later date when needed. It should not be forgotten, however, that the nomadic tribes have a symbiotic relationship with the settled peoples with considerable interchange of agricultural products and artefacts.

Since the 1930s the movements of nomads have been greatly restricted by both political actions and economic development (Stace Birks, 1981). In Iran, Reza Shah saw nomads as a challenge to his authority and forcibly settled them whenever he could. Elsewhere the growing populations in villages have meant that more land has been brought into cultivation, so restricting movement and grazing of the large moving herds. In other cases the pull of the oil fields has taken away many able-bodied menfolk. So great has been the reduction in nomadism in Iran that in many parts of the Zagros Mountains high summer pastures are today not being fully utilized. To some extent this waste of resources might be overcome by bringing in animals by lorry for the summer season, but many of the higher pastures still remain inaccessible to wheeled transport.

In any work on agriculture it is important to know the relative magnitude of cultivated areas in different countries; but at the same time it must be realized that early data on such matters were highly variable in quality. Looking at the arable and permanently cropped land, Turkey, Iran and Sudan dominate the region (Table 1.1) and account for 57 per cent of all cultivated lands in the Middle East. Algeria, Iraq, Morocco, Syria, and Tunisia provide another 32 per cent

AGRICULTURAL DEVELOPMENT IN THE MIDDLE EAST

Table 1.1   Cultivated and irrigated land in the Middle East in the early 1960s and in 1981

	Arable land and permanent cropped (1,000 ha) Early 1960s	Irrigated land (1,000 ha) 1961–65	Irrigated land as percentage of arable & permanent cropped Early 1960s	Arable land and permanent cropped (1,000 ha) 1981	Irrigated land (1,000 ha) 1981	Irrigated land as percentage of arable & permanent cropped 1981
Algeria	6,863	259	3.77	7,513	345	4.59
Bahrain	2	N.D.		2	1	50.00
Egypt	2,548	2,548	100.00	2,860	2,860	100.00
Iran	15,358	4,800	31.25	15,970	5,900	36.94
Iraq	(4,810)	1,030	21.41	5,450	1,750	32.11
Israel	401	142	35.41	418	205	49.04
Jordan	1,177	57	4.84	1,380	85	6.16
Kuwait	1	N.D.		1	1	100.00
Lebanon	276	49	17.75	350	85	24.29
Libya	2,509	123	4.90	2,085	225	10.79
Morocco	7,076	199	2.81	8,394	520	6.19
Oman	36	N.D.		41	38	92.68
Qatar	2	N.D.		3	N.D.	
Saudi Arabia	705	270	38.30	1,117	395	35.36
Sudan	6,180	952	15.40	12,448	1,850	14.86
Syria	6,523	579	8.88	5,759	567	9.85
Tunisia	4,406	74	1.68	4,673	163	3.49
Turkey	25,775	1,336	5.18	28,489	2,080	7.30
UAE	8	N.D.		13	5	38.46
Yemen AR	1,410	175	12.41	2,790	245	8.78
Yemen PDR	195	23	11.79	207	70	33.82
	86,261	12,616	14.63	99,963	17,390	17.40

Source: FAO Production Yearbooks, various dates.

of the total, leaving only 11 per cent among the other countries. Another point which the table clearly illustrates is that the size of the country is no guide to the importance of arable farming within it, as many of the largest countries, such as Saudi Arabia, have relatively small cultivated areas, while small countries, like the Lebanon can have large areas in proportion to their size.

In many countries of the Middle East irrigation plays a vital role in agricultural production and in some countries, such as Egypt all arable land has to be irrigated (Table 1.1). Perhaps surprising is the fact that Iran has almost double the irrigated land which Egypt possesses, although this only represents about

Table 1.2 Cereals and wheat — cultivated area and production

	Cereals area (1,000 ha) 1982	Cereals production (1,000 mt) 1982	Wheat area (1,000 ha) 1982	Wheat production (1,000 mt) 1982	Wheat area as percentage of total cereal area 1982	Wheat production as percentage of total cereal production 1982
Algeria	3,002	1,935	2,000	1,200	66.6	62.0
Bahrain	N.D.	N.D.	N.D.	N.D.	N.D.	N.D.
Egypt	2,022	7,768	577	2,017	28.5	26.0
Iran	7,805	9,189	6,000	6,500	76.9	70.7
Iraq	2,082	1,797	1,200	900	57.6	50.1
Israel	101	138	78	100	77.2	72.5
Jordan	152	47	100	20	65.8	42.6
Kuwait	N.D.	N.D.	N.D.	N.D.	N.D.	N.D.
Lebanon	24	31	18	23	75.0	74.2
Libya	586	235	303	160	51.7	68.1
Morocco	4,285	4,154	1,699	1,824	39.6	43.9
Oman	1	3	N.D.	1	N.D.	N.D.
Qatar	N.D.	N.D.	N.D.	N.D.	N.D.	N.D.
Saudi Arabia	605	538	200	400	33.1	44.3
Sudan	4,124	2,540	253	150	6.1	5.9
Syria	2,692	2,776	1,300	1,544	48.3	67.8
Tunisia	1,300	1,331	800	1,250	61.5	93.9
Turkey	13,577	26,387	9,250	17,650	68.1	66.9
UAE	N.D.	N.D.	N.D.	N.D.	N.D.	N.D.
Yemen AR	830	763	66	67	8.0	8.8
Yemen PDR	61	94	10	15	16.4	16.0
	43,249	59,226	23,854	33,821	55.2	57.1

Source: FAO Production Yearbook: 1982.

one-third of its total cultivated area. Other countries with large absolute amounts of irrigated land are Iraq, Sudan, and Turkey. Currently about 17 per cent of the total cultivated area of the Middle East is irrigated. Since the early 1960s almost all the countries of the region have recorded an increase in their irrigated lands, amounting overall to a gain of approximately 37 per cent. Details of irrigated areas before the 1960s cannot be obtained with any accuracy, making it difficult to distinguish what was happening in the immediate post-war period when population numbers were growing rapidly.

Cereals, including wheat, rye, barley, oats, maize, millet and sorghum, and rice (paddy), are by far the most important group of crops grown in the Middle

East and form the staple diet of most of the people of the region. In 1982 about 43 per cent of the total of arable and permanently cropped land of the Middle East was devoted to cereal production. Of the cereals the most important one is wheat, followed a long way behind by barley. Wheat alone covers 23.5 million hectares, or about 24 per cent of the total cultivated area of the Middle East (Table 1.2). In terms of tonnage wheat makes up 57 per cent of all cereals produced, stressing the pre-eminent nature of this crop. However, from country to country there are major differences in the relative importance of wheat *vis-à-vis* the other cereal crops (Table 1.2). Turkey is the largest single producer followed by Iran. Although production in any one country varies considerably from year to year, overall production in the early 1980s was relatively static between 33 and 36 million tonnes.

AGRICULTURAL SYSTEMS

Agriculture has been developed to provide man with food and raw materials for his livelihood. The aim is to put man at the end of as many food chains as possible so that he is able to benefit from the greatest amount of biological productivity. To achieve this end the natural vegetation from as much land as possible is removed and replaced by crops, which man can harvest. All non-useful plants are regarded as weeds and eliminated. Wherever possible animals are utilized to harvest low-quality natural vegetation and to convert it to edible protein and to wool and skins. When man acts as a herbivore he is able to utilize about 0.2 per cent of absorbed solar radiation in the form of edible vegetable matter. If he eats animal meat this figure drops to about one-tenth the value. However, in the drier areas animals are able to harvest sparse natural vegetation, which would have no value to man unless converted into animal products. Animal husbandry does, therefore, play a vital role in the agricultural economy of the Middle East.

Whether irrigation or dry farming is practised, the key to understanding traditional agriculture in the Middle East is the pattern of villages and their associated urban market centres. These areas were largely able to provide their own food and agricultural raw materials until very recently. In terms of staple crops, such as cereals and vegetables, these areas were totally self-sufficient except in times of drought. There would be trade in certain high-cost goods, such as metal cooking utensils and specialist crops, but the volume of this trade would be minimal compared with the volume of agricultural crops produced and used locally.

The society which evolved was one which made efficient use of the available scarce resources. Solar energy was captured through vegetation growth, to be subsequently harvested to feed man and his animals. All the energy consumed for agricultural activities was provided by human and animal labour. Houses were built using local materials such as stone or mud brick and local wood if available provided structural timber. Fuel for cooking and heating was obtained from animal dung, crop residues and natural vegetation. For the most valuable crops, such as vegetables and fruits, fertilizers would be provided in the form of animal dung and fire ash. Transport of goods was by large animals, of which the camel and the donkey were the most important. Trade between centres consisted mainly of materials in short supply in any one region and high-value goods, such as carpets, made of agricultural materials.

A point worth mentioning, but one that often is overlooked today, is that these village/agricultural complexes always utilized the best land in a region, that could be farmed with their available technology levels. This has meant that when agricultural expansion has occurred, from the mid-part of the twentieth century as a result of population pressure, it has been onto land which presents greater physical difficulties than the land already under cultivation. It is not surprising, therefore, that many of these 'new land' schemes have not proved as successful as originally hoped.

It is chiefly environmental constraints which determine the pattern of agricultural activity within the Middle East. Solar energy is the driving force of the system and this is converted by plants into different chemicals through the process of photosynthesis. This is termed primary production. The amount of solar radiation reaching the ground throughout the Middle East reveals considerable variations. The most important factors are the angle of the sun above the horizon, and day length, both of which are functions of latitude and season. Cloud cover can also greatly reduce the proportion of the sun's rays being utilized by plants. The highest values for solar radiation at ground level in the Middle East occur in summer in an east-west belt stretching from central Iran, through northern Arabia into Egypt. Here figures in excess of $300 \, MJ/m^2$ are recorded. However, at this time almost all of the rest of the region is receiving high values for radiation between 240 and $300 \, MJ/m^2$. In winter a much clearer north to south gradient is seen. Northern Turkey and Iran record values as low as $60 - 70 \, MJ/m^2$, while the southern parts of the Arabian peninsula and Egypt experience figures approaching $200 \, MJ/m^2$. Throughout the Middle East these variations in incoming solar radiation are not of sufficient magnitude to provide a constraint to vegetation growth by itself, as less than 1 per cent of this radiation is actually used for photosynthesis. It is, therefore, other environmental factors such as temperatures, and the availability of moisture and nutrients, that are the major limits to plant growth.

Although there are variations between species, most plants require a mean temperature of at least $6°C$ over a period of several days for successful growth. This means that average monthly temperatures for climatological stations can be used as a guide to the length of the potential growing season. In a Middle Eastern context though, outside of the highland areas, it is not only low temperatures, but also high ones which can limit crop growth. For example, the cereals and many temperate crops seem to attain optimum growth rates between about 25°C and 30°C and have difficulty in surviving when temperatures approach 40°C. Daily maximum temperatures above 30°C are common throughout the Middle East during the summer period. In contrast, many tropical plants require much higher temperatures. For example, melons and sorghum need minimum temperatures for growth of 15–18°C and optima from 31–37°C (Bayliss-Smith, 1982). They are also able to tolerate temperatures approaching 50°C. Maize, sugar and date palms can survive at even higher temperatures, though yields are depressed.

The effects of temperature on agriculture are, therefore, varied. In the highland areas the low autumn, winter, and early spring temperatures can severely limit the length of the growing season to less than 120 days, while the relatively low summer temperatures will mean that only a limited range of crops will be able to be grown. In the hotter lowland areas crops can be raised throughout the year. Here, it is the high temperatures which can check growth of crops, such as cereals. This difficulty is overcome by planting cereals early in the year so that they can be harvested in May and early June before the highest temperatures of summer are recorded. The crops which thrive on high temperatures will be planted to mature in June, July, and August.

One factor which it is very difficult to take account of in agriculture is climatic variability. All too often it is assumed that 'average' conditions are the best guide to a climatic regime, yet in semi-arid and arid regions the inter-annual variability of precipitation is considerable. In effect this means that every year is different from the previous one and that, therefore, successful agricultural management has to be based on accumulated experience of similar conditions which might have occurred many years previously. In dry farming areas, where precipitation provides all water for growth, crop production can reveal tremendous annual variations (Lomas, 1972). Even in years with similar precipitation totals yields can vary, dependent upon the timing of when precipitation occurs. Therefore, whenever agricultural systems are being examined it is essential that the inherent variability of local climatic conditions is appreciated. This could mean that 'drought' conditions, in which a crop might be expected to fail, may well occur in some regions with a frequency of more than once a decade.

Irrigated agriculture is also affected by precipitation variability, as this will control the total volume of water available for agricultural production. In such circumstances, what tends to happen during drought conditions is that water use/unit area is reduced, but if the drought worsens areas of cropland will be left unwatered so that successful crop growth can be maintained over at least part of the agricultural lands. Once again it is a matter of local experience which determines when watering of land ceases and what sort of area is involved. In some cases where new dams have been constructed or new wells sunk to tap deep aquifers it might appear for a few years that water resources are unlimited, and, therefore, that water poses no constraint on crop growth. This, however, is an erroneous view as it pays no attention to the fact that all fresh water resources are finite in nature. All that is happening in such cases is that water is being withdrawn from storage, whether from reservoirs or from aquifers; and once the storage is depleted it can take considerable time for the water levels to be replenished.

Any agricultural planning in the Middle East has to accept the constraints imposed by an unpredictable climate. Unfortunately, this tends not to happen and in most development plans there is an implicit assumption that environmental conditions, including climates, are constant, though minor fluctations around an 'average' value are accepted. It is this emphasis on the 'average' conditions which causes the most difficulty, as for many planners the average is regarded as synonymous with 'most frequently occurring'. This approach means that any deviations away from the 'average' value are regarded as being in some way exceptional, which in an arid or semi-arid climate is not the case. What planners ought to be doing is to accept climatic variability as the norm, with a range bracketed by recorded maximum and minimum precipitation values and make all their plans on this basis. If this view is taken fixed annual values for the production of a particular crop become meaningless and have to be replaced instead by a likely range of crop production figures.

Soils provide the medium in which plants grow, and are, therefore, a critical part of the agricultural environment. Over the Middle East as a whole little detailed information on soil types and their properties is available, though locally intensive surveys have been carried out for specific projects (Beaumont, Blake, & Wagstaff, 1976). From the agricultural point of view soils have to provide a medium in which the plant roots can live and develop. They have to hold an adequate moisture supply and should not be prone to problems such as salinity build-up. Many of the soils used for irrigated agriculture are depositional in origin, formed by fluvial sediments being washed from the uplands to lowland regions. Along river valleys, these sediments form flood plains and terraces, while adjacent to upland margins alluvial fans and alluvial plains predominate. Given the origin of the sediments on which these soils develop they tend to be coarse-grained near the upland margins, and become progressively finer-grained in a downstream direction. Along the major river valleys where gradients are gentle, such as along the lower parts of the Tigris/Euphrates and Nile, silts form the dominant sediments as all the coarser material has been deposited upstream. In general, clay-sized particles are flushed through the river system in suspension and so do not contribute greatly to the formation of these soils unless widespread flooding is a common feature of the river regime.

Agricultural activity on *in situ* soils takes place in the foothills adjacent to the main highland ranges. Here precipitation totals are greater and so rain-fed agriculture predominates. It is very difficult to generalize about soil types in these areas as a result of variations in parent material, topography, climate and vegetation. A wide range of soils is found, but many show evidence of truncated profiles as a result of erosion, both natural and man-induced (Beaumont & Atkinson, 1969). High stone contents are also common. Well developed Terra Rossa, Rendzinas, and brown forest soils are found in the Mediterranean belt of Turkey and the Levant, while in northern Turkey and Iran podsolic soils are common. As precipitation totals decrease towards the arid lowlands soil horizon development becomes less pronounced. Throughout the uplands, even at relatively low altitudes, soil cover is not continuous and large areas of bare rock are found. Agriculture has, therefore, to be confined to the pockets of deeper soil where crop growth is feasible.

Soil characteristics are of particular importance for irrigated agriculture, as they determine the water-holding capacity and in turn, therefore, the frequency with which watering has to occur. Coarse alluvial soils will hold about 50 mm of available water for every metre of depth, whereas with sandy loams the figure may rise to as high as 150 mm. Perhaps the greatest difficulty with irrigation is trying to ensure equal water application rates over a field or basin. This problem arises from the fact that not all parts of the field can be watered simultaneously. What usually happens is that water is let into a furrow or basin from one end and then a wetting front moves across the field/basin in a downslope direction. On very dry soils it will take a considerable time for the water to reach the lowest point in the area being irrigated. This means that the upper part of the field will have received considerable quantities of water before any has reached the lower point. In contrast, when the water flow is turned off the retreating front of water across the field is very rapid. The net effect is that the upper part of the field always receives more water than the lower part. The farmer, is therefore, faced with a dilemma. Does he put on enough water to ensure that the lower part of the field is properly wetted, and thereby waste water on the upper part of the field, or does he conserve water by providing the upper part with the correct amount and, at the same time, risk less than optimum yields in the lower part. In most traditional societies overwatering tends to occur as the farmer is not willing to risk losses to planted crops.

Actual water application rates vary widely, dependent on local conditions, but in general the farmer aims to put on the equivalent of about 100 mm water depth, which is the approximate water-holding capacity of the soil. The situation is further complicated by the fact that most crops will obtain about 70 per cent of their water needs from the top half of their total rooting depth. As a result it is the uppermost layers of the soil which have to be maintained at a high moisture level if crop growth is not to be checked.

On the alluvial soils, especially those which are irrigated, the greatest problem is caused by the build-up of soil salinity. When the soils are sandy or even gravelly in texture natural drainage is usually adequate and excess water drains down to the water table, carrying with it the unwanted salts. In contrast, the soils with high silt contents, such as are found on the major flood plains of the Tigris and Euphrates rivers, have very low percolation rates. As a consequence salinity build-up is an ever-present danger, which is very difficult to combat (Elgabaly, 1977). Although drainage ditches, tile or plastic drains alleviate the problem they rarely solve it completely.

As many irrigation projects in the Middle East have been developed on fine-textured soils a gradual deterioration in soil quality can often be witnessed. When first commissioned, new irrigation projects often show a period of variable length, up to a number of years, when crop production is good and difficulties appear few. However, during this time water tables on the project are usually rising despite the drainage systems which have been installed. A second period can be delimited, once again several years in length, when increasing problems, particularly with regard to salinity, are recognized by the project managers. Water tables rise sufficiently close to the ground surface for capillary action to allow water to move upwards and be evaporated from the soil leaving behind any dissolved salts (Schulze & De Ridder, 1974). Extra drainage schemes are sometimes installed, but these rarely solve the problems completely and extra costs are always high. However, during this time attempts are continuously being made to overcome the difficulties. A final stage is reached when the problems become so severe that economic production is no longer feasible and the project is abandoned. It should not be thought that all new irrigation projects follow the pattern outlined above. If soils have adequate natural drainage, irrigation can be practised on a long-term basis, provided that careful management procedures are followed. However, the fine-grained soils with impeded drainage do seem to pose particular difficulties for flood or furrow irrigation which are not easily overcome even with modern management techniques (Reeve & Fireman, 1967).

Soil erosion is another problem which has limited agricultural activity especially in foothill regions where moderate slopes are found. Sheet and gully erosion both occur and these have often been intensified as a result of man's activities. Turkey, Syria, Lebanon, Israel and Jordan all reveal evidence of truncated soil profiles. As natural soil profiles tend to become more indurated with depth, progressive erosion exposes poorly structured subsoils with lower infiltration rates and consequently produces higher runoff potential. One of the most significant soil properties as far as erosion is concerned is the lack of stability of the soil structural units under the combined influence of wetting by precipitation and the impact of raindrops. The addition of water to the soil breaks down the aggregates into their constituent particle sizes. These particles clog the macropores and form an impermeable surface crust, which in turn reduces surface infiltration and so promotes even more water erosion.

The main impact of human activity on soil erosion has been the widespread removal of the vegetation cover, whether by deforestation, overgrazing or the monoculture of cereals. Of these deforestation is the most difficult to combat, because forest rehabilitation is such a long-term process (Mikesell, 1969). Over the years woodlands have been cleared to provide additional agricultural land as well as timber and fuel supplies. Government attempts to control deforestation have met with only limited success. Overgrazing of pastures and woodlands has meant that regeneration of many species has been prevented and in some cases, particularly around the villages, all the vegetation has been removed leaving large areas of bare soils and rock. Although it is possible in theory to control animal numbers, it is difficult to do this in practice and it seems inevitable that overgrazing of certain areas will continue. Goats are a special problem as their very close grazing of vegetation and barking of trees effectively prevents regeneration. On the other hand the goat is an extremely efficient harvester of poor quality vegetation and hence a very valuable animal in a semi-arid environment.

This decline in biological productivity in arid zones has become known as desertification or desertization. It is usually associated with the spread of the desert margin into semi-arid regions where cultivation and livestock farming formerly took place. The causes of desertification are varied. In some cases physical factors, such as changes in rainfall patterns or movement of sand dunes are sufficient to render arable farming difficult or even impossible. However, in most cases it would seem that human factors are pre-eminent (Department of Environment, Iran, 1980). Overgrazing and deforestation often result in the breakdown of a particular vegetation type, as regeneration proves impossible. Soil erosion becomes a problem and topsoil is removed, so making future vegetation growth even more difficult.

In other cases man has been able to eke out a livelihood on the desert margin by labour-intensive, irrigated agriculture. Under such conditions man's task has been to keep the desert at bay. However, with the advent of modern communication systems even people in such isolated environments have become aware of the opportunities available in urban centres. Many young people are no longer willing to accept the drudgery of subsistence agriculture and have left their villages to seek employment in the towns. As a result the old and very young remaining in these remote areas are no longer able to adequately maintain the complex infrastructure of irrigated agriculture. Productivity declines and eventually the village is abandoned to the desert (Beaumont, 1981a).

WATER RESOURCES

In the Middle East the most important environmental constraint on agriculture is lack of water. Only along the lowlands of the Black Sea and the Caspian Sea does precipitation occur all the year round. Elsewhere a marked seasonal precipitation pattern is found, with a pronounced winter maximum. In the highlands of northern Jordan, 95 per cent of the precipitation falls between November and March inclusive, with 70 per cent of the total occurring in the three months of December, January, and February. This precipitation is produced by eastward moving frontal systems from the Atlantic and

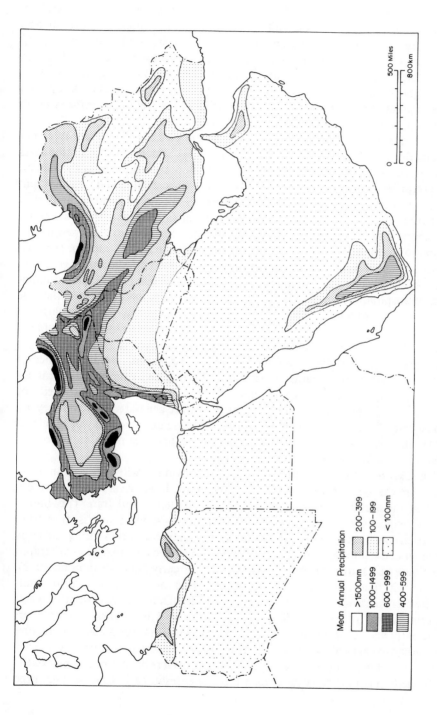

Figure 1.1   Precipitation in the Middle East

Mediterranean areas. Only in the extreme southern parts of the Arabian peninsula is there summer precipitation as a result of the monsoonal circulation from Ethiopia across to the Indian subcontinent. The highest precipitation totals, in excess of 1500 mm/annum, occur along the highland areas of the Black Sea and Mediterranean coasts of Turkey and the Caspian Lowlands (Figure 1.1). Throughout most upland areas totals in excess of 400 mm are found. Away from these highlands precipitation falls to less than 100 mm over much of the Arabian peninsula, central and southeastern Iran, and Egypt.

Precipitation totals alone, however, do not give a true picture of water availability, as this depends as well on prevailing temperature conditions. As a result of the high summer temperatures which prevail thoughout the region, evapotranspiration rates (the combined effects of evaporation and transpiration) exceed 570 mm almost everywhere except in the high mountain areas of Turkey and Iran. Indeed, in most of the lowlands figures in excess of 1140 mm are common. The importance of these data is that when crops are growing the evapotranspiration rates determine the amount of water which must be supplied to the fields by irrigation to ensure optimum growth conditions.

From the agricultural standpoint it is water surplus, that is precipitation minus evapotranspiration, which determines water availability and, therefore, the type of agriculture which can be practised. In the Middle East, water surpluses in excess of 100 mm/annum are confined to narrow belts enclosing the highland areas of Iran, Turkey, and the Levant (Figure 1.2). Even in these uplands the total area receiving more than 400 mm is very restricted. The importance of these water surpluses in terms of agricultural productivity in the region cannot be overstressed. In most cases these water surpluses are not utilized for agricultural purposes in the areas in which they are generated. Instead they are transported by rivers and aquifers to adjacent lowlands which may be hundreds of kilometres away and which may themselves experience precipitation of less than 100 mm/annum (Beaumont, 1981b).

The nature of the flow pattern of the rivers obviously will have an important effect on the agricultural calendar. Almost all the larger rivers in the northern part of the area rise in the high mountains of Turkey and Iran. These areas receive winter precipitation, most of which falls as snow. Water is, therefore, stored in snow-packs in the uplands until rising temperatures in spring and early summer cause it to melt. As a result, all these rivers possess regime hydrographs dominated by snowmelt peaks in April, May or early June (Beaumont, 1973). This means that there is abundant water available for irrigation in the lowlands at the early part of the crop-growing season, though the sheer volume of water pouring down the river valleys makes it difficult to harness these waters efficiently and so flood damage to diversion structures was a common feature of traditional agricultural societies. During the late summer, when many crops are approaching maturity, water volumes in the rivers have dropped markedly.

In the twentieth century the emphasis of water resource management has been

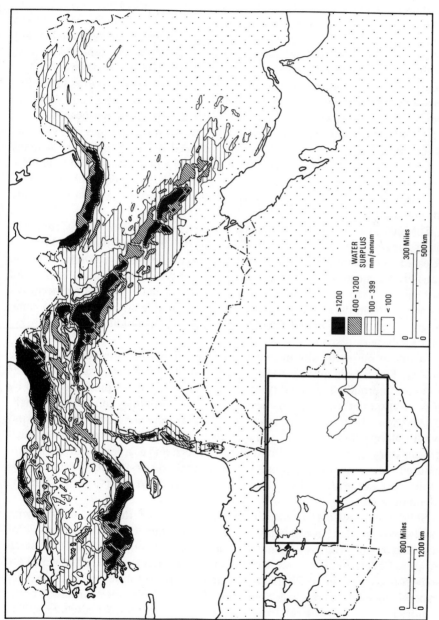

Figure 1.2   Water surplus in the Middle East

to concentrate on larger dam and reservoir schemes on the major rivers to provide increased amounts of water for irrigation and urban/industrial use. The basic objective of these projects has been to store the snowmelt peaks of early summer so that the water can be utilized later in the year. Unfortunately many of these projects have not been as succesful as was originally planned. The short periods of hydrological observations which were available when the dams were being constructed meant that calculations of available water resources have often proved over-optimistic and so water volumes available for beneficial uses have not been as great as was hoped for. An even bigger problem has been the gross underestimation of the amount of sediment flowing down the river systems (Vali Khodjeini & Mohamed, 1975). This has reduced the life-expectancy of many reservoirs from original estimates of in excess of a hundred or even a hundred and fifty years to in some cases as little as forty years. Under such conditions the original cost-benefit calculations which were used to justify a project are almost worthless.

Difficulties have also arisen with water distribution systems. New concrete-lined canals in Isfahan, Iran, built over unstable ground have cracked and released large quantities of water, whilst in Turkey the use of preformed concrete canalets has meant that when they have become damaged the local farmers did not have the necessary resources to repair them and had to wait for replacements to be sent from the nearest big city. The traditional hand-dug irrigation canal might have had large percolation losses, but at least it could be repaired quickly and easily with a spade.

Water surpluses are also moved by aquifer systems into areas of considerable aridity. Flood plain gravels make excellent aquifers with high water movement rates and normally high quality water. Such aquifers are rapidly recharged by precipitation and runoff, but their water storage volumes can be quite small. Throughout the more arid parts of the region productive aquifers are found within alluvial fans and underlying alluvial plains. These usually have much greater storage volumes than flood plain aquifers, and as a result provide a more assured source of water supply. The main recharge for these systems occurs at the apex of the alluvial fans and is achieved by floodwaters issuing from the highland zone.

Solid rock aquifers are also important in parts of the Middle East, with sandstones and limestones being the most important formations. In parts of the Zagros and Taurus Mountains, these are relatively small, but in Saudi Arabia, Egypt, and Libya some of the biggest aquifer systems in the world are to be found (Gischler, 1979). Natural recharge to these aquifers often takes place in highland areas where surface outcrops of the particular rock formation occur hundred of kilometres away from the point where the aquifer discharges. The most likely way in which recharge occurs is by runoff being concentrated into stream channels followed by percolation into the underlying rock aquifer. Dispute still exists as to the degree to which these large rock aquifers are being recharged under the prevailing environmental conditions. Some workers claim that most of the water within them is of fossil origin dating from about 16,000 to 35,000 years ago (Thatcher, Rubin, & Brown, 1961; Swailem, Hamza, & Aly, 1983); while others have suggested that where sand dunes overlie the aquifer recharge can occur in areas where precipitation totals are as low as 50 mm/annum (Dincer, Al-Mugrin, & Zimmerman, 1974).

In Saudi Arabia the Eocene aquifer system is the best known as its waters feed the oases at Hofuf and Qatif as well as the springs on Bahrain Island (Beaumont, 1977; Pike, 1983). This aquifer is largely composed of carbonate units which have been weathered during the Tertiary period and so greatly increased their water holding and transmission capacities (Burdon & Al-Sharham, 1968). Further west, in Egypt and Libya, the Nubian Sandstone forms the major rock aquifer. This formation attains a thickness of more than 3,500 m and covers an area of about 2.5 million km^2 (Hammad, 1970; Himida, 1970). Most of the recharge of the aquifer is thought to have occurred in the highlands of northeast Chad, the western uplands of Sudan, and the Ennedi and Tibesti plateaux. Groundwater movement occurs in a northeasterly direction and natural discharge from the aquifer takes place at the Siwa, Farafra, Bahariya, and Dakhla oases.

Obtaining groundwater for irrigation has always posed problems for a traditional society with limited energy resources. Water is a heavy substance and to provide adequate amounts to meet the demands of growing crops requires considerable expenditure of effort. For irrigating small gardens it is possible to lift sufficient water manually in a bucket or other receptacle from a well. With larger fields a device making use of animal labour has to be devised. Throughout North Africa an inclined plane was often used down which a camel or donkey walked. As it did so it lifted a large bucket of water to the surface which was then poured into irrigation channels to take it to the fields.

In Iran, but also elsewhere, an engineering construction known as the qanat was used (Beaumont, 1971; English, 1968). This is a sloping tunnel which was dug through alluvial material until it went below the water table. The gradient of the tunnel was such that water was always able to flow along it under the influence of gravity and so reach the surface. Although prodigious effort was needed to build these qanats, once constructed they produced water between 1 and 80 m^3/h, while lengths varied from 1 km to almost 50 km. A measure of how valuable water was is found in eastern Iran where some qanats were measured to be in excess of 250 m in depth. More usually, however, depths were between 10 m and 50 m. Today, however, many qanats are falling into disrepair as adequate maintenance is not being carried out (Birks, 1984).

From the Second World War onwards the use of goundwater for irrigation has been greatly aided by the introduction of electric and diesel water pumps. Initially these pumps were installed on existing wells in Iran and so supplied water to fields already under cultivation. However, new wells were sunk by urban entrepreneurs in areas where lack of water had previously prevented cultivation. An unforeseen result of the pumped well was that over-pumping led to serious declines in water tables during the late 1950s and early 1960s. In Iran, on the Varamin Plain near Tehran, newly installed wells in the late 1950s caused the water table to fall and as a result many qanats supplying water to the local villages ceased to flow or else produced much lower water volumes (Beaumont, 1968). This meant that the villagers had difficulty in producing corps at a time when growing population numbers were increasing pressures on available land resources. The introduction of new technologies has, therefore, often caused new problems rather than solved old ones.

CONCLUSION

In the final section of this chapter the overall effects of increasing fossil fuel subsidies on the agricultural system of the Middle East will be assessed and possible future problems identified. On the largest scale, at the regional level and above, the greatest impact has been on the manipulation of water supplies. Rivers as large as the Nile have effectively been turned into irrigation canals in their lower reaches by major dam construction (Beaumont, 1981b). For example, so great is the storage capacity of Lake Nasser, behind the Aswan High Dam, that all flood flows along the Nile can be contained for a number of years. Similarly, along the Euphrates the construction of dams has given much more control over the regime of the river and permitted the development of massive new irrigation projects (Beaumont, 1978). Similar projects are also planned for the upper Tigris (Özis, 1983). In their initial stages many of the large dam projects have created the illusion that water resources are unlimited. This has led to water pricing policies which have given rise to wasteful water use and in many cases caused salination of large areas of once potentially fertile land. With groundwater, the electric motor and the internal combustion engine have permitted the development of aquifers on a much larger scale than was previously possible when animal and human effort was all that was available for lifting water. Inevitably the safe yield of many aquifer systems has been exceeded, producing falling water tables, poorer quality water, and, in certain coastal locations, sea water intrusion.

As populations grow and as standards of living increase the domestic and industrial demands on available water resources will increase to the point where the only feasible way to obtain water at a reasonable cost will be to cut back on agricultural water usage. There is little doubt that urban/industrial water consumers can and are willing to pay more for a cubic metre of water than an irrigation farmer. This situation has already been reached in some of the smaller countries of the Gulf and it seems inevitable that even the larger countries will experience the same pressures in due time (Beaumont, 1977). This brings into focus the question concerning the role of irrigated agriculture throughout the region in the future. Where water resources are very scarce, in places like Bahrain, Qatar, Kuwait, and Israel, irrigation seems bound to decline in importance in the near future as water is switched to the growing urban complexes, even though agriculture has currently a high priority in terms of water demand (Hall & Hill, 1983; Annesley, Hall, & Hill, 1983). Even in large countries such as Iran the demands of cities such as Tehran and Isfahan, both sited in arid regions, are placing growing strains on available water resources. Within a decade or two, irrigated areas are likely to contract and, as all arable farming in these areas needs irrigation, the impact on agriculture will be considerable.

Further pressures on water resources will also come in terms of water quality changes. Sewage effluents from the larger cities, together with runoff from fertilized agricultural lands, mean that the nutrient contents of surface waters are rising and the dangers of eutrophication are growing (Beaumont, 1980; Saad, 1973). These waters, too, are often important recharge sources for local aquifers and so groundwater pollution, particularly by nitrates, will increase (Shuval, 1977). As agriculture has begun to use fertilizers, herbicides, and pesticides, a whole range of new chemicals are beginning to find their way into both surface and groundwater systems. Application rates for these chemicals are rarely scientifically assessed and the lack of training of operators often means that excessive quantities are used.

With the development of new industries in the urban centres, toxic chemicals are being introduced to sewer systems and they too will be discharged into water courses or groundwater systems. In most parts of the Middle East these discharges have not yet begun to cause serious problems, but the potential for pollution is very great and, of course, growing. In Saudi Arabia the threat of industrial pollution is widely recognized and the government has established a Meteorological and Environmental Protection Agency (MEPA) to deal with it. The agency has been granted powers to demand information on waste discharges and is able to make recommendations to control pollution. How successful it will be remains to be seen.

A feature of the Middle East which has already been alluded to is the growing use of fossil fuel subsidies at the farm level. This is represented by machinery, fertilizers, pesticides, and herbicides. Unfortunately, comprehensive data on these items are extremely difficult to collect for the present day, and almost impossible to obtain in an historical context. Therefore, individual items have to be selected which can be used as indicators of change. In this context one of the best is the number of tractors in a country. Data on tractors in the Middle East during the post-war period provide an indication of the growing dependence on Western technology and fossil fuels.

At the time of the Second World War very few tractors indeed are believed to have been in use in the Middle East, though it must be admitted that data are sparse. By the early 1950s, tractors were beginning to be of growing

Table 1.3   Tractors in the Middle East in the early 1960s and in 1981

	Tractors (number) 1963/64	Tractors per 1,000 ha of arable & cultivated land Early 1960s	Tractors (number) 1981	Tractors per 1,000 ha of arable and cultivated land 1981
Algeria	26,800	3.90	44,000	5.86
Bahrain	N.D.	N.D.	N.D.	N.D.
Egypt	10,994	4.31	25,500	8.92
Iran	6,000	0.39	59,000	3.69
Iraq	2,404	0.50	22,100	4.06
Israel	9,300	23.19	28,335	67.79
Jordan	1,432	1.22	4,570	3.31
Kuwait	N.D.	N.D.	37	37.00
Lebanon	3,556	12.88	3,000	8.57
Libya	2,812	1.12	15,000	7.19
Morocco	10,915	1.54	24,800	2.95
Oman	N.D.	N.D.	109	2.66
Qatar	N.D.	N.D.	N.D.	N.D.
Saudi Arabia	197	0.28	1,300	1.16
Sudan	1,761	0.28	11,600	0.93
Syria	6,698	1.03	31,387	5.45
Tunisia	11,594	2.63	35,500	7.60
Turkey	50,034	1.94	457,425	16.06
UAE	N.D.	N.D.	N.D.	N.D.
Yemen AR	N.D.	N.D.	2,050	0.73
Yemen PDR	N.D.	N.D.	1,265	6.11
	144,497		766,978	

Source: FAO Production Yearbooks, various dates.

importance in certain countries, particularly Morocco, Algeria, and Tunisia, where European settlement was strong. Tractors provide considerably more power than, say, a bullock team, but they are dependent on sources from outside the villages for spare parts and fuel. With such machinery, ploughing is often deeper than with the traditional pattern and soil structures can be seriously disturbed. Under certain conditions this can greatly increase the erosion potential. For example, the practice of disc-harrowing under very arid conditions can provide ideal soil conditions for wind erosion.

However, absolute numbers of tractors do not convey a realistic indication of the intensity of mechanization within a country. A much better impression is conveyed if the numbers of tractors are expressed per unit of arable and permanently cropped land. For example, in the early 1960s Israel had 23 tractors/1000 ha and Lebanon 13 (Table 1.3). Following a long way behind were Egypt with 4.3, Algeria with 3.9, and Turkey with 2.6. During the 1960s

Table 1.4   Fertilizer consumption in kg/ha in 1961–65, 1974–76, and 1981

	Consumption of fertilizer per ha of arable and permanent crops 1961–65	Consumption of fertilizer per ha of arable and permanent crops 1974–76	Consumption of fertilizer per ha of arable and permanent crops 1981
Algeria	7.2	20.6	26.2
Bahrain	N.D.	10.5	N.D.
Egypt	109.9	168.3	247.5
Iran	1.6	22.1	42.3
Iraq	0.5	6.9	14.1
Israel	90.3	169.1	199.6
Jordan	2.7	4.5	5.3
Kuwait	N.D.	N.D.	500.0
Lebanon	62.4	83.4	100.6
Libya	1.6	16.4	37.5
Morocco	5.7	21.7	23.9
Oman	N.D.	9.0	39.5
Qatar	N.D.	83.3	280.0
Saudia Arabia	6.3	7.4	60.2
Sudan	3.7	6.4	6.0
Syria	2.8	10.1	23.2
Tunisia	4.5	10.8	18.0
Turkey	3.9	30.7	45.4
UAE	N.D.	77.8	281.2
Yemen AR	N.D.	1.0	4.3
Yemen PDR	N.D.	4.0	8.8

Source: FAO Fertilizer Yearbooks, various dates.

and early 1970s massive investments were made in agricultural machinery in the Middle East, so that by the early 1980s the number of tractors had risen from 144,000 in 1963/64 to 766,978 (1981). Well over half these tractors were located in Turkey. In terms of the level of mechanization, Israel is seen to be by far the most important with more than 67 tractors/1000 ha. Kuwait comes second, followed well behind by Turkey. Many countries at this time, including Iraq, Iran, Jordan, Morocco, and Sudan still had fewer than five tractors/1000 ha, revealing a level of mechanization which is considerably less than one-tenth of that of Israel.

Fertilizer use also reveals a massive increase in the post-war period, and particularly from the mid-1960s onwards. As with tractors the most revealing figures are obtained if one compares kilogram of fertilizer use per hectare of arable and permanently cropped land (Table 1.4). In the early 1960s Egypt, Israel, and Lebanon are by far the most important consumers, with all other countries way behind. By the mid-1970s these three countries still remain in the vanguard, but have now been joined by small, rich oil states such as Qatar and the UAE. What is also significant at this time is that many countries including Algeria, Iran, Libya, and Turkey show major relative increases compared with the early period, and many other states reveal growing fertilizer usage. By the early 1980s the small, oil-rich states of Kuwait, UAE, and Qatar top the list of users, with Egypt, Israel, and Lebanon still in the top echelon. Some way behind, but still making important gains on the 1970s are Iran, Libya, Oman, Saudi Arabia, and Turkey, all of which, with the exception of Turkey, are large oil producers. At the bottom of list, with less than 10 kg/ha of fertilizer use, are Jordan, Sudan, Yemen AR, and Yemen PDR.

Using both indicators, namely tractors and fertilizers, it is clear that only Israel scores highly on both counts and is, therefore, in a class of its own in terms of fossil fuel consumption. What is significant, though, is that the rate of increase in the richer oil-producing countries is considerable and that within the next decade they may well attain standards directly comparable with Israel.

In conclusion, agriculture in the Middle East can be seen to be taking place in environments where conditions for optimum plant growth are often constrained by lack of water, high temperatures, poor soils, and lack of plant nutrients. Over the years man has successfully modified his agricultural practices to minimize these disadvantages, and so permit stable agricultural systems to develop. However, it must always be remembered that the Middle East environment, from the ecological point of view, can be easily damaged. With increased population pressures and the expansion of the cultivated area into ever more marginal lands the dangers of degradation of water and land resources continue to rise.

REFERENCES

Annesley, T. J., Hall, M. J., & Hill, N. A. (1983) 'A master water resources and agricultural plan for the state of Qatar. Part II: The systems model and its application', *International Journal of Water Resources Development,* 1, 31–49.
Barth, F. (1962) 'Nomadism in the mountain and plateau areas of South West Asia', *The problems of the arid zone,* Unesco, Paris.
Bayliss-Smith, T. P. (1982) *The ecology of agricultural systems,* Cambridge University Press, London.
Beaumont, P. (1968) 'Qanats on the Varamin Plain, Iran', *Transactions of the Institute of British Geographers,* 45, 169–179.
Beaumont, P. (1971) 'Qanat systems in Iran', *Bulletin of the International Association of Scientific Hydrology,* 16, 39–50.
Beaumont, P. (1973) *River regimes in Iran,* Occasional Publications (New Series) No. 1, Department of Geography, University of Durham.
Beaumont, P. (1977) 'Water in Kuwait', *Geography,* 62, 187–197.
Beaumont, P. (1978) 'The Euphrates river — an international problem of water resources development', *Environmental Conservation,* 5, 35–43.
Beaumont, P. (1980) 'Urban water problems', in Blake, G. H., & Lawless, R. I. (Eds), *The changing Middle Eastern city,* Croom Helm, London, pp. 230–250.
Beaumont, P. (1981a) 'Changing patterns of land and water use in the Isfahan Oasis, Iran', in Meckelein, W. (Eds) *Desertification in extremely arid environments,*

Stuttgarten Geographische Studien, Band 95. IGU Working Group on desertification in and around arid lands, 1981, pp. 29–63.

Beaumont, P. (1981b) 'Water resources and their management', in Clarke, J. I., & Bowen-Jones, H. (Eds), *Change and development in the Middle East*, Methuen, London, pp. 40–72.

Beaumont, P., & Atkinson, K. (1969) 'Soil erosion and conservation in Northern Jordan', *Journal of Soil and Water Conservation*, **24**, 144–147.

Beaumont, P., Blake, G. H., & Wagstaff, J. M. (1976) *The Middle East: A Geographical Study*, John Wiley, London.

Birks, J. S. (1984) 'The falaj: modern problems and some possible solutions', *Water Lines*, **2**(4), 28–31.

Brichambaut, G., & Wallen, C. C. (1963) *A study of agroclimatology in semi-arid zones of the Near East*, World Meteorological Organization, Technical Note No. 56, Geneva, 64 pages.

Burdon, D. J., & Al-Sharhan (1968) 'The problem of the Palaeokarst Dammam Limestone aquifer in Kuwait', *Journal of Hydrology*, **6**, 385–404.

Clarke, J. I., & Fisher, W. B. (1972) *Populations of the Middle East and North Africa*, University of London Press, London.

Department of Environment, Iran (1980) 'The Turan Programme', in Biswas, M. R., & Biswas, A. K. (Eds), *Desertification*, Pergamon Press, Oxford, pp. 181–251.

Dincer, T., Al-Mugrin, A., & Zimmerman, U. (1974) 'Study of the infiltration and recharge through the sand dunes in arid zones with special reference to the stable isotopes and thermonuclear tritium', *Journal of Hydrology*, **23**, 79–109.

Elgabaly, M. M. (1977) 'Water in arid agriculture: salinity and water logging in the Near East region', *Ambio*, **6**, 36–39.

English, P. W. (1968) 'The origin and spread of qanats in the Old World', *Proceedings of the American Philosophical Society*, **112**, 170–181.

Floret, C., & Hadjej, M. S. (1977) 'An attempt to combat desertification in Tunisia', *Ambio*, **6**, 366–370.

Gischler, C. (1979) *Water resources in the Arab Middle East and North Africa*, Wisbech: Middle East and North African Studies Press.

Hall, M. J., & Hill, N. A. (1983) 'A master water resources and agricultural plan for the state of Qatar. Part I: Physical setting and resources', *International Journal of Water Resources Development*, **1**, 15–30.

Hammad, H. Y. (1970) *Groundwater potentialities in the African Sahara and the Nile Valley*, Beirut Arab University, Beirut.

Himida, I. H. (1970) 'The Nubian Artesian Basin, its regional hydrogeological aspects and palaeohydrological reconstruction', *Journal of Hydrology, New Zealand*, **9**, 89–116.

Johnson, D. L. (1969) *The nature of nomadism: a comparative study of pastoral migrations in south western Asia and north Africa*, University of Chicago, Department of Geography, Research Paper No. 118.

Kenyon, K. (1969) 'The origins of the Neolithic', *Advancement of Science (London)*, **25**, 144–160.

Le Houerou, H. N. (1977) 'Man and desertization in the Mediterranean Region', *Ambio*, **6**, 363–365.

Lomas, J. (1972) 'Forecasting wheat yields from rainfall data in Iran', *World Meteorological Organization Bulletin*, **21**, 9–14.

Mabbutt, J. A., & Floret, C. (Eds) (1980) *Case studies on desertification*, Unesco/UNEP/UNDP National Resources Research No. 18.

Mikesell, M. W. (1969) 'The deforestation of Mount Lebanon', *Geographical Review*, **58**, 1–28.

Özis, Ü. (1983) 'Development Plan of the Western Tigris Basin in Turkey', *International Journal of Water Resources Development*, 1, 343–352.

Pike, J. G. (1983) 'Groundwater resources development and the environment in the Central Region of the Arabian Gulf', *International Journal of Water Resources Development*, 1, 115–132.

Population Reference Bureau Inc. (1983) *World Population Data Sheet 1983*, Population Reference Bureau, Washington, D.C.

Reeve, R. C., & Fireman, M. (1967) 'Salt problems in relation to irrigation', in *Hagan, R. M., Haise, H. R., & Edminster, T. W. (Eds), Irrigation of agricultural lands* American Society of Agronomy, Madison, Wisconsin, pp. 988–1008.

Saad, M. A. H. (1973) 'Distribution of phosphates in Lake Mariut, a heavily polluted lake in Egypt', *Water, Air and Soil Pollution*, 2, 515–522.

Schulze, F. E., & De Ridder, N. A. (1974) 'The rising water table in the West Nubarya area of Egypt', *Nature and Resources*, 10, 12–18.

Shuval, H. I. (Ed.) (1977) *Water renovation and re-use*, New York: Academic Press.

Stace Birks, J. (1981) 'The impact of economic development on pastoral nomadism in the Middle East: an inevitable eclipse?', in Clarke, J. I. & Bowen-Jones, H. (Eds), *Changes and development in the Middle East*, Methuen, London, pp. 82–94.

Swailem, F. M., Hamza, M. S., & Aly, A. I. M. (1983) 'Isotopic composition of groundwater in Kufra, Libya', *International Journal of Water Resources Development*, 1, 331–341.

Thatcher, L., Rubin, M., & Brown, G. F. (1961) 'Dating desert groundwater', *Science*, 234, 105–106.

Tomaselli, R. (1977) 'The degradation of the Mediterranean Maquis', *Ambio*, 6, 356–362.

Unesco (1979) *Map of the world distribution of arid regions*, MAB Technical Notes, No. 7.

Vali Khodjeini, A., & Mohamed, A. (1975) 'Étude du débit solide et de la sédimentation du barrage de Shah-Banou Farah', *Bulletin of the International Association of Scientific Hydrology*, 20, 223–231.

Wilkinson, J. L. (1977) *Water and tribal settlement in south east Arabia*, Oxford Research Studies in Geography, Oxford.

Agricultural Development in the Middle East
Edited by P. Beaumont and K. McLachlan
© 1985 John Wiley & Sons Ltd

Chapter 2

The Agricultural Development of the Middle East: An Overview

KEITH MCLACHLAN

INTRODUCTION

There are few areas of the world that have changed so profoundly as the Middle East in the period since 1945. Not only has there been the emergence of new nation states out of the dissolution of the British, French, and Italian colonial, including mandated or trucial, territories, but many traditional patterns of life have been irrevocably altered. Among the social revolutions that have transformed the region has been an explosion of population beginning in Egypt and later sweeping through many other states (Clarke & Fisher, 1972). Rates of population growth have occasionally exceeded 5 per cent annually and were of this order in Kuwait and the United Arab Emirates as late as the mid-1980s. The total population of the Middle East as reviewed in this volume was approximately 176 million in 1982 (IBRD, 1983) as against 44 million in 1950 (UN, 1962).

The burden of population on resources was always seen to be heavy (Warriner, 1948). In the early years after the Second World War the basic stress of providing food and employment was on agriculture and the rural sector. At that time, some three-quarters of people lived directly or indirectly on the land (Keen, 1946). The distribution of population has, however, changed radically. By 1981 the Middle East was organized predominantly within towns and cities. Averages can be misleading, but it might be hazarded that almost half of the people of the region became urban resident, and in some of the states of the Arabian peninsula proportions in towns rose to more than 80 per cent (Table 2.1). The economic causations of this rapid and fundamental realignment in the distribution of population are varied (Issawi, 1982) and possibly not yet entirely understood (Asad & Owen, 1983). It represents a major social and economic revolution that, together with rapid growth in population numbers, has had important implications for agricultural development of the region in the period since 1945.

Movement of people to towns has fortunately not brought with it an extinction of traditional values of the countryside. The point is well made that parts of

Table 2.1   Rural population in the Middle East, 1981

Country	Total population at mid-year estimate (millions)	Urban population (%)	Rural population (%)	Rural population (millions)
Sudan	19.2	26	74	14.2
Yemen AR	7.3	11	89	6.5
Yemen PDR	2.0	37	63	1.3
Egypt	43.3	44	56	24.5
Turkey	45.5	47	53	24.1
Syria	9.3	49	51	4.7
Jordan	3.4	57	43	1.5
Iran	40.1	51	49	19.7
Iraq	13.5	72	28	3.8
Lebanon	2.7	77	23	0.6
Saudi Arabia	9.3	68	32	2.9
Kuwait	1.5	89	11	0.2
UAE	1.1	73	27	0.3
Total	198.2	47	53	104.3

Source: IBRD, 1983, p. 148.

many cities have more the characteristics of rural society than of the more established urban systems to which they are attached (cf. Cairo in Egypt (Abu-Lughod, 1961) or Tehran (Costello, 1977)). The principal change is that those who live in extensions of the rural village communities on the fringes of towns rarely undertake work in cultivation for agricultural purposes. Their lack of skills may well tie many of them to manual and labouring jobs in their new locations but this will normally be for employment other than agriculture.

Yet, within the countryside, quite another situation has become common, in which a proportion, often substantial, of rural dwellers will move from agriculture to other employment. There are many variants to this theme. It has been noted in Iran by Salmanzadeh (1980), that former agricultural workers who moved to the towns for well-paid employment during the 1970s retained their village links by keeping their wives and children in their places of origin. In Saudi Arabia, indigenous farmers tend to gain the bulk of their income from non-farm activities, often in the form of disguised social subsidies (Chapter 10). In the 1960s in Libya, Mabro noted that government policies favoured keeping population in the rural areas, though not necessarily within agricultural employment (Mabro, 1973). Whatever the form of the phenomenon, the outcome of the trend was for a further weakening of the importance of agricultural employment, albeit at a much lower social cost than permanent drift from the villages.

Perhaps the most powerful and universal of factors making for revolution of the way of life in the Middle East in recent years was the rise in national and personal incomes that was stimulated by increased earnings from oil exports. Oil revenues paid to producing countries were comparatively small until 1950. After that date, a much improved share of profits of crude oil sales was realized by the producers (Johany, 1980). Moreover, an increasing number of states benefited from oil revenues. In 1946 only Iran could be classified as a major world exporter. By the mid-1950s Iran had been joined by Kuwait, Iraq, Bahrain, Abu Dhabi, Qatar and Saudi Arabia. Other countries had begun small-scale oil exports or earned income from oil pipeline transit fees, provision of labour or creation of services for the oil exporters themselves.

The impact of oil-sponsored expenditures was felt most immediately in the oil-producing countries, especially those with small populations and primitive economic structures. Effects were to be insidious, however, and, aided by the dislocation of wars, growing inter-state temporary migrations of labour (Seccombe, 1981), or failures in Middle Eastern countries without oil to establish productive economies, most of the region was destined to experience forms of rising personal incomes. This trend was greatly enhanced following the oil boom of 1973 (Table 2.2), after which spending by the oil states grew extremely quickly in current, if not real, terms (McLachlan & Ghorban, 1979).

Effects of this apparently fast growth in incomes were not all positive even in the broadest of economic contexts (McLachlan, 1979). The strategies that lay behind what was almost recklessly breakneck expenditures by the oil-rich

Table 2.2 Incomes per capita in the Middle East

Country	GDP/head in 1981($)	Average annual growth rate 1960–81(%)
Egypt	650	3.5
Iran	1,902	5.0
Iraq	1,467	4.5
Jordan	1,620	—
Kuwait	20,900	− 0.4
Lebanon	—	—
Saudi Arabia	12,600	7.8
Sudan	380	− 0.3
Syria	1,570	3.8
Turkey	1,540	3.5
UAE	24,660	—
Yemen AR	460	5.5
Yemen PDR	460	—

Source: IBRD, 1983.

countries of the region were called into question (Mabro, 1977; Sayigh, 1982). It was unfortunate that precipitate growth patterns were only abandoned, and then reluctantly, after oil income fell first during the mid-late 1970s and more markedly from 1980. By that time, considerable social and economic damage had been done in both oil-exporting and adjacent states. In the long term it may be that no worse damage was done than to agriculture.

Adverse influences arose mainly from the side-effects of over-rapid economic expansion rather than from deliberate neglect. Government spending in the oil states was of necessity augmented by rising oil income since payments by the international oil companies were made not to individuals but to the national treasury. The mechanism for turning oil income in foreign exchange to domestic account was state-sponsored development projects, some concerned with investment in productive schemes, others designed to improve the welfare of the people through expenditures on health, education, and housing. While the overt aims of 'development' were laudable, their achievements could be attained only through imported real resources (Penrose & Penrose, 1978) since the local base in skilled personnel and materials was poor throughout the region. The more rapid the rate of development adopted, the more the need grew to import goods and services. In practice, the flow of resources could not keep pace with demand and severe inflation beset virtually all the oil economies in the Middle East with particular effect after 1973.

But, of all shortages, that of labour turned out to be among the most pernicious. Wage rates soared in those areas such as services and construction which were first and longest affected by government programmes for expansion. Much of government concern was with central services and was essentially urban in expression. Agricultural wages fell increasingly behind those in other sectors and the rural areas lost ground to the towns. A wage gravity was brought into being that condemned agriculture to being not simply badly paid as compared with other occupations, but so inferior in the rewards it offered as to leave it the resort of only the very young, the very old and the unemployable. This gravity effect was noticeable very soon after the inception of high state spending in the countries with small populations such as Kuwait and the states of the former Trucial Coast. It was evident in the larger states including Iran, Iraq, and Oman once they became recipients of enhanced oil income following the oil boom of 1973, as noted by Salmanzadeh (1980) in the Khuzestan province of Iran.

Attitudes often supported trends engendered by differential rural – urban income levels. In all the shaikhdoms of the Gulf, small population size enabled governments to pass on welfare to their citizens by granting guaranteed employment to nationals (Al-Moosa & McLachlan, 1985). Burgeoning civil services, too, offered great scope for placing people in jobs that were more often sinecures than real contributions to national wealth. In such a situation, it was scarcely surprising that few citizens of the Arab states of the Gulf opted to stay in farming where conditions were poor and returns low from a difficult physical environment.

It must be pointed out in fairness, that a number of countries did attempt to develop their agricultures. In Part II of this volume a variety of authors discuss the very varied styles of agricultural development programmes put into effect. What is abundantly clear from these later studies is that few development plans as a whole and only a handful of individual schemes were successful in the sense that they consistently added extra output at costs that made them genuinely competitive with imports. Few could be expected to survive once the oil income that provided the initial funds to set them up and subsidies to sustain them afterwards had ended (McLachlan, 1984). The differences between the oil- and non-oil-exporting states are less than might at first be hoped in this area of activity. Turkey, Syria, and Sudan have performed better than most in implementing new developments in agriculture, but not without notable failures. In the oil-exporting states few schemes have stood the test of time. Initially, for example, Sadiyat Island farm in Abu Dhabi showed great promise (Durham University, 1974), but has stopped short of becoming a model for intensive greenhouse vegetable farming as was originally hoped.

More surprising than the slow and viable expansion of the Sadiyat Island venture was the abortive attempt to set up agri-business enterprises in southern Iran. Given the known skills in farming techniques of the Iranian farmer and the high levels of managerial talents of the many foreign companies that took up contract areas in the Khuzestan irrigation scheme (including Shell-Mitchell Cotts and US participants), it was depressing to note the degree of failure there (Salmanzadeh, 1980). Worse, the new farms took over lands formerly worked efficiently under traditional systems and, on the demise of the agri-businesses, there was a double loss to agriculture.

Iran illustrates yet another feature of the modern period of developments in agriculture instituted under economic plans devised to utilize oil revenues. Here, in addition to all the problems of land reclamation and settlement (FAO, 1967) that attend new development schemes set up by governments and their agencies, state activities in bringing hectares into use have failed to keep pace with the loss of land from cultivation arising from spoilage through alkalinization and salinization of the soil within existing settled areas.

The catalogue of failures in new state developments in agriculture is long and is offset by only a short list of projects that may be termed successful, even when the difficulties of economic viability are ignored. Despite, therefore , the often much-acclaimed government programmes for agriculture, together with a considerable propaganda reinforcing belief that agriculture holds a priority role within the development plan (cf. Iraq, (EIU, 1980)), the reality in most countries has been quite different.

Spending on agriculture has often run at much lower levels than that declared in the national development plans as noted in the case of Iraq. Dedication of other resources than capital has often had an even poorer record. The vital manpower required to maintain agricultural activities has been ignored by the planners in most states and has been permitted to be deployed elsewhere in the economy as the forces of economic revolution have penetrated into the countryside. In the Middle East, cultivation skills and knowledge of the vagaries of the environment are important attributes of the long-term farming community. The handling of water for irrigation, the protection of crops from pests, and the application of special skills for getting the best from individual crops are important techniques that can make farming possible in arid and physically marginal conditions. Loss of proficiency in managing cultivation in local microclimates has been underrated as a negative aspect of the movement of population from the land.

In effect, once marginal areas have been abandoned as a result of mass migrations or the movement of individual farmers from their holdings, they are unlikely to be reclaimed for intensive farming. Sophisticated skills are stripped from the land and later attempts to replace them by mechanized farming can prove fruitless, as in the case of Khuzestan, Iran (Samanzadeh, 1980). A further dimension of the same problem is visible in the oil-rich states, where loss of any kind of population from villages distant from the cities has been irreparable. Libya illustrates the difficulty. Here, the Saharan territories of Fezzan and southern Al-Khalij have static-to-declining populations since all gains in population numbers are transferred to the coastal districts of the north. Migrants have often not moved willingly but only in response to strong economic forces generated by government policies inimical to private agricultural enterprise. Simultaneously, official programmes have concentrated on developing new state-controlled estate farms that are socially alien to the local people (Alawar, 1979) and which have had no success in drawing in Libyan labour from the north. Short-term Sudanese and other emigrant labour has been required to service the labour needs of the expensive new farms. The irony of displacing locals desirous of remaining on the land by others that have no permanent interest, at high cost and without the justification of economic success, is not confined to Libya.

The several revolutions that have affected the Middle East in the last three decades have brought deep changes in rural life. The status of agriculture has been relegated from principal to minor importance despite government promises that it would be protected as a source of food supply and a basic area for employment. The speed of change has been rapid and there are few signs in the mid-1980s that either the pace or direction of progress will alter significantly in the immediate future.

INDICATORS OF AGRICULTURAL CHANGE

Although reliable statistical data are notoriously hard to come by for the current period and are almost entirely lacking for earlier times, there are a number of broad conclusions that can be drawn from existing materials. First, development onto the extensive margin of cultivation has been slow. Rates of reclamation of new irrigated land have been negligible, even in those states that have apparently lavished large sums on such developments. The Iranian experience in the most intensive years of land and irrigation development was typical. Gains of reclaimed land amounted to approximately 765,000 ha in the area under full irrigation and some improvement was effected on lands under supplementary irrigation in the years 1962–83. Taking into account comparatively small areas brought under the plough in the dryland zone, less than one million hectares was added to the cultivated domain. However, some reclamations were not permanent. The much-advertised irrigation scheme in Khuzestan, taking in more than 70,000 ha, collapsed immediately prior to the revolution of 1979. In other areas, such as Jahrom in the east of the country, a mere 10 per cent of reclaimed land has remained under cultivation. At the same time, there were losses of land through abandonment, both in the upper mountain valleys of the Zagros and Alburz ranges and in the more distant villages, especially of the east and south. The consequences of the gains and losses of cultivated land in the past two decades were that the total cropped area remained at some 16.6 mn ha but that, of this total, 40 per cent was actually under the plough in any given year against only 30 per cent twenty years previously. Measured by the costs of water provision and land reclamation, such an improvement was expensively purchased, granted that the base level was initially low and the system so open to betterment (Murray, 1950).

Table 2.3   Changes in the area of irrigated land (ha. mn)

Country	1940s	1950s	1960s	1970s	1980s
Iraq	1.50[a]	2.91[b]		3.68[c]	
Iran	2–3[d]	2–3[e]		4.10[f]	4.80[g]
Syria	0.32[h]	0.40[a]		0.62[c]	0.60[c]
Egypt	2.65[h]			3.00[c]	3.70[c]
Saudi Arabia				0.18[i]	0.23[i]
Other Gulf states				0.04	0.04

Sources: [a]FAO, 1959. [b]Salter, 1955. [c]Gischler, 1979. [d]Hadary & Sai, 1949. [e]Plan Organization, 1344. [f]Kayhan, 1978. [g]EIU, 1984. [h]Warriner, 1962. [i]Bowen-Jones & Dutton, 1983.

Similar situations can be found in other Middle Eastern countries such as Egypt, where net additions to the cultivated area have been small and where early optimism for the outcome of irrigated and dryland reclamation projects such as those expected to arise from the construction of the Aswan High Dam, has not been matched by performance. Such conclusions are supported by official reports of changing land use in the region. Syria, where Lake Assad had been created in recent years and the potential area commanded by irrigation much increased, has experienced fluctuating fortunes. The actual area farmed under irrigated culture each year has varied greatly in a range between 300,000 and 650,000 ha and, while the average area has increased slightly over the years, the net permanent gain has been slight (Table 2.3). In Iraq, the irrigated area was reported by Lord Salter in 1955 as 2.9 mn ha. Despite enormous expenditure on hydraulic engineering on the major river systems since that time, the total grew to only 3.7 mn ha by 1978 (Gischler, 1979).

The relative decline in the importance of agriculture as an employer has been rapid and sustained, as noted earlier. By the mid-1980s only Sudan, Saudi Arabia, and North Yemen had a clear majority of their labour forces engaged in farming, in contrast to the situation twenty years previously, when Lebanon, Jordan, and Kuwait were the sole countries with less than half of their active labour forces in agriculture (Table 2.4).

Table 2.4   The labour force in agriculture
(per cent)

Country	1960	1980
Egypt	58	50
Iran	54	39
Iraq	53	42
Jordan	44	20
Kuwait	1	2
Lebanon	38	11
Oman		63
Saudi Arabia	71	61
Sudan	86	72
Syria	54	33
Turkey	79	54
Yemen AR	83	75
Yemen PDR	70	45

Sources: IBRD; 1983; Sayigh; 1978.

The position is probably worse than the figures indicate. In Egypt, for example, underemployment and unemployment in the countryside is considerable (Sayigh, 1978). Many economies, especially those based on oil exports, have become even more characterized by part-time employment in agriculture and, while this is not necessarily entirely an adverse trend (Fuller & Mage, 1976), it must diminish the absolute nature of some of the statistics that purport to show high levels of employment in agriculture, as in the case of Saudi Arabia. The reality is that many of those recorded as being in agricultural employ in countries such as Saudi Arabia are maintained by income from the state through retainers for participating in regional militias, national guards or similar types of undemanding security work.

There are small numbers but in some cases relatively important proportions of immigrants working in agriculture. Moroccan, Egyptian, and Sudanese have been brought into Iraq, for example, while Omanis, Yemenis, and others are employed as agricultural labour in the Gulf oil states. In Libya, Sudanese, Tunisians, and Moroccans undertake much of the hard manual work on state farms in the south of the country. Even in Iran, migrant agricultural labour was drawn in during the period of the oil boom in the 1970s, with possibly as much as a quarter of the one million Afghan workers absorbed by agriculture, of which a residue still remained in the mid-1980s. The effect of use of foreign labour in agriculture has been to release yet more local farmers from the land while sustaining an apparently high ratio of the total labour force in the sector. Taking the problems of unemployment, underemployment, part-time farming, and the substitution of temporary foreign labour for indigenous workers, it can be appreciated that agriculture has undergone an even more precipitate relative decline as an employer than is shown in official estimates.

It might be added that the quality of the labour force has exhibited few improvements. Productivity has generally gone up if real prices are used. One or two gains have been remarkable, such as that in Syria, where the value added in agriculture for each member of the agricultural work force climbed from $50 to $311 in the period 1960–81 (Table 2.5). Elsewhere gains were steady but less dramatic except for the oil-rich states of the Gulf region, where rapid rises were achieved from very low initial levels as a result of high state expenditures on development in the sector. In one or two cases the gain in productivity per head was attained with the assistance of an absolute decline in the number

Table 2.5   Value added in agriculture —
per capita ($)

Country	1960	1981
Egypt	140	393
Iran	189	582
Iraq	142	480
Saudi Arabia	—	391
Sudan	134	391
Syria	50	311
Turkey	300	886
Yemen AR	—	267
Yemen PDR	—	157

Source: Calculated from IBRD, 1983.

employed in agriculture, Syria again being an instance of this. In almost every other country of the region, improvements in productivity came about despite increasing numbers on the land (Table 2.5), though rates were sluggish in most of the larger states such as Turkey, in which numbers rose from approximately 12.04 mn in 1960 to 14.0 mn in 1981, and Iran, where those in agriculture rose from 6.33 mn to 8.13 mn over the same period. Egypt, with a steady growth in the farming group from 8.32 mn in 1960 to 12.34 mn in 1981 was only slightly more subject to expansion.

Other studies of productivity of labour in agriculture (Weinbaum, 1982) suggest similar conclusions. Using a ratio of agriculture's contribution to GDP to its share of the total labour force, it was suggested that the best achievements were slight improvements by Egypt (from 0.52 to 0.57) and Jordan (from 0.36 to 0.41). All other states saw a decline in efficiency measured by this ratio, with Iran (down from 0.54 to 0.2) and Iraq (from 0.32 to 0.19), the worst performers.

The pattern of change in agricultural production over the last two decades would appear to have been complex. Overall, the tendency has been for rates to be fairly buoyant during the years 1960–70 but for the subsequent decade to be less so. But great divergencies occur between countries. Iran and Iraq did badly. Their growth rates halved or worse as the rest of their economies expanded rapidly. Egypt maintained a steady average growth rate, assisted in part by the long-term storage at the Aswan High Dam, while Turkey markedly improved its rate of increase in output on an already large base.

Political factors were not without importance in depressing agricultural production. Civil war in Lebanon was disruptive for agriculture as communities were compelled to abandon their fields and exports were curtailed. First the revolution and later war with Iraq did much to reduce production in Iran in the late 1970s and early 1980s. The Gulf war also had a profound impact in Iraq.

The main influence on growth rates in agriculture remained weather conditions and particularly the amount of rainfall in any given season. From this point of view averages can be as misleading now as in the past, except for Egypt, where water provision is generally reliable on a year-on-year basis. Wheat production in Iraq has varied in the last ten years between a low of 696,000 tons in 1977 and a high of 1,492,000 tons in 1979. Iran shows less pronounced but similar characteristics with a low of 5,526,000 tons in 1978/79 and a high of 6,610,000 tons in 1981/82. Even irrigated crops such as rice have fared little better. In Iraq 1978 was a poor year with output at 172,000 tons in contrast to highs of 250,000 tons in 1981 and 1982. Rice production in Iran has ranged between 1,181,000 tons in 1980/81 and 1,624,000 tons in 1981/82 (Table 2.6).

Weather affects production on both dry and irrigated areas, but is more damaging on the former when rains are light or badly timed from a crop growth point of view. Much agricultural development activity in the arid zone of the Middle East has been directed towards reducing dependence on rainfall by increasing the proportion of land fed by irrigation water. New cereal varieties

Table 2.6   Annual changes in wheat and rice output in Iraq and Iran — 1976 to 1982 (%)

	1976	1977	1978	1979	1980	1981	1982
IRAQ							
Wheat	+ 55	− 47	+ 30	+ 64	− 13	− 15	− 18
Rice	− 32	+ 22	− 14	+ 65	− 23	+ 14	0
IRAN							
Wheat	+ 9	− 8	0	+ 8	− 4	+ 15	—
Rice	+ 10	− 11	+ 9	− 17	− 7	+ 37	—

Source: EIU, *QER Annual Supplements 1984, Iraq & Iran.*

and cultivation techniques have been tried on the drylands as a means of adapting to problems of limited and unreliable rainfall. Some gains have been made in both areas but only at the margin in stabilizing levels of production from year to year. The region remains subject to violent fluctuations in output of crops, despite improvements in irrigation and dryland technologies, and the loss from cultivation through land abandonment by those farmers in the most risky and least rewarding areas of cultivation as part of the process of rural depopulation.

It is notable that the growth of production in the region has been assisted partially and inconsistently by the introduction of improved crop varieties, particularly in cereals. Average annual rates of change have been poor when measured against those achieved elsewhere in the world. Worse, few countries other than Turkey appear to have been able to sustain their rates of increment in yields so that there is no immediate prospect for the region making up its failure to expand the acreage under cultivation by intensifying output on the existing agricultural area.

Turkey is the outstanding instance of a substantial and successful agricultural country in the region. In almost every area of cultivation, it is the regional leader, often producing as much or more than all other states of the Middle East together. By the early 1980s Turkish output of cereals totalled more than 25 mn tons/year, while Egypt and Iran, the next largest producers, each recorded some 8 mn tons/year (Figure 2.1(a)). This generalization only fails in the case of rice, where Egypt, with more than 2.5 mn tons/year and Iran, more than one million tons/year, stand above Turkey's 250,000 tons/year (Figure 2.1(b)). For wheat (Figure 2.1(c)) and barley (Figure 2.1(d)), Turkish leadership is unchallenged.

Production of fruit and vegetables has remained unfortunately static in recent years, despite the obviously fast-growing market that developed as incomes burgeoned under the impact of the oil boom following 1973. Once again, Turkey produces more and has a faster rate of growth than other regional states for both vegetables (Figure 2.1(e)) and fruit (Figure 2.1(f)). Indeed, some states have displayed absolutely no apparent response to improvements in demand

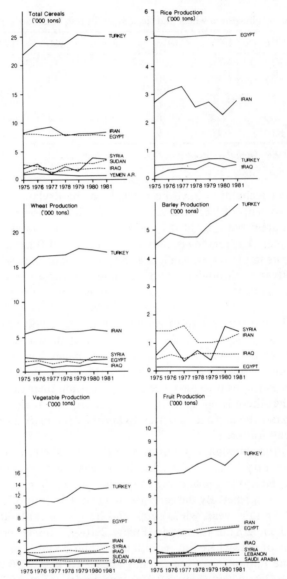

Figure 2.1 Crop production in the Middle East, 1975–81

or methods of cultivation and there has been no conversion of land formerly under annual cultivation to orchards as population pressure has fallen off in rural areas during the 1970s and 1980s.

The dismal performance by most states of the region in increasing output simultaneously with a decline in numbers of those employed in agriculture is

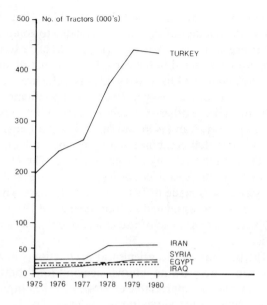

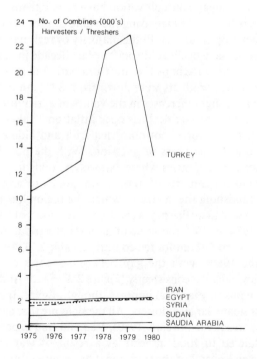

Figure 2.2 Numbers of tractors and harvester-threshers in the Middle East

explained in part by the low rate with which machinery has been substituted for labour. Absolute numbers of tractors have tended to go up within the region but with neither urgency nor consistency (Figure 2.2(a)) except in the case of Turkey, where growth faltered in the early 1980s. A not dissimilar position is found in respect of combined harvesters-threshers (Figure 2.2(b)). Ranked on the basis of the number of tractors per thousand of population, Turkey retains its lead with 9,567 and the others scarcely register on any normal scale Syria having 2,962, Iraq 1,644, Iran 1,446, Jordan 1,329 and Lebanon 1,111 (FAO, 1980). Only Iraq with 400 combined harvesters per million population and Turkey with 300 had relatively significant holdings. The efficiency of use of machinery is dubious in several countries, where care and maintenance is inadequate or servicing is made difficult by lack of spare parts or absence of trained mechanics. It was estimated in Iran during the 1970s that almost half of the country's tractors were out of use at any one time as a result of failures in maintenance.

In view of the poor levels of agricultural production, the region has been unable to keep pace with growing internal demand for foodstuffs. It is clear, too, that there has been no appreciable expansion in the output of commercial crops such as cotton or other agricultural raw materials. Innovations such as sunflower, soya and sugar cane cultivation have been attempted, but these have had limited effects by way of increasing the value of production or stimulating radical changes in crop rotations. In the majority of regional states, traditional agricultural exports have declined during the last decade in real if not absolute terms, even for Turkey. Perhaps the most dramatic fall was in Egypt, where exports of agricultural products were down from $982 mn in 1974 to $610 mn in 1979. For the rest, slight increases in the values of agricultural exports (Table 2.7) disguised greater or lesser declines once inflation is taken into account. In the case of Iran, for example, cotton, dried fruit and hide exports gradually fell by volume during the years in question. With the possible exception of Turkey, there were no countries where exports of foodstuffs and agricultural raw materials paid for imports of agricultural goods to any marked extent.

The key to establishing the degree to which the region has failed to meet its own targets for food self-sufficiency is change in the import of food commodities. In the six years 1974 to 1979, the value of imports of agricultural products rose from $5,416 mn to $10,597 mn for ten countries (Table 2.7). This trend remained effective into the 1980s, with the gap between imports and exports of the agricultural group widening consistently (Figure 2.3). Saudi Arabia has played an important role in pushing up total imports, though, Egypt, Iran, and Iraq have also become persistently large importers. All the signs are that agricultural imports rose more or less steadily in the early 1980s, Iraq and Iran in particular becoming increasingly addicted to food imports as a result of war conditions. More worrying in its way, several of the traditional basic agricultural economies such as Egypt, with severely constrained foreign exchange incomes, were beginning to

Table 2.7 Imports and exports of agricultural commodities ('000 $)

Imports Country	1974	1975	1976	1977	1978	1979
Egypt	1,201,359	1,413,774	1,358,949	1,524,416	1,973,227	1,643,881
Iran	1,270,185	2,011,023	1,473,418	1,932,307	2,097,218	1,994,710
Iraq	704,051	776,582	591,579	802,964	1,003,593	1,378,887
Jordan	151,729	165,568	266,603	251,766	316,020	389,216
Kuwait	276,102	408,890	473,893	574,457	670,394	798,119
Lebanon	325,837	351,146	323,501	381,544	437,120	543,595
Saudi Arabia	561,827	617,755	995,257	1,498,087	2,259,023	3,087,920
Sudan	158,182	179,498	129,870	126,705	168,089	167,255
Syria	351,327	363,550	349,193	334,628	444,835	456,859
Turkey	416,010	399,327	192,775	150,810	110,487	136,175

Exports Country	1974	1976	1976	1977	1978	1979
Egypt	982,088	782,089	733,609	822,823	663,775	610,111
Iran	239,381	301,146	175,872	326,140	302,985	287,929
Iraq	52,430	58,326	63,668	59,777	59,255	67,655
Jordan	37,748	39,959	72,766	87,887	92,522	152,482
Lebanon	129,306	99,487	72,329	70,473	153,029	211,020
Saudi Arabia	26,667	21,818	24,177	20,160	29,630	47,864
Sudan	419,931	404,082	556,530	603,635	486,994	552,356
Syria	264,434	191,937	256,864	302,226	271,193	297,380
Turkey	964,087	897,962	1,332,123	1,155,765	1,627,483	1,468,704

Source: FAO *Yearbook 1980*, pp. 301–343.

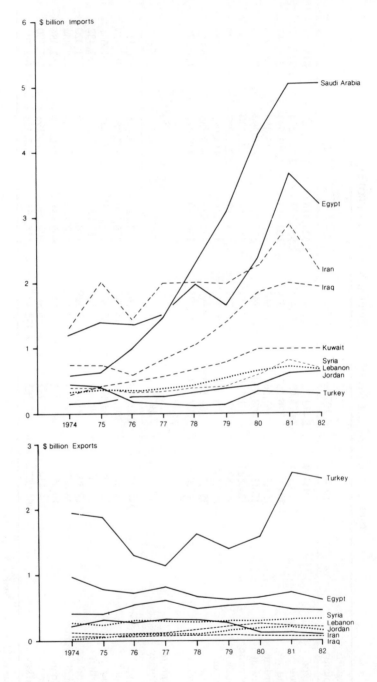

Figure 2.3   Imports and exports of agricultural commodities, 1974–82

follow a similar route. In 1981, Egypt was in receipt of food aid in cereals amounting to 1,865,000 tons and was importing no less than 7,287,000 tons of cereals.

Oil-rich states of the region in possession of long reserves of crude oil and natural gas can face the prospect of external food dependency with few qualms. Saudi Arabia, Kuwait, the United Arab Emirates, and Iraq were firmly oil-based economies by the close of the 1970s with reserves of hydrocarbons sufficient to last them for between 40 and 100 more years on then rates of output. Countries with oil but with the problem of fast-depleting reserves or high rates of domestic oil consumption or undergoing rapid rates of population increase were less well placed. Iran and Oman suffered from problems associated with declining oil industries. Yet in both cases, agriculture had been badly neglected over a decade and more. Traditional agricultural systems were permitted to be overwhelmed by the forces of change let loose by high oil revenues during the 1970s but were not replaced with viable modern farming systems adequate to feed expanding populations (Bowen-Jones & Dutton, 1983). For those nations that have chosen or have drifted inexorably to import-dependence but lack the financial strength to procure imports—a situation that applies to Egypt above all others—the only serious question that requires answering is from what source will food aid be forthcoming in the future?

In a situation where only Turkey and Sudan have a positive balance of trade in agricultural products, the preoccupation of Middle East states with food self-sufficiency strikes a hollow note. There are few states that do not claim to be working towards a new era of food self-sufficiency either as a declared policy of the national development plan or as a general political aim. Present trends indicate that that there are few exceptions to the trend by which countries of the region are becoming more rather than less reliant on imports (Figure 2.3). Iran since the revolution of 1979 has put a great deal of emphasis on the spiritual as well as the economic need for the nation to return to farming. Agriculture was elevated to the position of priority sector for allocations of state funds and the government promised that production from agriculture would rise. Even with strong religious, political, and economic motivations, the Iranian authorities have failed to stimulate significant new developments and by 1983 it was estimated by the US Department of Agriculture that Iranian food imports would continue to rise for the foreseeable future (US, 1983). Given the wealth of Iran in soil and water resources, the agricultural skills of its farming populations, and the unusual dedication of the authorities to agriculture, Iran's failure to diminish its food dependency on imports must give little hope that other, less advantaged states, can turn back from a slide into permanent dependency on the outside world for their food supplies.

AGRARIAN STRUCTURES

If physical developments of irrigation schemes, crop innovations or introduction of the green revolution have failed to produce a brave new world for agriculture in the Middle East and have left the region among the least successful in the world *vis à vis* productivity and growth rates in agriculture, it might be wondered whether changes to the agrarian structure might not be more rewarding. Certainly, the systems of land holding on traditional lands have attracted criticism on a growing scale since the Second World War, beginning with the Middle East Supply Centre, which attempted to sustain local crop production during the war years (Keen, 1946). Rural poverty and low agricultural production were felt to be among the results of maldistribution of land ownership (Warriner, 1948).

The skewed nature of land ownership is well documented. In Egypt, in 1952, 0.1 per cent of owners held 20 per cent of agricultural land, while less than 6 per cent of owners held 65 per cent of the land (Saab, 1967). Similar if slightly less dramatic situations existed in other counties with traditionally cultivated lands, whether Arab (Warriner, 1962), Turkish (Bonné, 1945), or Persian (McLachlan, 1968). Most large land owners operated their holdings through traditional, labour-intensive tenant systems or, more often, through share-cropping. The combination of uneven distribution of land and a farm management structure, that frequently divorced owners from day-to-day operations, kept cultivation techniques at an apparently primitive level, and maintained the bulk of the farming population in medieval conditions of subservience, was much criticized by Western and Western-trained observers (UN, 1951) on grounds of poor productivity and humanitarian considerations.

Share-cropping practices varied greatly throughout the Middle East region depending on the cultural regime, whether dryland or irrigated cultivation was in question, or the type of landlord. The land systems of the Levant were among the most complex in the region, showing changes in practices of crop division, rights of peasants to cultivation, and liability to servitudes by the peasants to their landlords (Granott, 1952). Perhaps the share-cropping arrangement was seen to be at its worst in Iran (Denman, 1973). Here, peasant holders of rights to cultivation, who were among the most fortunate of village dwellers, often took as little as one-fifth of the crop they cultivated, depending on how many factors of production—land, water, seed, animal power, and labour—they provided. Their lot was unenviable by modern standards and the impositions upon them severe (Lambton, 1969).

The key to solving problems of rural economic distress and low levels of production in agriculture was felt to lie in the introduction of land reform by which the cultivated area could be redistributed or consolidated and the sector opened up to modernization. Often, too, political objectives were aimed at, including destruction of the power of the landlord classes and the imposition of central control on the farming community. The process was far from new to the region. The Ottoman Land Codes of the mid-eighteenth century, which *inter alia* freed cultivators from remaining in communal ownership, were the precursors of later reforms in the area of the Ottoman Empire, while, in Persia,

Reza Shah Pahlavi's distribution of *khaliseh* or crown land in the 1930s represented an early move in that country towards albeit modest improvement in the system of land holding. In Turkey, comprehensive restructuring of agriculture that took in land redistribution and the mass provision of extension, credit and marketing services together with improvement in water supply and environmental protection measures, began in the 1940s.

But radical change came as part of the reform measures introduced by the independent and frequently revolutionary Arab states, beginning with Egypt in 1952 and following by Syria and Iraq in 1958. Other Arab states undertook land reforms of varying effect in subsequent years or brought in improvement in land law and agrarian support services that stopped somewhat short of full reform. Despite its many critics, the land reform in Iran, begun in 1962, was among the most far-reaching in the region, penetrating ultimately into each of the country's 50,000 villages.

Agrarian reform, however, meant very different things not only in each of the individual countries in which laws governing land ownership were introduced, but also changing in its implications over time. In Egypt, land reform affected only limited areas and few farmers, mainly in the less populated areas of the Delta (Saab, 1967). A great deal of land that was 'reformed' was consolidated within large management units, in which rotations were governed by state officials. Some gains were made in production during the period of the reform in Egypt, cotton output rising by 21 per cent between 1952 and 1969, but claims that this could be attributed to changes in land ownership (Marei, undated) are difficult to sustain since the major gains were made for most crops in the late 1960s, long after the end of land distribution.

Performances of the land reform agencies in Syria and Iraq, where motivations were above all political and their bureaucracies altogether inadequate to handle the enormous burden of administration, were far from satisfactory. The pre-reform agrarian systems were deficient in many ways, as noted by Warriner and others. But research has shown that land reform was no cure, and reform activities probably diminished security of tenure, reduced peasant initiative, undermined private farming enterprise, and inhibited growth in output from the cultivated area in Syria (Andrews, 1981) and Iraq (Gabbay, 1978). The two main features of implementation of the reforms were inability of governments to redistribute lands so readily expropriated from former owners and yet a commitment to creeping bureaucratization of the countryside. On the one hand, farmers were deprived of freedom of action or reasonable access to normal farm inputs such as fertilizers and seed or participation in marketing yet, on the other hand, they were not provided with effective new management structures and supporting services by the government. Constrained in Syria by strong development of political vested interests by the ruling Ba'ath party in the cooperative organization and in Iraq by constantly changing government policies towards agriculture, private farmers have reduced their commitment to cultivation. In both countries, the rates of rural depopulation are high and rates of change in farm output erratic or falling in the wake of continuing land reforms.

The Iranian experience of land reform (see Chapter 8) has been only slightly less bitter than in the Arab states. Early successes for the first phase of reform, in which the beneficiaries were new, land owning, small peasant farmers, gave way to a more temporizing solution in 1963 and 1964 (Lambton, 1969) in which most land owners were able to retain their best or indeed all their remaining lands. By 1967/68, further modifications were introduced in the law for farm corporations, a move to consolidate the smaller agricultural holdings in farm corporations, producer cooperatives, and (later) agri-businesses. All these structures appeared to promise greater mechanization, improved management, and high production. In fact, they too often caused severe disruption. Worse, the threat that consolidation would take in their villages put a blight on the greater area of Iran's agricultural land as peasant farmers reacted to the new wave of insecurity by leaving the land or abstaining from new investment in their holdings (Stobbs, 1976). Taken together with the damaging impact of growing urban wealth on the countryside during the years of the oil boom following 1973, the reform lost its potency and in its ultimate pre-revolutionary character failed to satisfy the peasantry (Hooglund, 1982).

The revolutionary government experienced no greater success in land reform than the old regime. Two land reform measures were attempted after 1979 but religious, legal, and administrative problems made them at best partially effective in areas under the control of the central government. Revolutionary events in general, insecurity in rural areas, and further land reforms in view all tended to drive out the large and medium land owners and farm managers, leaving behind considerable confusion allied to low production of crops in many areas.

In all, the coming of politically inspired land reforms to the Middle East was surrounded by the highest of aspirations for social and economic improvement in the agricultural sector. It achieved remarkably little in either area. With the benefit of hindsight, it might be speculated that the old system had more merits than it had been given credit for. Many of its attributes, despised by the modernizers as primitive and demeaning, had the virtue of holding rural society together through climatic and political vicissitudes. Absentee landlordism is a case in point. In the situation before land reform, landlords would spend long periods in provincial or national capitals in what was assumed by Western writers to be unproductive consumption of leisure (Denman, 1973; Weinbaum, 1982). Both the observers cited tend to overlook the vital role of the landlords as agents of the village community as a whole during their stay in the town. Large land owners would be required to raise capital when needed in the urban suks or bazaars. In Iran this was particularly the case when major works were needed on *qanats*, the underground water channels that fed agricultural land on much of the Iranian plateau. They would also defend their own and their peasants' position on taxation, conscription, and the corvée against encroachments by the regional and central authorities. For many areas of the Middle East, the centralized nature of political power and the land grant system or *'iqta* ensured that landlords who wished to survive spent at least part of the year attending to their affairs outside their villages. In those parts of the region where forms of nomadic livestock herding were practised, landlords were also owners of herds and travelled with them on the annual migration, as happened until recently in Iran and is still the case in Afghanistan. Whereas there were absentee landlords who never visited their villages and in some cases lived in Western Europe or shrine towns far removed from their lands, it was far more common for landlords to be vitally concerned with the well-being of their prime sources of income and political power — their farms.

Once the landlords had been removed by the land reforms in the general movement to modernization that affected the region in recent decades, it became clear that their roles had been much undervalued. Rural leadership, protection of the peasantry from the extortions of the bureaucracy or security services, and access to funds for agricultural development were found to be lacking. Substitution of these activities by the state either fell far short of requirements, as in the case of Iraq, or was transmuted to a political tool of rural control, as in Syria, or introduced insecurity in the countryside, as in Iran.

Other aspects of traditional agriculture were equally misunderstood or ignored by the idealists who urged or implemented the land reforms of the 1950s and 1960s. Dedication of rural peoples to their villages, however small and miserable, was very powerful and was allied to linkages of kinship, family, and tribe. Insofar as land reform and modernization meant the uprooting of peasants, consolidation of farm holdings, and imposition of state, collective, or cooperative management, peasant communities were gratuitously disturbed and their social structures undermined. Once this first step had been taken, cohesion was less effective and the process of migration to the towns made easier, especially when taken in parallel with the other important economic, social and demographic upheavals that affected the region, noted earlier.

CONCLUSIONS

The story of change in Middle Eastern agriculture is of necessity both complex and frustrating, though not lacking in interest. In Chapter 1, it was made apparent that the agricultural potential of much of the physical environment in the region is very limited. This second chapter has, it is hoped, demonstrated that those who have developed special skills and structures with which to manage and cultivate this difficult area have left their villages and tents in increasing numbers in the modern period, leaving agriculture to decline relatively against other sectors and especially urban occupations. In some instances, such as Iran and Iraq in the 1980s, decline may well have been absolute.

Government policies towards agriculture have proved singularly ineffective. Stated intentions have been universally favourable to the sector but real dedication by way of allocation of the best human and technical resources has been lacking. Intervention into agriculture by the state has been remarkably unproductive. Indeed, much that has been done in changing land ownership, reshaping other aspects of agrarian structure, and centralizing marketing has been destructive, even if undertaken in the name of economic and social modernization.

Farming and herding in the Middle East have been eclipsed above all by the oil revolution and its side-effects. Possibly, the passing of traditional agricultural society would have come about, regardless of government action and inaction, as a result of the profound changes in economic structure wrought by the short period of high oil revenues, beginning in the early 1950s but rising to dominating prominence in the 1970s. The growth of petroleum and service sectors meant the inexorable decline in agriculture as principal employer and generator of national wealth. It seems less inevitable that farming should have lost all momentum for growth and have failed by so wide a margin to respond to growing demand for foodstuffs and agricultural raw materials in what were booming economies. Watson made the point (Watson, 1983) that population growth was a cause of rising agricultural productivity in the past, and there is no reason to assume that a similar dynamic could not have been in evidence in the contemporary period.

If environmental and economic determinism are put aside, much of the blame for the poor state of Middle Eastern agriculture must be placed on the ineptness of the various political establishments in the region, which, with more financial and technical resources at their disposal than at any time in the past, failed to meet the admittedly severe challenges of modernization in the post-independence era. The chemistry of agricultural success that characterized the region in the early centuries of Islam (Watson, 1983) and the ability at least to keep the skills of that period alive in the service of farm production was a comparatively simple formula of innovation and imaginative elaboration of existing techniques in land and water use. Ingenuity of this order has thus far evaded the authorities in the twentieth century, except in Turkey, and certain limited areas elsewhere in the region, where local initiative has triumphed over central government policies.

In the chapters that follow, this volume will review trends in specific sectors of agricultural activity and in each of the main agricultural countries of the region to throw further light on the paradoxes that surround modern agricultural developments in the Middle East.

REFERENCES

Abu-Lughod, J. (1961) 'Migrant adjustment to city life: the Egyptian case', In *American Journal of Sociology*, **57**, 22–23.

Alawar, M. A. (1979) *An examination of agricultural development projects in Fezzan Region, Libya*, Unpublished, Ph.D. thesis, University of Northern Colorado.

Al-Moosa, A. A., & McLachlan, K. S. (1985) *Immigrant workers in Kuwait*, Croom Helm, London.

Andrews, R. J. (1981) *Agrarian reform in Syria, 1958–73*, unpublished Ph.D. thesis, University of London.

Asad, T., & Owen, R. (Eds) (1983) *Sociology of developing societies: the Middle East*, Macmillan, London.

Bonné, A. (1945) *The economic development of the Middle East*, Kegan Paul, London, p. 41.

Bowen-Jones, H., & Dutton R. (1983) *Agriculture in the Arabian Peninsula*, Economist Intelligence Unit, London, p. 3 & p. 160.

Clarke, J. I., & Fisher, B. W. (Eds) (1972) *Populations of the Middle East and North Africa*, London University Press, London.

Costello, V. F. (1977) *Urbanization in the Middle East*, Cambridge University Press, Cambridge, p. 48.

Denman, D. R. (1973) *The King's vista*, Geographia, Berkhamsted, p. 7.

Durham University (1974) *A memorandum on agricultural development in Abu Dhabi*, Durham, pp. 19–20.

EIU (1980) *Iraq: a new market in a region of turmoil*, Economist Intelligence Unit, London, pp. 54–56.

EIU (1984) *Iraq, quarterly economic review: annual supplement 1984*, Economist Intelligence Unit, London. P. 8.

FAO (1959) *FAO Mediterranean Development Project*, Food and Agriculture Organization of the United Nations, Rome, p. 72.

FAO (1967) *Land policy in the Near East*, Food and Agriculture Organization of the United Nations, Rome.

FAO (1980) *Production Yearbook*, Food and Agriculture Organization of the United Nations, Rome.

Fuller, A. M. & Mage, J. A. (Eds) (1976) *Part-time farming: problem or resource in rural development*, University of Guelph, Guelph.

Gabbay, R. (1978) *Communism and agrarian reform in Iraq*, Croom Helm, London, p. 178.

Gischler, C. (1979) *Water resources of the Arab Middle East and North Africa*, Menas Press, London. 95–118.

Granott, A (1952) *The land systems of Palestine*, Eyre & Spottiswoode, London.

Hadary, G., & Sai, K. (1949) *Handbook of agricultural statistics of Iran*, Tehran, p. 9.

Hooglund, E. J. (1982) *Land and revolution in Iran 1960–1980*, Texas University Press, Texas, p. 98.

IBRD (1983) *World development report 1983*, Oxford University Press, Oxford.

Issawi, C. (1982) *An economic history of the Middle East and North Africa*, Methuen, London.

Johany, A. D. (1980) *The myth of the OPEC cartel*, Wiley, Chichester, pp. 4–5.

Kayhan (1978) *Iran's 5th plan*, Kayhan Associates, Tehran, p. 56.

Keen, B. A. (1946) *The agricultural development of the Middle East*, HMSO, London.

Lambton, A. K. S. (1969) *The Persian land reform 1962–68*, Oxford University Press, Oxford, p. 22 & p. 25.

Mabro, R. (1973) 'Employment and wage rates', in Allan, J. A., McLachlan, K. S., & Penrose, E. T. (Eds), *Libya: agriculture and economic development*, Cass, London, p. 165.

Mabro, R. (1977) 'Development—defects in OPEC's fast growth strategy', in *Middle East Annual Review 1977*, pp. 23–30.

Marei, S. (undated) *Egyptian agriculture and agrarian reform*, Ministry of Agriculture and Agrarian Reform, Cairo, pp. 19–21.

McLachlan, K. S. (1968) 'Land reform in Iran', in Fisher, W. B. (Ed.) *Cambridge History of Iran*, Vol. 1, Cambridge University Press, Cambridge, p. 687.

McLachlan, K. S. (1979) 'The disaster of the oil boom', in *Middle East Annual Review 1979*, pp. 21–28.

McLachlan, K. S. (1984) 'The agricultural potential of the Arab Gulf States', in Al Azhary, M. S. (Ed.) *The impact of oil revenues on Arab Gulf development*, Croom Helm, London, pp. 107–137.

McLachlan, K. S., & Ghorban, N. (1979) *Economic development of the Middle East oil-exporting states*, Economist Intelligence Unit, London.

Murray, J. (1950) *Iran today*, Tehran, pp. 75–98.

Penrose, E. T., & Penrose, E. F. (1978) *Iraq: international relations and national development*, Benn, London, pp. 476–488.

Plan Organization (1344) *Outline of the third plan 1341–46*, Tehran.

Saab, G. (1967) *The Egyptian agrarian reform*, Oxford University Press, Oxford, p.–9.

Salmanzadeh, C. (1980) *Agricultural change and rural society in southern Iran*, Menas Press, London.

Salter, Lord (1955) *The development of Iraq*, The Iraq Development Board, London.

Sayigh, Y. A. (1978) *The economies of the Arab world*, Croom Helm, London, p. 359.

Sayigh, Y. A. (1982) *The Arab economies: past performance and future prospects*, Oxford University Press, London.

Seccombe, I. J. (1981) *Manpower and migration: the effects of international labour migration on agricultural development in the East Jordan Valley, 1973–80*, Durham University Centre for Middle Eastern and Islamic Studies, Durham.

Stobbs, C. A. (1976) *Agrarian change in western Iran: a case study of Olya sub-district*, unpublished, Ph.D. thesis, University of London.

United Nations (1951) *Land reform: defects in agrarian structure as obstacles to economic development*, United Nations, New York.

United Nations (1962) *Statistical Yearbook 1962*, United Nations, New York.

US (1983) *Bulletin*, United States Department of Agriculture, Washington D.C.

Warriner, D. (1948) *Land and poverty in the Middle East*, RIIA, Oxford University Press, London.

Warriner, D. (1962) *Land reform and development in the Middle East*, RIIA, Oxford University Press, London, p. 16 & p. 73.

Watson, A. M. (1983) *Agricultural innovation in the early Islamic world*, Cambridge University Press, Cambridge, pp. 2–3. Professor Watson, writing of the period suggests the reasons behind the active advance of agriculture in the region in the following terms:

> The productivity of agricultural land and sometimes of agricultural labour rose through the introduction of higher-yielding new crops and better varieties of old crops, through more specialised land use which often centred on new crops, through more intensive rotations which the crops allowed, through the concomitant extension and improvement of irrigation, through the spread of cultivation into new or abandoned areas and through the development of more labour-intensive techniques of farming.

Weinbaum, M. G. (1982) *Food Development and Politics in the Middle East*, Westview Press and Croom Helm, Boulder and London.

Agricultural Development in the Middle East
Edited by P. Beaumont and K. McLachlan
© 1985 John Wiley & Sons Ltd

Chapter 3

Irrigated Agriculture in the Middle East: The future

J. A. ALLAN

INTRODUCTION

The Middle East's food economy was in deficit in the early 1960s and in the 25 years since then demographic factors and changes in the patterns of food consumption have further exacerbated the situation (Allan, 1981). Population growth rates of over 3 per cent per year are the rule and, though these have been matched on occasions by growth in agricultural production, the boosts given to consumption by the surges in oil revenues in 1973 and 1979 have pushed consumption in both volume and value terms to levels way beyond the capacity of the farming and marketing systems extant in the region in the mid-1980s. Happily the recession of the 1980s has tempered the rates of increase in food consumption as a consequence of the fall in oil revenues.

In earlier chapters it has been shown that, while recognizing the very real constraints upon irrigated and dryland farming in the region, these activities had not yet reached their full productive potential. The purpose of this chapter will be to evaluate the major resources upon which further agricultural development will depend and then briefly to examine the rural and marketing institutions through which such development would have to be organized.

The scarce factor of production in all agricultural activity in all Middle Eastern countries is water, and water will be examined most carefully of all the factors affecting production. In the region water must be supplied in depths of one metre per year (for two seasons of cultivation) on the heavy silt soils such as those of the river lowlands of Egypt and Iraq, and at depths of 1.6 metres or so for the sandy soils in which many Middle eastern governments and farmers have recently shown a strong interest. Some specialists would argue for higher rates of water utilization (Aboukalad et al., 1975), but the above figures are generally in currency. It is recognized that water utilization will vary according to climatic, soil and cropping conditions. The region is characterized by high rates of potential evapotranspiration of about two metres depth of water per year, rising to global extremes of over three metres per year in areas with climates approximating to that of Aswan in Egypt. Water utilization will also vary according to cropping practices and crop cover. Optimal wheat yields have been obtained with applications of 0.49 metres depth in Delta Egypt, while in southern Egypt 0.83 metres depth was found to be needed (Aboukalad *et al.*, 1975, p. 49).

WATER RESOURCES FOR AGRICULTURE IN THE MIDDLE EAST

A number of studies have included reviews of the water resources of the region (Clawson *et al.*, 1971; Gischler, 1979; Beaumont, 1981) but none of the authors have had the courage to quantify systematically the surface and groundwater of the individual countries of the region. Reference has been made to a number of additional sources concerning groundwater (Burdon, 1982; Khouri, 1982; Pallas, 1980; Wright *et al.*, 1982) in an attempt to establish the approximate extent of national water resources and then on the basis of the above estimates of water utilization (1.0–1.6 metres depth of water per year) to estimate the potential irrigable area and then the carrying capacity in terms of a modest set of national food demands. It will be argued that this is a logical approach in that there are some recent precedents for a regional agricultural economy to transform itself from being in deficit and imbalanced (with respect to specialization as well as the balance of payments), to being in overall surplus if still somewhat imbalanced with respect to specialization. This last is the story of the European Economic Community, and at a smaller scale Israel has pursued a similar policy. Careful management of scarce resources, and the amelioration of the constraints implicit in their scarcity, through high agricultural investment and subsidy together with the deployment of new and appropriate technology, have brought increases in production greater than in any previous agricultural revolution and self-sufficiency in agricultural production in terms of value. The Middle East is as advantageously placed as another arid and semi-arid region, namely the southwestern United States, in terms of the availability of timely solar energy, and its modest water resources are no worse disposed than those of southern California and Arizona with respect to major markets. By the 1970s the southwestern United States provided a significant and sometimes dominant proportion of the nation's specialized agricultural needs. Californian agriculture turned over US $13 billion in the early 1980s on an area much smaller than that which Middle Eastern water resources could safely irrigate. Figure 3.1 shows some official estimates of recent national irrigated areas; and, from a variety of sources, the water resources of the respective countries have also been estimated. A number of important features emerge, with respect both to the volume and the distribution of water resources, and Figure 3.1 has been prepared to emphasize some of these features. First, though the region has both large and small political units, it is characterized by countries with very small agricultural areas. Though there are giant states in terms of area, such as Algeria, Sudan, Libya, and Iran (all over 150 million hectares), only Turkey, Iran, and Sudan have over 20 million hectares of arable land. Only Iraq, Jordan,

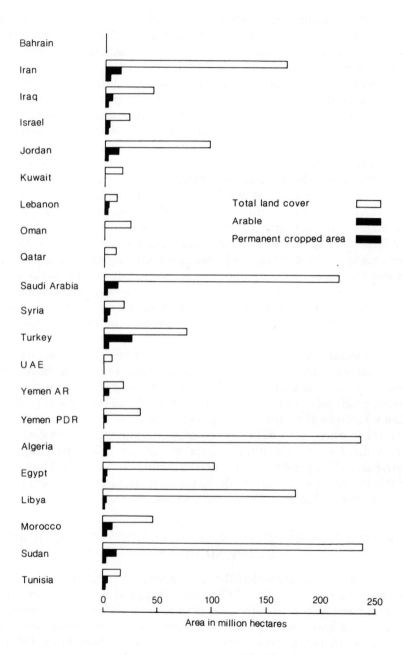

Source: FAO Production Yearbook, 1982 (Data for 1981 or estimates)
Figure 3.1   Total land area, arable area and permanent cropped area of Middle Eastern
and Northern African countries

Saudi Arabia, Morocco, and Algeria have cultivated areas greater than half a million hectares and the rest, including Egypt have 250 million hectares or less, and usually considerably less (250 million hectares is a diminutive 160×160 kilometres, approximately, or 100×100 miles).

The second feature of great importance with respect to the water resources of the region is that most countries of the region are dependent on groundwater or on surface flow mainly deriving from precipitation outside the region. Only Turkey, Iran, Yemen A.R., and southern Sudan have relatively reliable rainfall, albeit seasonal, while Syria, Lebanon, Israel, and, to a lesser extent, Iraq and parts of the Maghreb have precipitation upon which cultivation can be wholly or partially based. The utilization of rainfall from cultivation has for the most part been carried as far as is consistent with sustainable practices for some decades and some would regard the extension of dryland farming in many tracts, for example in northern Iraq and in Jordan, as resource degrading. Meanwhile the utilization of surface flow resources has been advanced steadily in the past three decades so that Syria, Iraq, Iran, Morocco, Egypt, and Sudan have extended their cultivated and cropped areas significantly and it could be argued that only Sudan has surface water resources for major future utilization since it had not by the mid-1980s taken up more than two-thirds of its 18.5 cubic kilometres allocated by the Nile Waters Agreement of 1959 between Egypt and Sudan.

In other words, the region has little in the way of potential new water resources for development and increases in production must come from increases in productivity and, above all, through higher returns to water. Egypt, for example, would need 20 cubic kilometres to develop the 1.2 million hectares (3 million feddans/acres) targeted for new irrigation schemes, which compares with the 37 cubic kilometres (or more in some estimates (Gischler, 1979 p. 98)) estimated to be utilized in the agricultural sector in the late 1970s and early 1980s (Waterbury, 1979, p. 240ff; Allan, 1983a; Ikram, 1980, p. 387). Most of this 20 cubic kilometres would have to derive from a reorganization of existing water resources rather than from their augmentation.

WATER AS A BASIS FOR SELF-SUFFICIENT AGRICULTURE IN THE MIDDLE EAST

Self-sufficiency is the avowed goal of a number of governments and national leaders in the region (Allan, 1983b). And despite their strength in capital resources many Middle Eastern states have found it difficult to combine such capital successfully with land and water resources in productive and cost-beneficial agriculture (Stevens, 1981; USAID, 1982, pp. 8–9; Allan, 1985). In looking to the future it is appropriate to assess to what extent it is realistic to aim at self-sufficiency if the constraining resource is inadequate or unreliable, and so there follows a crude evaluation (crude because of the inadequacy of

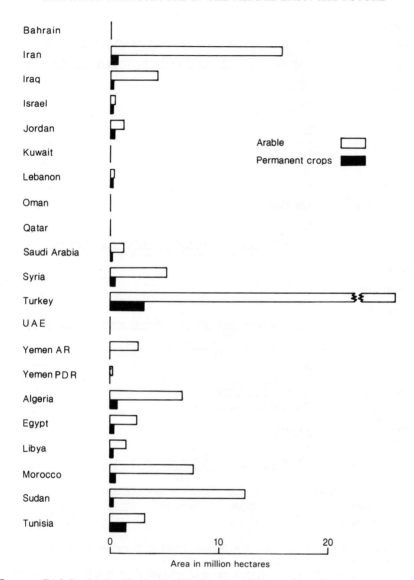

Source: FAO Production Yearbook 1982 (Data for 1981 or estimates)
Figure 3.2   Total arable area and permanent cropped areas of Middle Eastern and
North African countries

water balance data) of the mismatch between water resources and water demands
for national and regional self-sufficient agriculture.

Figures 3.2 and 3.3 are attempts to present a version of the water resources
of the countries of the region together with an assessment of the potentially
irrigable area based on such resources. In addition there are estimates of current

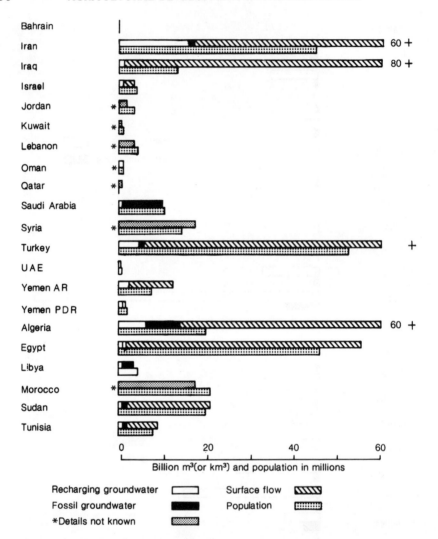

Source: Clanson *et al.*, 1971, Grischler 1978, Beaumont 1981, p. 55, authors' estimates; population from FAO Production Yearbook, 1982

Figure 3.3   Available water resources and potential irrigable area, showing *ca.* 1980 population and population supported by currently available water.

population as well as of the populations which could be supported by the potentially irrigable areas assuming modest levels of food consumption. (It is assumed that each individual could be supported by 0.13 hectares of very efficiently irrigated land producing two crops per year.) It should be emphasized that these calculations only take into account agricultural production from currently available, or imminent, surface flow and groundwater resources. The discussion will relate initially to those countries which are potentially self-sufficient and then to those that are marginally self-sufficient, and finally to those which have a permanent and deteriorating deficit.

THE POTENTIALLY SELF-SUFFICIENT COUNTRIES

Predictably the potentially self-sufficient group of countries includes those in the northern tier which enjoy levels of precipitation unknown elsewhere in the region, with the exception of the Maghreb countries. However, a close look at the estimates reveals the following position with respect to self-sufficiency based on irrigated agriculture alone:

Countries potentially self-sufficient based on irrigated farming:

Iraq

Countries potentially non-self-sufficient from irrigated farming, but potentially self-sufficient with dryland and rangeland outputs:

Iran	Yemen A.R.
Lebanon	Algeria
Syria	Morocco
Turkey	Sudan

Countries potentially not self-sufficient:

Israel	Qatar	Egypt
Jordan	Saudi Arabia	Libya
Kuwait	UAE	Tunisia
Oman	Yemen—P.D.R.	

Only Iraq is potentially self-sufficient with respect to water resources alone and this only if the water were to be managed carefully and applied to appropriate soils. Another eight countries are potentially self-sufficient, but only because in addition to their water resources for irrigation they have dryland and rangeland areas which provide additional agricultural production. Included are Turkey and Iran with their substantial rains in the north, as well as the countries of the long-recognized fertile crescent, namely Lebanon and Syria. Interestingly, the rest of the desirable 'fertile crescent', over which the peoples of the Middle East and elsewhere have fought so frequently over so many millennia, and which is variously called Israel, Jordan, and Palestine, had by the mid-1980s had its population boosted far beyond a sustainable level through natural increase and immigration. In the far south, Yemen A.R., although seriously deficient in groundwater, has substantial rains which would allow its present population to approach self-sufficiency. And Sudan has rains sufficient to make up its deficit from potential irrigated output, although as discussed elsewhere there are institutional as well as capital constraints which inhibit the realization of Sudan's agricultural potential. Algeria is also potentially self-sufficient although its agricultural development has been plagued by organizational problems (Lawless, 1984, p. 173).

THE POTENTIALLY NON-SELF-SUFFICIENT COUNTRIES

Only Egypt of the non-self-sufficient countries has a large population and therefore the problem of deficiencies in food production in regional terms is a minor one, even taking into account the rapid increase in population, usually over 3 per cent per year. It is not generally on such a basis, however, that the problem is judged by the governments of the countries in deficit. These governments are grievously, and some would say overly, aware of the strategic disadvantages of the economic dependency which is the consequence of agricultural inadequacy.

The majority of the food-deficit states are oil-enriched and for them the anxiety over food deficits must be something of a luxury. Kuwait, Oman, Qatar, Saudi Arabia, and the United Arab Emirates can for the foreseeable future finance their food needs and simultaneously engage in costly experiments to produce food in unlikely circumstances with expensive water. Libya and possibly Oman of this group of countries could have real food problems in a couple of decades if they do not marshall their water resources with care, for in such a period their energy resources, the source of their revenue, would make the financing of food imports difficult. However, if Libya does manage to engineer its water from its southern aquifers to the coast, then with careful management Libya could approach self-sufficiency for its current level of population on the basis of irrigated production supplemented by rain-fed output from the Gefara Plain (near Tripoli) and in the Jabal Akhdar in the east (Allan, 1985). Unfortunately, Libya's population will rise quickly, at least to the end of the century, and it should be emphasized that the approach to self-sufficiency identified here would be as measured by the value of specialized output rather than in the provision of the nation's food needs from indigenous production (Allan, 1984).

The case of Egypt is special from a number of points of view. First Egypt has a major proportion of the region's water resources (55 cubic kilometres per year at least at Aswan plus Sudan's currently unused share of about six cubic kilometres) but secondly of the major countries with large deficits in river and groundwater resources (Iran, Turkey, and Algeria) it is the only one that does not have supplementary rainfall. Thirdly it has a large population which continues to grow rapidly. Fourthly the scale of its food deficit is of global importance as it is one of the world's major importers of food staples and especially of flour. Finally its strategic position makes it of inescapable importance to the super-powers and it has been sequentially the client of both the Soviet Union and the United States; and since it is the latter which can alone provide food staples as well as the finance to fund Egypt's economic development, Egypt is committed, if albeit unwillingly, to a close dependence on the United States.

Even if Egypt could mobilize the capital and institutions to reclaim the 1.25 million hectares which it is the national goal to cultivate, it would still not be self-sufficient agriculturally even in value terms, and it never could be self-sufficient in commodity terms. Many would question whether there is water to irrigate the proposed additional 1.25 million hectares of sandy soil; sandy tracts are the only soil types available now that the accessible heavy silt soils have been utilized. The subject of water duties is a vexed one in that Egypt's Ministry of Irrigation assumes a water depth of one metre to be sufficient to irrigate for a year while international opinion suggests a minimum of 1.6 metres on sandy soils. Therefore the 1.25 million hectares would require either 12.5 cubic kilometres or 20 cubic kilometres of water depending on the assumption used. Additional water could be engineered from the system to the extent of 10 cubic kilometres by, for example, using drainage water judiciously and utilizing Delta groundwater to a greater extent than at present. In due course 2 cubic kilometres of water should come to Egypt from the Jonglei Scheme Phase 1, and subsequently two more; but it is clear that Egypt could still not satisfy its water needs even with a decade of extraordinary water engineering and agronomic achievement. In the event Egypt's population would almost double by the time such achievements had materialized. (Assuming 45 million in 1985 and a population increase of 3 per cent per year and two decades of engineering lead time.) Thus Egypt's predicament is extreme and a solution would require much more than irrigation and agricultural engineering.

A country with a much more manageable problem and with a much more manageable environment in which to seek a solution to agricultural production deficiencies is Tunisia. Its rainfall, surface flow and groundwater are not equal to the task of producing sufficient agricultural production for food self-sufficiency, but by careful use of water and of dryland productive resources then Tunisia could provide agricultural revenues from specialized crops to purchase complementary food staples. However, in order to achieve this Tunisia will need to assure markets for its agricultural output. In this, unfortunately, Tunisia is in a similar position to a number of Mediterranean countries and is peripheral to the European Economic Community. And this problem of marketing is one which must be addressed by all Middle Eastern and North African countries if they are to optimize their production in terms of value. This is seen as a necessity in the long term, even for the currently oil-enriched. A further necessity for all countries is the abandonment of the goal of food self-sufficiency. Nevertheless, there will be scope for continual experimentation, but mainly amongst the oil-enriched, with a view to testing the potential of unlikely environments for agricultural production. With the examples of the United States and the EEC to hand, it is clear that extraordinary increases in agricultural production can be achieved if sustained subsidies are made available to the agricultural sector. Rain-fed yields have been more than doubled in three decades for food staples in Europe, and during the same period technologies have emerged in the United States which have enabled the utilization of previously unusable sandy soils.

THE FUTURE OF IRRIGATED FARMING
IN THE MIDDLE EAST AND NORTH AFRICA

In the absence of reliable water balance data (Beaumont, 1981, p. 55) it is difficult to estimate the future for agriculture in the region, and in any event the volume of water available is by no means the most 'sensitive' in the equation of agricultural resource utilization. Just as important are the two matters mentioned at the end of the preceding section. The structure both of marketing and of subsidies within individual countries and, possibly more important, within the region, will determine or not, an environment in which Middle Eastern countries could emulate the performance of industrialized nations in agriculture. These nations are currently competing to supply grain and livestock products to feed countries as diverse in factor endowment as Egypt and Saudi Arabia.

Figure 3.3 provides an analysis of the countries of the region in terms of their water resources and their capacity to invest private and public capital in their initial or further development. Predictably, they fall into three categories, one of which is the oil-enriched, all of which countries have very poor water resources before hefty capital investments have been made. For example, it would require between one and two years of Libya's GDP to make water available (in terms of volume of water per head of population per year equivalent to that of Egypt (Allan, 1984)). Only the oil-rich can envisage the deployment of such pro-portional allocations; and even they find it difficult in periods of world recession such as the early 1980s.

A second group of countries, those with substantial water resources, large populations, and, consequently, low levels of GDP per head, includes Egypt, Iran, Turkey, and Algeria. These are countries which can through their very size of their economies mobilize critical volumes of investment capital, except in the case of Egypt, where only outside capital would be sufficient to initiate the necessary engineering infrastructure.

The third group of countries includes a number which have made significant and sometimes spectacular strides in developing aspects of their agriculture. They may, therefore, serve as models for the future. Amongst them are Israel, Jordan, Morocco, Syria, and Tunisia (in alphabetical order). FAO production indices are not everywhere recognized as reliable, but it is interesting to note that they reveal these same five countries to be special in comparison with others in the region in that, despite their indifferent renewable natural resource endowments, they advanced their agricultural production and productivity substantially in the decade 1971–1982. In the case of Syria the rate of increase was very much greater than in those countries with the advantages of oil revenues; and in the cases of Israel, Jordan, Lebanon, and Tunisia, the performance of their agricultural sectors was at least as good as their oil-enriched neighbours. The moral would certainly seem to be that unlimited supplies of capital do not stimulate the most effective combination of the factors of production relevant to agriculture. The energy-, and therefore capital-, deficient countries such as Turkey have also done as well, or slightly better, than their better-endowed neighbours, Iran and Iraq.

So much for recent performances; future growth in production will depend on the appropriate deployment of technology and the creation and adjustment of rural institutions. The technologies are available and proved to transform the management of water distribution so that specialized tree crops can be raised with water applications of less than one metre depth of water per year. The adoption of such systems is essential if Middle Eastern countries are to pursue water-using strategies which lead to high returns to water. Calculations of returns to water should be regarded as much more important than returns to capital in a world in which 'economic' agriculture is the exception and where soil (land) and capital are not generally serious constraints. It is true that some nations have a poor capital endowment, such as Egypt, Yemen A.R., Morocco, and Tunisia, but even in these countries the consideration of returns to water should still always come first. No longer will it be possible to assume that the agricultural sector can provide a surplus to initiate development in other sectors. It is only with the effective combination of oil revenues with other factors, or with capital investment deriving from remittances from oil-rich countries that the region can move from the serious deficit position currently endured, a position determined by the disposition of climate, water, and soil resources in the Middle East and North Africa and the level of management so far mobilized to utilize them.

REFERENCES

Aboukalad, A. *et al.* (1975) *Research on crop water use, salt-affected soils and drainage in the Arab Republic of Egypt*, FAO, Cairo.

Allan, J. A. (1981) *Libya: the experience of oil*, Croom Helm, London.

Allan, J. A. (1981) 'Renewable natural resources in the Middle East' in Clarke, J. I. & Bowen-Jones, H. (Eds), *Change and development in the Middle East*, Methuen, London, pp. 24–39.

Allan, J. A. (1983a) 'Some phases in the extension of cultivation in nineteenth and twentieth centuries in Egypt', *Middle Eastern Studies*, **19**, 470–481.

Allan, J. A. (1983b) 'Natural resources as national fantasies', *Geoforum*, **14**, 243–247.

Allan, J. A. (1984) 'Reorganising Libya's water resources', *Libyan Studies*, Vol. 15, (in press).

62   AGRICULTURAL DEVELOPMENT IN THE MIDDLE EAST

Allan, J. A. (1985) 'Should Libyan agriculture absorb more capital?', in Buru, M. M., Ghanem, S., & McLachlan, K. S. (Eds), *Economic development of modern Libya*, forthcoming.
Bakiewicz, W., Milne, D. M., & Noori, M. (1982) 'Hydrogeology of Umm Er Radhuma aquifer, Saudi Arabia', *Quarterly Journal of Engineering Geology*, **15**, 105–126.
Beaumont, P. (1981) 'Water resources and their management', in Clarke, J. I. & Bowen-Jones, H. (Eds), *Change and development in the Middle East*, Methuen, London, pp. 40–72.
Burdon, D. J. (1982) 'Hydrogeological conditions in the Middle East', *Quarterly Journal of Engineering Geology, London*, **15**, 71–82.
Clawson, M., Landsberg, H. H. & Alexander, L. T. (1971) *The agricultural potential of the Middle East*, Elsevier, New York.
Gischler, C. E. (1979) *Water resources of the Arab Middle East and North Africa*, Middle East and North African Studies Press, Wisbech, UK.
Ikram, K. (1980) *Egypt: economic management in a period of transition*, Johns Hopkins/World Bank, Baltimore and London.
Khouri, J. (1982) 'Hydrogeology of the Syrian steppe and adjoining arid areas', *Quarterly Journal of Engineering Geology, London*, **15**, 135–154.
Lawless, R. (1984) 'Algeria: the contradiction of rapid industrialisation', in Lawless, R. & Findlay, A. (Eds), *North Africa*, Croom Helm, London, pp. 153–190.
Pallas, P. (1980) 'Water resources of the Socialist Peoples Libyan Arab Jamahiriya', in Salem, M. J., & Busrewil, M. T. (Eds), *The geology of Libya, Vol. II*, Academic Press, London and New York, 539–596.
Stevens, J. H. (1981) 'Irrigation in the Arab countries of the Middle East', in Clarke, J. I., & Bowen-Jones, H. (Eds), *Change and development in the Middle East*, Methuen, London, pp. 73–81.
USAID (1982) *Egypt: strategies for accelerating agricultural development—summary*, Washington.
Waterbury, J. (1979) *The hydropolitics of the Nile*, Syracuse University Press, Syracuse.
Wright, E. P., Benfield, A. C., Edmunds, W. M. & Kitching, R. (1982) 'Hydrogeology of the Kufra and Sirte basins, eastern Libya', *Quarterly Journal of Engineering Geology, London*, **15**, 83–103.

Agricultural Development in the Middle East
Edited by P. Beaumont and K. McLachlan
© 1985 John Wiley & Sons Ltd

Chapter 4

The Use of Land and Water in Modern Agriculture: An Assessment

MARTIN E. ADAMS and
JAMES M. HOLT

INTRODUCTION

Early civilizations in the Middle East developed systems for the allocation and use of communal resources. The value of water for survival and for irrigation has required the establishment of clearly defined water rights. In the Middle East many have received the stamp of religious approval to become traditional law. In modern times, attempts have been made to extend such agreements across political boundaries to achieve a more rational use of scarce, common resources. As users place increasing demands on supplies, governments seek ways of using existing sources more efficiently and developing new sources which for physical, chemical, or bacterial reasons were previously considered too costly to harness.

Table 4.1, based on data published by FAO, shows the relative suitability of the land area of selected Middle Eastern countries for agriculture and the relative importance of irrigation. The FAO record all the cultivated area of Egypt as being irrigated, while in Sudan 86 per cent of the land under permanent and arable crops is rain-fed, most of it lying south of the Middle East as normally defined. Within the Arabian Peninsula, the Yemen Arab Republic has the most significant area for arable agriculture. By contrast, in Saudi Arabia, UAE, and Kuwait, a very small percentage of the total land area is cropped but, as high-income oil exporters (see Table 4.2) they can afford to spend large sums on irrigation development. Of the countries of the Mediterranean littoral, Syria, Jordan, and Lebanon have about the same GNP per capita and employ a similar level of irrigation technology. But Israel with a much higher GNP per capita and higher labour costs has been a pioneer in the field of irrigation technology, putting to use land and water which would go unused elsewhere in the Middle East. Iraq and Iran have similar proportions of cultivated land under irrigation but experience different conditions. In Iraq the major irrigation areas comprise extensive alluvial plains supplied by major rivers. In Iran many schemes use smaller rivers and more steeply sloping land. Both countries employ comparable production techniques both in rain-fed and irrigated areas.

Table 4.1 Cultivated area in selected countries of the Middle East (in '000 ha)

Country	Land area[a]	Arable and permanent crop land	Percentage of land area	Irrigated land[b]	Percentage of arable and permanent crop land irrigated
NILE VALLEY					
Sudan	237,600	12,417	5.2	1,750	14
Egypt	99,545	2,855	2.9	2,855	100
ARABIAN PENINSULA					
Saudi Arabia	214,969	1,105	0.5	395	36
Oman	21,246	41	0.2	38	93
UAE	8,360	13	0.2	5	38
Kuwait	1,782	1	0.1	1	100
Yemen AR	19,500	2,790	14.3	245	9
Yemen PDR	33,297	207	0.6	70	34
MEDITERRANEAN LITTORAL					
Israel	2,033	413	20.3	203	49
Jordan	9,774	1,380	14.1	85	6
Lebanon	1,023	348	34.0	85	24
Syria	18,405	5,684	30.9	539	9
IRAQ AND IRAN					
Iraq	43,397	5,450	12.6	1,750	32
Iran	163,600	15,950	9.7	5,900	37

Source: FAO *Production Yearbook* 1981, FAO, Rome, 1982.
Notes: [a]Land area refers to total area excluding area under water bodies (major rivers and lakes).
[b]Irrigated land refers to areas purposely provided with water, including land flooded by river water for crop production or pasture improvement, whether this area is irrigated several times or only once during the year stated.

Table 4.2 Costs of new irrigation development

Country	Population (millions)	GNP per capita 1980 (US$)	Labour force in agriculture (%)	Share of agriculture in GNP (%)	Marginal costs of irrigation (US$/ha, 1980)
NILE VALLEY					
Sudan	18.7	410	72	38	4,800[a]
Egypt	39.8	580	50	23	9,600–14,400[b]
ARABIAN PENINSULA					
Saudi Arabia	9.0	11,260	61	1	14,000–56,000[b]
UAE	1.0	28,850	n.a.	1	
Kuwait	1.4	19,830	2	0.5	
Yemen AR	7.0	430	75	29	
Yemen PDR	1.9	420	45	13	
MEDITERRANEAN LITTORAL					
Israel	3.9	4,500	7	5	
Jordan	3.2	1,420	20	8	
Lebanon	2.7	1,450	11	n.a.	
Syria	9.0	1,340	33	20	
IRAQ AND IRAN					
Iraq	13.1	3,020	42	7	
Iran	38.8	3,000	39	8	

Source: World Bank Development Report (1982) and various feasibility/pre-investment reports for irrigation development in the Middle East.
Notes: [a] Open channel.
[b] Automated sprinkler and drip systems.

The irrigated land listed in Table 4.1 covers a range of systems and cropping intensities. Spate irrigation in the foothills of the Arabian Peninsula may leave a field without water for several years. Sophisticated drip irrigation in controlled environments in the Gulf States permits the same piece of ground to be cropped more than three times in one year. These differences arise from the widely varying nature of the resources of land, water, and capital available for development. In Sudan where available Nile water provides scope for further development of open channel irrigation on the clay plains, the capital costs of new schemes are still relatively low (US$ 4,800 per ha in 1980). In Saudi Arabia, where conditions are more difficult, costs of automated sprinkler and drip systems have reached US$ 50,000 per ha. In the oil-rich countries of the Arabian Peninsula these schemes are financially profitable only with major subsidies to cover well construction and input and output prices. The choice of sophisticated, automated systems is dictated not only by the unfavourable physical conditions, but also by the high cost of labour.

RAIN-FED CROPPING

It can be calculated from the data in Table 4.1 that less than 30 per cent of the crop land in the Middle East is irrigated. Rain-fed cultivation makes an important contribution to the agricultural production of the region.

The traditional form of land use throughout the winter rainfall areas of the uplands of the Middle East has been the cultivation of winter wheat and barley and the grazing of sheep and goats, cattle, and camels. The area cultivated varied considerably from year to year. In a year of good early rainfall, the land was cultivated to the limit of the population's resources, but in a bad year a smaller area was sown. Mechanization of tillage has transformed land preparation for rain-fed crops since the 1950s, but the system of alternate husbandry, long practised in the Fertile Crescent, is still recognized as the best means of water conservation. This conserves moisture from one year to the next, thereby producing a crop on the combined precipitation of two seasons. Although a major part of the moisture stored in the soil during the fallow period may be lost, the additional 80 to 100 mm may make the difference between success and failure the following year. In the low-potential rain-fed areas of Iran, a rotation of crop–fallow–crop–fallow together with moisture conservation techniques on fallow land are recommended. These techniques are designed to leave the soil receptive to winter rainfall and snow melt-water during the fallow season. They involve autumn stubble cultivation with chisel ploughs and tine cultivators and grazing or weeding of early growth in the spring so as to keep the surface bare and minimize loss of moisture. In the following autumn, the seed is furrow planted. FAO have promoted a formal integration of crop and animal husbandry in Iran based upon techniques developed in Australia. The system consists of a ley rotation of two or more years of annual, self-seeding legumes with one or more years of cereals. In good years the legumes and the cereal stubble produce year-round grazing and at the same time good grain yields are obtained. In poor years the relatively low production of the legumes is supplemented by grazing the cereal fields which have no prospect of maturing. In this way a total loss of an exclusively grain-producing system is converted into an economic yield of forage even when rainfall is insufficient for grain production.

These dryland farming techniques and the introduction of early maturing varieties can do much to increase the reliability of cereal yields in semi-arid areas but they are unlikely to extend rain-fed cultivation beyond existing limits. Cereal production in the Middle East, as elsewhere, is greatly influenced by world grain prices. Economic margins have been reduced by competition from more favourable cereal-producing areas in Europe and North America which also enjoy price support. Barley used to be a major export crop from Iraq, but the country is now a net importer. Successful rain-fed production in marginal rainfall areas requires that production costs are kept to a minimum. The Australian system has the advantage that input costs are relatively low. The annual legumes are self-seeding. Only phosphatic fertilizers are needed and the legumes provide all the nitrogen required by the subsequent cereal crop. In a good rainfall year, nitrogen may be top-dressed. The cereal can be sown into the legume by means of a chisel plough drill without further tillage.

The integration of cereal and animal production can be satisfactorily achieved only when the stock are controlled by the cereal farmer. Too often pastoralism and mechanized crop production are in direct competition for land resources. This is so in the savannah belt of Sudan, both to the east and the west of the White Nile where there is no formal system of land allocation between the Baggara cattle nomads and the mechanized sorghum production schemes of the clay plains. These cut across the north-south migration routes of the pastoralists and lead to crop damage and conflict. Thus the struggle between 'farmers and cowboys' has also been enacted in Sudan and has led to the concentration of the pastoralists on more and more marginal land (Adams, 1982).

Elsewhere in the Middle East overstocking by pastoralists has led to the complete removal of the vegetation cover and severe erosion. There has been much concern at the rapid rate of silting of reservoirs behind expensive dams on large rivers. For example, the benefits of the great Dez dam in Iran would be much enhanced if the life of the reservoir could be increased. This could be achieved only by a comprehensive soil and water conservation programme throughout the catchment, which would restrict stocking rates, control access, and thus reduce erosion. The implementation of such a programme would require close co-operation from the pastoral tribes who would have to be adequately compensated for loss of traditional grazing rights.

IRRIGABLE LANDS

Seasonal rainfall is limited and highly variable over much of the Middle East, whilst extensive tracts of land are naturally watered by river floods. Irrigation has been developed both to supplement rainfall and as the total source in the more arid areas. The areas of supplementary irrigation include the winter rainfall zones of Syria and Iraq; the summer rainfall areas of central Sudan; and the summer monsoon regions of southern Arabia. Areas totally dependent on irrigation comprise the Nile Valley downstream of Khartoum and the Tigris and Euphrates plains of central and southern Iraq. Irrigation systems have been progressively developed to distribute water to both the natural flood plains and to the higher lands. Rivers have been dammed to provide water throughout the crop growing periods, both seasonally and perennially, and to improve the reliability of the supply. Groundwater resources have also been tapped by wells, often to supplement wadi spates.

The areas with the best land do not necessarily possess the best water resources and vice versa. All over the Middle East the most favourable opportunities for irrigation development have already been implemented. For example, in Iran fertile soils of the Zagros remain unirrigated over extensive areas because of the lack of water. Where irrigation supplies are available, attempts have been made to extend agriculture onto more and more marginal soils.

Before the nineteenth century, Egyptian agriculture was restricted to winter cropping in the wake of the Nile flood. Some summer crops were irrigated from shallow wells. Widespread perennial cropping in the Delta had to await the final heightening of the old Aswan dam in the mid-1930s. This brought the total capacity for perennial irrigation to about 15 billion[†] cubic metres, a quarter of the annual mean flood. Irrigation was confined to the Nile alluvium until the 1960s when the Aswan High Dam lake was filled. This allowed over-year storage, protection from potentially disastrous flood and drought, and irrigation of the sandy and calcareous soils of the western and eastern deserts. With each successive step in controlling the Nile discharge, the margin of cultivated land has been extended.

Sandy Soils

Irrigation development (or 'land reclamation' in Egyptian parlance) beyond the narrow confines of the Nile Valley and the Delta was Egypt's top agricultural priority in the 1950s. A target was set of 1.5 million acres of new land which was to receive the additional 7.5 billion cubic metres of water provided annually by the High Dam. The major part of the area for development was in the desert to the west of the Delta. Whereas previous expansion of irrigated agriculture had been on relatively flat expanses of clay reclaimed from Delta lakes, expansion of agriculture on higher land could be achieved only by pumping from the Nile onto sandy soils of uneven topography. On the first highland scheme, South Tahrir, lined canals were

†Billion = 10^9 throughout.

constructed and water was applied using both overhead sprinkler and surface methods. Later schemes followed a variety of approaches but traditional surface methods predominated. By 1980, less than a third of the 1.5 million acres was under crops. The reasons for the shortfall are complex, but among the technical problems that of inappropriate irrigation methods was crucial. Excessive water losses on the coarse-textured soils of West Nubariya resulted in very high operating costs, poor yields and a rapid rise in the salinity of groundwater. In the schemes with surface irrigation, the water table rose more than 20 m, causing waterlogging, resalinization and crop losses. Drainage water re-entered supply canals and caused serious contamination (HTS, 1980).

The irrigation of sandy soils was pioneered in countries where water was a more scarce commodity than in Egypt, but research work on sandy soils came remarkably late. Even in Israel, for example, an ambitious research programme was initiated in 1954 to obtain water production functions for all important crops and to develop a sound irrigation technology. By 1974 about 90 field experiments on 17 crops had been conducted throughout the country. For each crop, information was presented on crop production function (yield vs. water application), optimum irrigation frequencies, water uptake distribution within the root zone, and the daily evaporation rates for the optimum treatment. Regrettably, none of the experimental soils was sandy (Shalhevet & Bielorai, 1978). The reasons for this conservatism lay in the systems of surface irrigation available at the time. But once the breakthrough was made, soils considered unirrigable have been brought into production on a large scale using drip irrigation.

The United States Bureau of Reclamation (1953) standards for classifying land suitability for irrigation, widely used in feasibility studies for irrigation development in the Middle East until the 1970s, were evolved for surface irrigation and have been rendered obsolete by the new irrigation technology. Under the old classification, sands (i.e. soils consisting mostly of coarse and fine sand and less than 10 per cent clay) were placed in class 5 and 6 (non-irrigable) because of their low water-holding capacity (about 10 per cent, compared with about 40 per cent for clays) and infertility. They were downgraded because the incorporation of organic matter (by manuring, clover, etc.) in the soil over a number of years was considered essential prior to cropping. In addition to the direct costs involved, the traditional system of reclamation postponed the start of cash-cropping. Further, the effects of manuring were shortlived, owing to the rapid breakdown of organic compounds under conditions of high temperature and alternate wetting and drying.

However, with the advent of automated, mechanized overhead and sub-surface irrigation systems, the cultivation of these soils involves no special difficulties. Indeed, because of their high hydraulic conductivity, sands are more easily leached of accumulated salts than clay soils. Nutrients, pesticides and herbicides can be delivered via the irrigation system which can be controlled mechanically according to the crop, the soil texture, and the prevailing weather conditions. Trees and vegetables can be successfully grown in soils with a high gravel content if irrigated with a drip system. The drip lines can easily be arranged to permit the irrigation of sites with finer material which is an important consideration in variable desert soils. Modern systems of land classification for irrigation potential take account of these alternative irrigation methods. For example in the feasibility study of the Wadi Mujib scheme in Jordan (Binnie & Partners, 1979), overhead (sprinkler), surface, and sub-surface (drip) systems were considered. The net return to each soil subclass was estimated and expressed as a proportion of the highest return obtainable.

Clay Soils

Much of the more successful land reclamation in Egypt has been in the north of the Nile Delta, particularly in Behera Province where Lake Mariut and Lake Idku have been reduced to about a quarter of their former size; but to the east of the Delta the northern boundary of cultivated land has changed little since the beginning of the century. Towards Lake Manzala, the saline water table lies progressively nearer the surface and the costs of pumping to maintain the water table 1.0 m below the ground surface eventually exceed the returns from cropping.

No technical revolution has occurred in the reclamation of clay soils as it has for sandy soils. In Egypt, the methods currently used differ little from those described by Willcocks and Craig (1913) when large capacity drainage pumps were first introduced for the purpose of regional drainage. Following the construction of main drainage and irrigation canals, a close network of field drains (e.g. 25 m intervals) is excavated. The land is then flooded to a depth of about 20 cm in order to leach the soluble salts from the soil profile. Because evaporation may exceed seepage, it may be necessary to periodically dry out the soil so as to encourage cracking and increase infiltration and drainage flows. At this stage rushes may be planted both as a test crop and to speed drying out and leaching. Finally rice (in summer) and berseem or barley may be planted. The whole process of leaching, test cropping, and further leaching until a reasonable harvest can be gathered can take between 3 and 6 years in the more marginal areas of the Delta. A feasibility study of such a reclamation scheme covering 3,000 ha on the shores of Lake Manzala concluded that at least nine years would be required for the US$ 14,000 per ha scheme to break even and that by normal economic criteria such a scheme was unlikely to attract funds (HTS, 1980).

Drainage of areas such as Lake Manzala is economic only if carried out on a large scale. Accordingly, plans are going ahead for the construction of the 4.4 billion cubic metre per year Al-Salaam Canal in the eastern Delta which is to irrigate *ca.* 75,000 ha west of the Suez Canal, mostly saline alkali clays, and 400,000 feddan (*ca.* 165,000 ha) in the Sinai (Government of Egypt, 1980).

Saline Soils

There are two main causes of saline soils in the Middle East. The first is upward seepage from a naturally high water table. It gives rise to the spongy salty surface known as 'sabkha' which occurs in coastal regions and in certain enclosed basins where the rainfall is insufficient to leach the salt left after evaporation. The second cause is irrigation. Either insufficient water is applied to leach the salt or too much water is applied without adequate drainage and excessive evaporation occurs from the resulting high water table. This latter phenomenon is common in Iraq and Iran where a summer fallow is usual. Salinity caused by insufficient water has been observed in orchards irrigated by drip; in Dubai a citrus orchard was destroyed by a freak rainstorm which washed into the root zone salts which had accumulated near the surface.

Gezira Clays

The Blue Nile is the only major river valley remaining in the Middle East with a large reserve of fertile, potentially irrigable land. The clay plain extends from near Ed Damazin in the south almost to Khartoum, a distance of about 400 km. At the latitude of Sennar the plain is about 200 km broad.

The clay content of the soils decreases from 70 per cent in the far south to 40 to 50 per cent in the Gezira irrigation scheme. With this high clay content the soils have a low permeability and the downward movement of water is restricted. The soils are alkaline with an exchangeable sodium percentage of 5 to 25 per cent and, in the early days of the Gezira, the low permeability of the soils gave rise to doubts about the long-term practicability of irrigation without artificial drainage. However, this has now continued for over 50 years with no evidence of deterioration due to an increase in alkalinity or salinity. The montmorillonitic clays shrink markedly when dry and the cracks so formed allow irrigation water to spread through the uppermost 50 to 60 cm of the soil, counteracting the harmful effect of high exchangeable sodium normally associated with the irrigation of clay soils.

Large-scale irrigation development began in the 1920s with the construction of Sennar Dam and the creation of the Gezira scheme. During the 1950s and 1960s expansion in the Managil and other areas took place which more than doubled the scheme's total area. This now amounts to more than 850,000 hectares. A further 80,000 hectares are supplied by pumps downstream of Sennar Dam. Further irrigation development became dependent on the provision of additional storage which was provided by the completion of Roseires Dam in 1966. Full exploitation of Roseires is being achieved with the construction of the Rahad scheme (125,000 ha), intensification in the Gezira, and other smaller projects. This was expected to bring the net cultivated area to 1.12 million ha by 1981 (Blue Nile Study Consultants, 1978), when less than 30 per cent of the basin would have been developed for agriculture. Most of the undeveloped land lies to the south and has potential for rain-fed farming but considerable areas still remain within the command of the Blue Nile. The 100,000 ha Kenana sugar scheme is developed on similar soils but irrigation water is pumped from the White Nile. Further development of irrigation on the clays of the Blue and White Nile will be dependent on the construction of new conservation works in the Nile basin.

Urban Encroachment

No discussion of land and water use in modern agriculture can ignore the loss of agricultural land to urban encroachment. This competition is particularly serious in Egypt where it has been estimated that 20,000 ha is lost from agricultural use every year (Righter, 1982). The Egyptian Master Plan for Water Resources (Government of Egypt, 1980) included a more conservative figure of 6,250 ha (15,000 feddans). A significant portion is accounted for by the mud-brick industry. Every urban community is expanding, mainly on to adjacent irrigated land. To reduce this loss the government is encouraging the siting of new towns on desert land outside the Delta, despite the exceptionally high infrastructural costs involved.

WATER RESOURCES

Generally, an increase in the application to the land of a controlled supply of good quality water is the key to raising the agricultural production of the Middle East. There are three aspects to the problem; first the potential water resource which is variable in its incidence; secondly the extent to which the water supply is regulated or stored and its transmission in the main system controlled; thirdly the efficiency with which the water is used from the point of main supply to the farm level.

Surface Sources

By far the most important source of irrigation water in the Middle East is from the rivers listed in Table 4.3. Sudan, Egypt, Syria, Iraq, and Iran all derive their major share of irrigation water from surface sources. For resource planners, rivers have three important characteristics; the flood, the average flow, and the annual and over-year variability.

The floods of the major rivers have done much to shape the human environment of the Middle East. The annual overflowing of the Nile nourished the valley and the Delta since before the dawn of civilization. The Tigris flood fostered the evolution of the city protected by its flood banks and built up as an island above the plain. Flood control was the priority in times past and bunds were

Table 4.3 Principal rivers of the Middle East

Rivers and tributaries	Recharge location	Catchment area (000 km²)	Downstream countries	Mean annual discharge before use (km³)	Water storage capacity (km³)	
White Nile	E.C. African uplands and S. Sudan		Sudan	12.56	3.5	Jebel Auliya
Blue Nile	Ethiopian Highlands		Sudan	52.92	0.9	Sennar
					3.0	Roseires
Main Nile	as above	3,100	Sudan	89.0	1.2	Khashm el Girba (Atbara)
			Egypt		5.3	Old Aswan
					164.0	Aswan High Dam
Euphrates	Mountains of S. Turkey and N. Syria	350	Turkey	24.0	25.0	Keban Dam
			Syria	31.82	11.9	Tabqa Dam
			Iraq			
Tigris	Mountains of S. Turkey and N.E. Iraq	435	Turkey	44.4	7.5	Dokan (Lesser Zab)
			Syria		5.0	Darbandikhan (Diyala)
			Iraq			
Karun	Central Zagros, Khuzestan, Iran		Iran	22.0		

raised, first around the settlements and then along vast reaches of the river bank. In the Nile Valley bunds were constructed to impound flood water in large basins prior to cultivation. This self-regulatory system replenished nutrients and washed away salt. In contrast, the Euphrates and the Tigris flooded when crops were in the ground and massive diking operations had to be organized to protect them. In both the Nile valley and Mesopotamia, the danger of flood precluded settlement on the flood plain. An unfortunate side effect of river regulation has been the encroachment of settlement on fertile agricultural land.

The amount of a river's flow that can be extracted depends on its variability. The best streams are those which discharge throughout the year and provide a usable base flow. To irrigate the land, man has devised many ingenious ways of lifting water from the river bed. The huge water wheels of the Euphrates and the Orantes in Syria are examples, but for large-scale diversion of water for irrigation the weir or barrage has been used for thousands of years to raise the level of water. In recent times, this function has been combined with large-scale flood control and river regulation. The Sammara barrage on the Tigris was built to divert the flood peak flow into the Tartar basin to protect Baghdad from floods such as those of 1954. In later years, the Sammara barrage has been used to divert irrigation water to the Ishaqi area.

To exploit a river fully, the whole of its flood must be saved, but this is not possible without constructing a reservoir that is significantly larger than the average flow. However, regulatory reservoirs lose water by evaporation. Therefore there is a trade-off between saving water from flowing to waste and losing water from evaporation. Such a problem confronted the designers of the Aswan High Dam. The mean annual discharge of the Nile at Aswan (1900 to 1959) was estimated at 84 billion cubic metres. This mean concealed annual variations of up to 30 per cent which annually exposed the downstream areas to threat of flood or drought. A guaranteed and predictable supply during the summer months was judged to outweigh evaporation losses of about 10 billion cubic metres from the 300 km long reservoir in the Nubian Desert (Agreement for the Full Utilisation of the Nile Waters, 1959). In practice the evaporation has been estimated to exceed 13 billion cubic metres (Howell, 1984), a rate of about 3,000 mm per year from the surface of the lake.

The quality of river water in an arid country deteriorates as it passes downstream. It is contaminated by irrigation drainage and natural seepage. The concentration of dissolved solids in the Blue Nile at Roseires in Sudan is about 150 parts per million (ppm). By Cairo it has increased to about 300 ppm. When this water is used to irrigate cotton in the Delta, some 70 per cent evaporates, raising the concentration of salts in the drainage water to about 1,000 ppm. Re-use of this water for the same purpose raises the salinity to 3,300 ppm at which stage it is of little further use for agriculture. The Egyptian Master Plan for Water Resources (Government of Egypt, 1980) estimated that drains discharge 16 billion cubic metres of Nile Water into the sea. It is planned to capture and re-use 5.4 billion cubic metres of this by diluting it with fresh canal water. The remainder must flow to the sea to carry away unwanted salt. The Al-Salaam project involves the recycling of drainage water, which is the cheapest way for Egypt to augment her water resources.

Disposal of saline drainage water into rivers, canals and, to some extent into natural depressions in the desert are short-term solutions, as in the Tigris Euphrates plain. In the lower reaches of the Euphrates, salinity doubled in the period 1970 to 1979 and it is now not possible to use the river as a source of drinking water. The long term solution is to stop all but a very limited amount of saline drainage water from being discharged into the river and to construct an outfall drain which would collect the drainage water from some 1.4 million ha of irrigated land and discharge it directly into the Shatt el Basra (Robson *et al.*, 1984).

Groundwater Sources

In the Middle East groundwater is of utmost importance because of high variability of the climate and high evaporation rates. Israel's annual water potential is estimated to be 1.7 billion cubic metres, of which 60 per cent comes from groundwater, with the Jordan River, storm runoff, and reclaimed sewage making up the remainder (International Irrigation Information Centre, 1978). In Iran the surface flows have been estimated to average about 82 billion cubic metres per year and allowable groundwater extraction 20 billion cubic metres per year (Bookers, 1975), but the distinction between surface water and groundwater is not always clear. The same river that is drawing on groundwater to maintain its flow long after the flood has passed away may be losing water by seepage in another reach. Because of the nature of groundwater flow, increase in any extraction inevitably involves a lowering of the water table and a depletion of storage. This can be an advantage where lowering of the water table can induce additional recharge to groundwater of river flows which would have gone unused otherwise or reduce groundwater flowing to waste.

Irrigation from wells in the Middle East dates from earliest times (Beaumont *et al.*, 1976). Nowadays the equipment most favoured by the small farmer is the slow-speed diesel motor with a shaft-driven turbine pump delivering 10 to 20 litres per second. Where the water table is not more than 20 metres deep, this equipment is used in a simple hand-dug, lined, shallow well to serve up to 5 ha. In the Arabian Peninsula, where flash floods are common, many ephemeral streams (wadis) are exploited in this way. The occasional torrent infiltrates rapidly into the stream bed and the farmers use the stored water to carry them over dry periods. In 1974, the Wadi Najran had more than 300 wells with pumps for about 1,200 ha of cultivation. The average farm size was about 1.5 ha. Some wells were shared (Binnie & Partners, 1974).

Nowadays, the tubewell can exploit groundwater much more efficiently because it penetrates far below the water table. The depth from which it is profitable to pump depends on the cost of inputs, including fuel oil, and the value of outputs. Some feasibility studies of groundwater development for irrigation have concluded that water is too deep for economic crop production, especially in areas remote from markets. However, the ease with which wells can be constructed in shallow aquifers has led to groundwater being depleted with serious social and financial effects. Groundwater can be mined like any other mineral, which is acceptable provided the dependent community recognizes the temporary basis of its wealth, but this is rare. In the 1960s, Mashad in northeast Iran was a city of pilgrimage. The market for fruit and vegetables was good and many local landowners put down wells into the shallow aquifer. Recharge was substantially less than extraction for irrigation and by the end of the decade, many wells were drying up. Those farmers who were unable to invest in a deeper well were forced to withdraw.

Water-bearing strata are not infrequently a source of salt, which can lead to high mineralization of groundwater and affect the quality of surface streams, hence causing serious salinity of the soil. Soil salinity is generally defined in terms of the concentration of salt in the saturation extract. Russell (1973) suggests the following limits: at 3,000 ppm no crop is likely to suffer; above 5,000 ppm only salt-tolerant crops will grow and above 10,000 ppm no crop will be economic. However, the salt concentration in the moisture in the soil profile is constantly changing as the soil is wetted and dried. As the roots extract moisture, the concentration increases. For this reason the safe limits for salt levels in irrigation water are far lower than those set for soils. Table 4.4 gives water salinities for salt-tolerant crops on coarse-textured soils as suggested by one study in Saudi Arabia, but it should be noted that the US Department of Agriculture lists all water with an electrical conductivity of 2,250 micromhos/cm as having a very high salinity hazard, to be used only under special circumstances. Water of extraordinarily high salinity (e.g. 5,000 ppm TDS) can be tolerated if it is used in large quantities and on very permeable soils.

Table 4.4   Groundwater quality classification for irrigation of salt-tolerant crops[a]

Salinity/water quality	ppm	Electrical conductivity at 25°C (Micromhos/cm)[b]
Satisfactory	<1,500	<2,300
Marginal	1,500–2,500	2,300–3,500
Poor	>2,500	>3,500

Source: Umm Er Radhuma Study, Saudi Arabia (G.D.C. Ltd. 1980).
Notes: [a]In rough order of decreasing tolerance: date, barley, wheat, ryegrass, fodder sorghum, alfalfa, onion, tomato, potato, carrot, grape, melon.
[b]Approximate EC values; for guidance only.

Other Sources

In the absence of surface and underground supplies of fresh water, desalinization of sea water for irrigation has been investigated, but the technology has not reached the stage where sea water offers a practical alternative to conventional sources of water supply except in special circumstances. Perhaps the best-known example is on Sadiat Island in Abu Dhabi where fresh water is derived from the sea by flash distillation. Vegetables are grown in plastic greenhouses in a controlled climate that reduces the transpiration to 2 mm per day. This is achieved by blowing air into the greenhouses through screens wetted with sea water. The crop is grown in sand and the distilled water is dosed with fertilizer and fed through a drip system.

Numerous proposals have been made for shipment of supplies to water-deficient areas for irrigation, but none of them have so far proved practicable on any scale. Probably the most promising proposal has emerged from the necessity to use the former oil terminal facilities at Milford Haven in Wales, which is encouraging oil tankers delivering oil in Northern Europe to return with fresh water from Wales rather than take on sea water as ballast.

Urban waste water may be recovered for agricultural use after treatment provided that it is free of heavy metals and toxic substances. Because of the risk to workers, disease transmission by flies and direct contamination of food crops, raw sewage is not suitable for irrigation, but subject to ethical objections, treated effluent can be safely used for certain crops, e.g. cotton, cereals and forest plantations such as the Khartoum shelter belt. In the plans for constructing the main sewage system of Cairo, consideration is being given to the re-use of treated effluent for the irrigation of desert soils east of the city. Until now, the mostly untreated effluent has flowed in a 200 km drain to Lake Manzala, a large, shallow lagoon which is rich in fish close to the outfall. The costs of waste water treatment are not competitive with the cost of developing other incremental supplies (e.g. agricultural drainage), but in Egypt the costs of waste water treatment are of the same order of magnitude as the economic or shadow price of water for agriculture, which suggests that treatment will be economically feasible once other incremental supplies are exhausted (Government of Egypt, 1980).

MANAGEMENT OF WATER RESOURCES

Most governments take the view that measures are needed to achieve efficient allocation of water among users. The land area suitable for irrigation is greater than that which the total resources will be able to supply and there is a high degree of interdependence between users. Firm control must be maintained through an adequate water law and a pricing system for different uses which reflects their true profitability. Further, where a river or aquifer impinges on two sovereign states, the use of the common resource should be coordinated. In practice, however, few governments have been able to effectively implement such a policy. Probably only Israel has achieved a full measure of control over water use and then only within the confines of the state. The Jordan waters are also used by Syria and Jordan. In the 1950s attempts to reach agreement on joint use failed.

International Competition

Egypt and Sudan have honoured the 1959 agreement for 25 years. This defined the main stream discharge as 84 billion cubic metres of which 55.5 were allocated to Egypt and 18.5 to Sudan. The remaining 10 were allocated to evaporation and seepage from the High Dam reservoir. Waterbury (1979) has predicted that the 1959 agreement will not stand up to the impending water crisis occasioned by the combination of rising population and rising demand for water in both countries during the 1980s. However, ambitious plans for new schemes are unlikely to be implemented in the short term and the scope for more intensive use of the available resources is probably underestimated. Both countries are likely to pursue their joint interests by upholding the 1959 agreement and by a united front should Ethiopia at the head of the Blue Nile and the countries surrounding Lake Victoria on the White Nile insist on being parties to the agreement (Adams, 1983).

Other riparian states in the Middle East have been less successful in obtaining agreement on coordinated use. Abstraction from the Euphrates and the Tigris is estimated to exceed 60 per cent of the annual discharge and an agreement for the sharing of the river is long overdue. In years of low flood, the discharge has been known to fall from an average of 84.5 to 25 billion cubic metres. So far, hydro-power development by Turkey on the Euphrates has been to the advantage of downstream states because it delays the flood and releases it in the dry season; but proposals for irrigation within Turkey are being considered. Syria is also in direct competition with Iraq for irrigation water from the Euphrates. Iraq is responding by building its own regulating reservoirs, but the evaporation from these reduces the total water supply and water quality.

Competing Demands for Water

Most if not all states have made an inventory of existing and potential water supplies and have attempted to plan their allocation between different sectors: domestic, industrial, agricultural, power generation, and navigation. The bases for the estimation of supply and demand are often uncertain. Supply projections may relate to a brief series of discharge readings. Records of water diverted for irrigation, irrigated area, return flows, etc. are usually shaky and detailed computations of actual water use can rarely be made. Where there are competing needs, the official data are often a source of controversy. Different ministries, for example Irrigation and Reclamation in Egypt, may publish different data. As in international negotiations, existing users claim priority over potential users and the former tend to inflate their requirements to improve their negotiating position.

It is nonetheless possible to draw some general conclusions regarding the relative proportions allocated to different sectors and the priority governments normally accord them. Demand for domestic and industrial water is relatively price-elastic and urban users are able to outbid farmers for scarce supplies; irrigation, by comparison, requires huge quantities. In Israel the water quota for municipal authorities is 80 cubic metres per person per year, about one per cent of the amount required to irrigate a hectare of cotton over a growing season. Cotton requires 8,000 tonnes of water per tonne of lint, wheat about 1,500 tonnes per tonne of grain.

Fish farming can take two forms: extensive pond and intensive tank production. In the former, water is consumed by evaporation from the open-water surface and biomass yields per unit of fresh water cannot compete economically with intensive irrigation. In Egypt, future ponds will be confined to the saline-alkali soils in the low-lying coastal areas and will use saline drain water which would otherwise have passed to the sea. Similarly opportunities exist in the marshes of the lower Tigris and Euphrates. Intensive tank production, on the other hand, gives very much higher yields per unit of water surface and the water can be re-used for irrigation after treatment.

In Sudan and Egypt, power generation competes with crops for Nile water. Power generation itself does not consume water, but the fairly constant need

Table 4.5   Water use in selected countries of the Middle East (km³ per year)

Country	Agriculture	(%)	Domestic	Industry	Navigation and spills at Idfina	Evaporation etc.
Egypt[a]	45.4[b]	(81)	1.8[c]	0.3[c]	3.8	2.7
Saudi Arabia	2.33	(58)	0.83	0.15		
UAE	0.331	(91)	0.031			
Kuwait	1.15	(39)	1.73	0.05		
Bahrain	0.166	(83)	0.020	0.013		
Jordan	0.375	(97)	0.004	0.006		
Lebanon	0.647	(87)	0.94			
Syria	6.0	(93)	0.4			

Source: All countries with exception of Egypt based on Gischler (1979); Egyptian data from Water Master Plan (Government of Egypt, 1980).
Notes: [a] 1980 demand.
          [b] Includes drainage to Mediterranean of 15.0 km³ but excludes some releases for hydro-electric generation and navigation.
          [c] Net consumption; i.e. excludes drainage water re-utilized by agriculture.

for water to drive the turbines does not match the wide variation in the seasonal water requirements of crops downstream. In January and February, for example, irrigation and other demands require a flow much less than the turbines at Aswan can use effectively. At such times the need to satisfy load requirements leads to water surplus that must be spilled into the Mediterranean at Idfina. However, the indirect benefits from power generation may be substantial for, without the potential benefits from the sale of electricity, the cost of dam construction solely for irrigation can be prohibitive. Also, hydro-electric power often provides a relatively cheap source of energy for lifting water for irrigation and drainage.

Table 4.5 shows that despite these competing demands, many countries in the Middle East, allocate more than 80 per cent of their water supply to agriculture. These proportions are not expected to change very greatly in the foreseeable future since there is scope for achieving increased efficiency in both urban and agricultural use.

Government Control over Water Use

The control exerted by Middle Eastern governments over existing and potential irrigation supplies varies from country to country. In Egypt all farmers are ultimately dependent on a single source of water controlled at Aswan. The organization of water distribution is highly centralized in the Ministry of Irrigation. In Israel the sources of supply are more dispersed, but an all-inclusive water law provides for state ownership of all water sources and for centralized planning of its distribution and control. A single, interconnected grid distributes supplies according to a national plan, which provides for the complementary use of groundwater and surface water.

Management of a watershed may require the integrated development of surface water and groundwater. Groundwater can supplement the irrigation supply obtained from surface water during the dry season and surplus surface water can be used to replenish the aquifer. Underground storage eliminates evaporation losses and can substantially reduce the capital and maintenance costs of surface storage. This joint use of surface and underground sources requires that they are controlled by the same authority.

Some form of state control and monitoring of groundwater is essential. The regulation of the number of wells by licensing is normally the means of control, but this is not always effective as experience at Wadi Dhuleil in Jordan shows (Carruthers & Clayton, 1975). In Qatar, over-pumping of the Umm er Radhuma aquifer is causing deterioration by salt water intrusion. In order to ensure that the water table is stabilized at a particular level, it is proposed that groundwater extraction for agriculture is rationalized and controlled and that the groundwater balance is maintained by artificial recharge with sea-water distillate using natural gas as a source of energy in the distillation process (Gischler, 1979).

Another area for state intervention may be the construction and maintenance of an effective regional drainage system which will remove saline water from the root zone, lower the water table so as to prevent resalinization when the soil is fallow, and ensure that drainage water does not enter irrigation canals. The successful implementation of a regional drainage programme usually demands large capital sums and strong government intervention since construction can disrupt cropping in the short term and take some land out of production altogether. The largest regional drainage project in the Middle East commenced in 1969 with the installation of tile drains in the Nile Valley and Delta with IBRD assistance. By the year 2000 about 1.6 million hectares are expected to benefit. The rising water table and salinity in the Old Lands followed the widespread replacement of basin irrigation by perennial irrigation made possible by the construction of barrages across the Nile, a process which began in the 1860s with the Delta Barrage and culminated with the Aswan High Dam a century later.

The overall objective of water management is the efficient and equitable use of water resources and the disposal of drainage water. Efficiency and equity are not necessarily compatible and careful balances must be struck between the two. Crop consumptive use is potentially much higher in Upper Egypt than in the Delta for climatic reasons, but it would be unacceptable to deny supplies to farmers in Upper Egypt on grounds of economic efficiency. 'Irrigation efficiency' is normally defined as the proportion of water diverted from the source (e.g. a barrage or well) which is used by the crop. It recognizes that losses occur due to evaporation and seepage from canals and spillage into drains.

Table 4.6 Water use efficiencies in Egypt

Field efficiency					
Irrigation method	Surface			Sprinkler	Drip
Soil type	Sand	Clay		Sand	Sand
Water used by plant	0.55	0.65		0.7	0.9
Farm to field					
Irrigation method	Unlined			Lined	Piped
Situation[a]	Highland	Lowland		Highland	Highland
Soil type	Sand	Sand	Clay	Sand	Sand
Water reaching field	0.8	0.8	0.85	0.9	0.9
Main canal to farm					
	0.8	0.8	0.85	0.9	0.9
	0.35	0.42	0.47	0.57	0.73

Source: Hunting Technical Services (1980).
Notes: [a]Highland = pumped supplies, gravity drainage, New Lands. Lowland = gravity supplies, pumped drainage, Old Lands.

Water losses in the field result from evaporation and percolation beyond the root zone. Estimates of irrigation efficiency obtained in Egypt are shown in Table 4.6. The amount of water needed by a crop for evapotranspiration varies by crop type, season and geographical location and may be calculated according to standard formulae, but the estimation of water duty must take into account more nebulous items: the crop mix, cropping intensity, the proportion of land cropped, and efficiency factors at different levels. In Egypt, the official water duty for surface irrigation in the Old Lands is $12,000 \, m^3/ha$, but the actual duty probably exceeds $17,000 \, m^3/ha$ (Waterbury, 1979). In Israel, where there is supplementary rainfall and where most of the water is delivered through sprinkler or drip systems, the average duty is $6,000 \, m^3/ha$. Such comparisons serve only to demonstrate the wide divergence between different agricultural systems. The adoption of capital-intensive irrigation technology would be inappropriate in densely settled, traditionally irrigated, lands of the Middle East where substitution of capital for labour would be difficult to justify on economic and social grounds.

It has been customary in Egypt, as elsewhere, to put the blame for poor water management on farmers, despite the fact that canals are normally designed to maintain a water level that requires them to lift water to their fields using a saqia or Archimedes screw. Yet, the main causes of wastage are found not on farms, but in the operation of main canals and watercourses (Bottrall, 1980). Wastage results from failure to regulate discharge as well as from the physical characters of the system.

There is a growing body of evidence that the improvement of existing irrigated infrastructure and its management offers the most economic means of increasing agricultural production in the water-scarce countries. A pre-condition for this is an understanding of farmers' management practices at field and watercourse levels as well as the operation of the main water distribution system. Investment in improved operation and maintenance is likely to be more cost-effective than investment in major civil engineering works such as the Jonglei Canal on the White Nile or large-scale reclamation and settlement projects in sandy deserts using sophisticated irrigation equipment. Such projects are highly dependent on factors beyond the control of individual farmers (e.g. pumps, equipment, spare parts, energy supplies, and other farm inputs controlled by officials) the failure of which can result in rapid desiccation of the crop. Such equipment is better suited to small-scale irrigation of high-value cash crops by entrepreneurs who retain control over their water and energy supply.

Any major improvement in standards of operation and maintenance in the major riparian states of the Middle East, who account for over 90 per cent of the irrigated land, is unlikely to be achieved in the short term. Efficient operation of irrigation systems is as much a consequence as it is a cause of economic development. Nor are prospects good for international management of resources. In the present climate, more urgent issues dominate the attention of political leaders. Nation states are likely to continue to insist upon their sovereign rights to exploit trans-national rivers to their own advantage.

REFERENCES

Adams, M. E. (1982) 'The Baggara problem. The future of nomads in Africa', *Development and Change*, **13** 259–289.
Adams, M. E. (1983) 'Nile water; a crisis postponed?', *Economic Development and Cultural Change*, **31** 639–643.
Beaumont, P., Blake, G. H., & Wagstaff, J. M. (1976) *The Middle East*, Wiley, London.
Binnie & Partners (1974) *Wadi Najran Water Development Study*.
Binnie & Partners (1979) *Wadi Mujib Irrigation Scheme Feasibility Study*.
Blue Nile Study Consultants (1978) *Blue Nile Waters Study*.
Bookers Agricultural & Technical Services Ltd & Hunting Technical Services Ltd (1975) *National Cropping Plan*.
Bottrall, A. F. (1980) *Irrigation in Egypt Past and Present*, Network paper 1/80/3, Overseas Development Institute, London.
Carruthers, I., & Clayton, E. S. (1975) *Ex-post Evaluation in the Agricultural Sector—A Case Study of a Groundwater Project in Jordan*, mimeo, Overseas Development Institute, London.
Gischler, C. (1979) *Water Resources in the Arab Middle East and North Africa* Middle East and North African Studies Press, Wisbech, Cambridge.
Government of Egypt (1980) *Master Plan for Water Resources Development and Use* UNDP/EGY/73/024. Ministry of Irrigation & Sudanese Affairs.
G. D. C. Ltd (1980) *Umm Er Radhuma Study*.
Howell, P. P. (1984) *The Jonglei Project* public lecture, Cambridge University African Studies Centre, 1 November, 1984, quoting John Waterbury.
HTS (Hunting Technical Services Ltd) (1980) *Suez Canal Region Integrated Development Study*.
Righter, R. (1982) 'Muddy hands are spoiling Egypt's dream of plenty', *Sunday Times*, 3 January 1982.
Robson, J. F., Stoner, R. F., & Perry, J. H. (1984) 'Disposal of drainage water from irrigated alluvial plains', *Water Science and Technology*, **16**, 41–55.
Russell, E. W. (1973) *Soil Conditions and Plant Growth*, Longman, London.
Shalhevet, J., and Bielorai, H. (1978) 'Crop water requirement in relation to climate and soil', *Soil Science*, **125**, 4, 240–247.
United States Bureau of Reclamation (1953) *Bureau of Reclamation Manual*, Vol. 5, *Irrigated Land Use*, Part 2, *Land Classification*, Dept. of the Interior, Washington.
Waterbury, J. (1979) *The Hydropolitics of the Nile Valley*, University of Syracuse, Syracuse.
Willcocks, W., and Craig, J. I. (1913) *Egyptian Irrigation*, Spon, London.

Agricultural Development in the Middle East
Edited by P. Beaumont and K. McLachlan
© 1985 John Wiley & Sons Ltd

Chapter 5

Livestock in the Middle East

CHARLES GURDON

INTRODUCTION

This chapter is concerned with the production and trade of livestock in the Middle East. The area under examination is the Arabian Peninsula, the Levant excluding Israel, Iran, Egypt, and Sudan, and Turkey which is by far the largest producer. The development of the livestock sector of the economies of these countries is analysed from 1965 to the present day. By doing so it is possible to see long-term trends in both the production and trade of all livestock products. It should be borne in mind that the statistics are not entirely accurate. While the trade figures can easily be assessed, the production statistics of live animals were little more than guesswork until recently. An illustration of this is that a livestock census of Sudan in the mid-1970s discovered that there were about two million additional animals than had previously been thought, even once growth rates had been accounted for (Watson, 1975). It is for this reason that the statistics appear to be so erratic, particularly in those countries with large proportions of nomadic herds.

This review will attempt to show trends which have occurred in the Middle East in the past twenty years. They include the dramatic rise in the level of production of dairy and poultry products, particularly in the richer oil-exporting countries of the Gulf. At the same time, the production of beef and mutton has not kept pace with demand. This has led to a substantial rise in the level of imports of both live animals and frozen meat to all but a few countries. Since countries such as Iran and Syria, which used to be major regional exporters, today have to import a large proportion of their livestock, only Turkey and Sudan are net exporters. This chapter will also examine the way in which both net importers and exporters are now seeking to develop their livestock sector and to what extent they have been successful.

TRADITIONAL SYSTEMS OF LIVESTOCK PRODUCTION

Livestock have always been an integral part of life in the Middle East where there are two distinct systems of production—nomadic pastoralism and sedentary farming. The largest sources of animals in the region are the herds

85

of the nomadic tribes which use the natural rangelands outside the cropped area for animal feeding. Because of the nomadic way of life of the herding groups this major rangeland resource has not been developed for other, possibly more beneficial uses. Although efforts are being made to settle the nomads of the Middle East, they remain largely outside the day-to-day control of their respective governments and can appear to contribute little to national economies. By contrast, the sedentary livestock farmers are usually situated near major urban centres and production is often oriented towards domestic and export markets. There are increasing numbers of specialist livestock farms being set up in the region, but the main suppliers of meat and other animal products are still the nomadic and sedentary farmers in the traditional sectors. Many of the settled farmers perceive their livestock to be secondary sources of income to field cropping, though this is far from universal. Haseeb (1964) noted that approximately half of all value added in agriculture came from livestock. Given the strong predilection of people in the Middle East for local mutton in preference to imports, it is likely that livestock rearing remains a principal interest of farmers in many parts of the region.

There has grown up a strong belief that the nomadic and pastoral livestock traditions have evolved separately, a notion fostered by analyses that emphasize the dichotomous nature of tribal and village societies (Braudel, 1972). In reality, the cleavages between the nomadic pastoralists and the settled population are less wide than imagined. There is often a close economic and social linkage. Tribal chiefs can be large land owners in village lands, as were the Bakhtiari *khans* in Iran until recent years. Villagers in Western Afghanistan have for long been dependent on acquiring animals from passing nomadic groups as a means of providing meat during the winter and of replenishing their herds in the spring. Likewise, herding groups have found it to their advantage to be associated with villages on their migration routes to ensure access to pasture. Many nomads rely on work in the harvest field to earn cash or cereals, while their presence at the peak in demand for agricultural labour is an important aid to the farmers (Ercon, 1974). In a number of Arab-occupied areas, however, conflict between the two groups is not infrequent and integration is weakly developed (Thesiger, 1965).

Nomadic Pastoralism

The size of the nomadic population has always been difficult to estimate with certainty. Definitions of the nomad tend to vary between sources, depending on whether semi-nomads and transhumant peoples are taken into account or not. The very mobility of the true nomad makes assessment of both human and livestock population numbers an unreliable business. At the beginning of the 1970s, it was suggested that approximately 1.4 per cent of all the peoples of the region were nomads (Beaumont *et al.*, 1976). In all, it was thought that there were some 2.2 million nomads at that time, with the largest absolute total, 600,000, in Turkey. Saudi Arabia was seen as second in ranking, reporting approximately 400,000 followed by Iraq and Iran, both with 300,000. Only Saudi Arabia and the peninsula states had more than 3.5 per cent of their total populations in nomadism and many countries had small numbers involved such as Egypt, where 50,000 nomads made up a mere 0.1 per cent of the population as a whole.

The 1970s estimates noted above exclude Sudan. Yet Sudan is a particularly interesting case, where two distinct forms of pastoralism survive. The northern desert and savanna regions contain as many as two million nomadic tribesmen, who utilize the area's limited pasture resources in sheep, goat, and camel herding (Gurdon, 1984). A further million transhumant cattle owners also move into this region to take advantage of seasonal availability of pasture (Lebon, 1965). The transhumant herders come from the nilotic Dinka and Nuer tribes of southern Sudan (Evans-Pritchard, 1940), although there are some Arab tribes that are cattle-owning semi-nomads to be found circulating in this area. The former follow the rains northwards to new pastures in the savanna lands, later returning to their home territories around the Sudd swamps, where they live out the dry season.

Reasons for the success of nomadism on so large a scale in Sudan are not hard to find. Sudan is comparatively rich in water and pasture to a degree unmatched elsewhere. It is also favoured by the sheer scale of the country, with 2.5 million square kilometres of land within national frontiers. Land rights have been vested in the state rather than private hands until recent years, inhibiting the division of pasturelands. The fruit of the unique situation in Sudan is widespread cattle herding, seen in few localities elsewhere in the Middle East, though sheep and goats, the basic constituents of animal herds throughout the region, are also present.

Despite the efforts of several of the national governments of the Middle East (Tapper, 1983; FAO & Government of Libya 1967), many nomadic groups have remained outside effective state control. Constant movement between pastures and a certain element of economic self-sufficiency has made them elusive of the machinery of government. Nomadic herdsmen use their flocks to provide milk and meat as basic foods and also can convert wool, hides, and skins into household utensils such as rugs for sale to outsiders. There are severe limitations on the amount of goods than can be carried by migrants and the acquisition of consumer goods is kept to a minimum, even when they are available. For many nomadic groups their only contact with government officials is when they are pressed into having their animals inoculated under disease-control programmes or when they use government-controlled well and water reservoirs during the dry season.

There are many reasons why governments seek to settle their nomadic populations. First, the nomadic herds constitute in many cases the most important source of livestock production within the domestic economy. There are attractions for governments to bring nomadic herds within the commercial sector to save on imports of meat and products from abroad, especially since the region as a whole has undergone a profound increase in urban populations during recent decades. Efforts to draw pastoral herders within the monetized economy have been hindered by social as much as technical considerations. Nomads are often unwilling to sell their animals which are regarded as forms of wealth on the hoof that reflect personal, family, or tribal prestige. It can be the number of animals rather than the quality of each animal that dictates a man's standing among his peers, a factor that militates against the effective conversion of nomadic herds into useful instruments for solving problems of urban food supply, even when nomads are prepared to release them for purely commercial considerations.

Other social attributes of livestock holdings are their utility as currency within dowry or similar negotiations for weddings in nomadic communities (Evans -Pritchard, 1940). For a number of social reasons, therefore, reinforced by practical factors such as a need to keep high animal numbers in order to survive drought, many nomads are reluctant to dispose of livestock except when compelled to do so to pay for water, veterinary services, goods from urban markets, or other vital items.

A significant but possibly overstated motivation for governments to attempt sedentarization of their nomadic communities is that these highly mobile peoples are extremely difficult to tax. Once settled, they are more vulnerable to inclusion in the tax system. Since livestock holdings are often valuable and represent an appreciable proportion of national wealth in some states without oil, it seems an acceptable target for government revenue raising. Cattle, for example, can each be worth up to $450, camels about $350 and sheep upwards of $115 equivalent (Bank of Sudan, 1983). Herd numbers averaging 100 head are sizeable assets in the poorer states of the region and seem, at least in theory, to be potential sources of tax revenues. There are one or two countries still where low income from customs and excise duties could be augmented by taxes levied on livestock. It must be doubted, however, whether creation of settlement schemes for nomadic peoples will assist this objective, since the first casualty of such projects has inevitably to be herds, the size of which cannot be accommodated in an essentially field cropping system and which are better liquidated in order to ensure the continuing settlement of the nomads (Fabietti, 1982). The outcome of many settlement schemes for pastoral nomads has been diminution of the livestock holding leading to a reduced taxable asset and, more importantly, a poorer supply base for animal food products.

The nomadic pastoralist way of life has been largely misunderstood and its economic wishes are not appreciated adequately. Throughout large areas of the Middle East, the land is unsuitable for producing crops and can only be utilized efficiently by animal herding. Nomadism is a special response to harsh environments and should not be considered as primitive or inferior to other systems of agriculture. In some areas there are serious problems of over-grazing which have been caused by a combination of government action and traditional attitudes towards livestock. In trying to develop the livestock sector and assist their nomadic populations, governments of the region have all undertaken programmes of water well construction. This has enabled the herds to find sufficient water throughout the year and has greatly reduced the animal mortality rate. Large areas around newly installed government wells have been over-grazed and range deterioration has become an increasing problem. Some of the richer countries are now also supplying herders with heavily subsidized fodder at wells against payment in animals. This enables increasing urban demand to be met in part. Access to new supplies of water and government encouragement to nomads to sell their animals has perpetuated the emphasis on quantity rather than quality in livestock production. The result has been to increase the animal population in the past twenty years without necessarily improving its quality. It has also severely damaged the already fragile environment as over-grazing has exacerbated trends towards desertification. Unless measures are taken to conserve and improve the rangelands and to change the emphasis of livestock production from quantity to quality, the nomadic pastoral system as constituted at present will eventually be insupportable. Alternatively, it could lead to the development of an expanding and valuable sector of the economy, reducing the need for imports in some countries and greatly expanding exports in others.

Sedentary Livestock Farming

The nomadic pastoral tradition of the Middle East is long but is far from being exclusively important at the present time. Livestock holdings of the sedentary agriculturalists have tended to become increasingly important in recent decades. In the settled sector, animal husbandry is practised either within or on the fringes of the cultivated zone as an activity complementary to that of crop production. Arable farming and livestock rearing can be separate occupations. Professional shepherds look after the livestock while the owners of the animals concentrate on their interests in the village or single tribal groups can divide, one following the herds and the other sedentary occupations as in the case of the Shammar tribe of Saudi Arabia.

This is not the norm, however. Livestock are generally worked from the individual farmstead and taken to their pastures each day or looked after by women or children as they scavenge in and around the village. Most holdings are small and mixed, comprising draught animals in the poor communities of, for example, Egypt, together with sheep and goats to provide milk, meat and other products for the household. Fodder growing can be an extremely important part of the crop rotation, including berseem, clovers, alfalfa, and barley produced entirely to feed animals kept on farm. Libya is typical of many countries, where the largest crop by area cultivated is fodder (Allan *et al.*, 1973).

In recent years the livestock resources of the Middle East have come under considerable strain. On the one hand, numbers engaged in nomadism have tended to fall as alternative occupations have opened up in the expanding oil economies of the region. The point is well made that nomads, despite the social and psychological advantages of the herding life, will always move to more comfortable and rewarding livelihoods when the opportunity arises (Fabietti, 1982). In the Arabian peninsula especially, this inevitably implies the demise of nomadic pastoralism (Birks, 1981). On the other hand, there has been a veritable explosion in human population numbers (Hill, 1981), with growth to a large extent concentrated in urban areas (Clarke, 1981). The combined effects of poor performance in production of animal food products and the much enhanced levels of domestic demand for them has opened up a gap that has been filled by imports in the richer states and by forms of rationing in those unable to fund imports. A further response has been for governments to foster development of large-scale commercial livestock enterprises, an activity examined in more detail later in this chapter.

Sheep and Goats

Sheep are the most numerous of farm animals in the Middle East and are raised in most indigenous systems of agriculture. With the exception of the better-watered Sudan, sheep are more suited than cattle to the arid conditions of the region. In addition, they produce milk and wool as well as a conveniently small carcass, which is important in a hot climate. Mutton is also the favoured meat for celebrations and feasts on Muslim holy days and during the fasting month of Ramadan. Indeed, the import of live animals for Muslim pilgrims at the Haj in Saudi Arabia is a very important element in livestock trade to the Middle East. Goats are usually kept with flocks, particularly in very marginal areas where they thrive better than sheep. They are also more environmentally destructive as their close cropping of all green plant life leads to greater vulnerability of soils to erosion in many areas of the Middle East. Goat hair, which is water-resistant, is favoured for making cloth for nomad tents.

Cattle

With the exception of Sudan, where pastoralism predominates, cattle tend to be kept in the cultivated areas because rangeland food and water supply is generally inadequate. Cattle tend to be used for milk rather than beef since mutton and lamb are far more popular as meat. They usually rely on stubble pasture from cereal crops and on uncultivated land, although fodder growing is becoming more widespread. The newer commercial dairy farms near urban areas sometimes stall-feed their animals. Because of their higher value than sheep and goats, cattle are more likely to be vaccinated against diseases, although tsetse fly and rinderpest remain problems in Sudan.

Draught Animals

As mechanization becomes more widespread in the rural areas the number of draught animals is declining. Asses are the most common draught animal in the Middle East because they are hardy and more suited to the climate of the region than mules or horses. The cost of keeping horses limits their normal use to that of riding and racing, although they are used as beasts of burden in cities such as Cairo and Istanbul. Camels are primarily kept by nomadic pastoralists for their milk and sometimes their meat. Their role in long-distance travel has largely been replaced by mechanized transport except in the poorer countries where fuel is expensive and roads impassable. In the oil-rich Gulf states large stables of camels are kept for both prestige and racing. While the single-humped Arabian camel is most common in the region, the two-humped Bactrian species, which is more tolerant of the cold winters, is found in Iran and Turkey.

Poultry

Traditionally poultry production was limited to the laying hens kept by farming families for their own use. Surplus eggs were sold in the small markets and the only meat was from old hens and surplus cocks. The feed for poultry was confined to food scraps and what they could find around the house and farmyard. As a consequence, the laying rate is often as low as 100 eggs per hen per year and they are very scrawny birds with little meat. A dramatic rise in poultry and egg production in recent years has been a direct result of growing urban demand throughout the Middle East. To meet this increased demand, commercial poultry production has been substantially expanded in the past decade, particularly in the Gulf countries.

CONCLUSIONS

As nomadic pastoralism declines throughout most, but not all, of the region and urban demand for high quality livestock animals and products increases, the traditional systems of production are gradually being superseded. They are being replaced by more modern, capital intensive, and possibly more efficient, methods, particularly in the dairy and poultry sectors. The demand for live animals and particularly for sheep and goats will, however, ensure that the traditional livestock sector remains important in the region. In the poorer countries, and especially in Sudan, livestock are amongst the most important natural resources and it will be decades before pastoralism is completely replaced by modern ranchland farming techniques. At the same time, the oil-rich Gulf states are preserving the nomadic and sedentary small-scale producers by subsidies for water, fodder, and transport.

THE DEVELOPMENT OF THE LIVESTOCK SECTOR

It has been seen that the traditional systems of animal production are being increasingly supplemented by a modern livestock sector. As the urban centres in the Middle East have expanded, both through natural growth and rural-urban migration, the demand for livestock products has increased. The traditional sector has been unable to meet this demand for a variety of reasons. As already noted, livestock is kept by pastoralists for their own requirements and for prestige rather than for commercial reasons. Animals tend to be of low quality in countries such as Sudan because of the emphasis on the number of animals a family owns. The long distances between pasturelands and urban centres makes transportation of animals to market a major problem. For these and other reasons, the offtake rate is often not much higher than 5 per cent per year. It has, therefore, been necessary to develop a modern commercial livestock sector while also encouraging the traditional sector to market a larger proportion of its animals.

Saudi Arabia

The assistance given to the traditional sector varies from country to country. In some Gulf countries it appears that little effort has been made and they seem willing to expand imports rather than develop indigenous production (Moliver & Abbondante, 1980). In the case of Saudi Arabia, agricultural policy has always had a large measure of political reasoning. As early as 1912 the government made concerted efforts to settle the pastoralists and thereby reduce tribalism and potential opposition to the ruling family. At the same time, it used its increasing oil revenues to subsidize the livestock sector and thereby increase tribal dependence on the central authority. It was the successive laws on land distribution, however, which affected the pastoralist system most profoundly. The law of 1968 particularly gave individual title, and therefore market value, to both land and water and also broke down the traditional tribal homelands in favour of the individual. A new class of land owners who were generally members, friends, and allies of the royal family was created (Barker, 1982). This increased the pace of settlement and the degree of government control. The most famous and ill-fated plan was the King Faisal Bedouin Settlement which aimed at settling the al-Murrah tribe in the Empty Quarter. While government subsidies are still given for water, fodder, and transport, pastoralism is gradually disappearing in Saudi Arabia and is being replaced by tenant agriculture and modern commercial farming.

Sudan

By contrast, the Sudanese Government has recognized that livestock represents one of its most important natural resources. It is, therefore, making substantial efforts to develop its pastoral system to benefit both its domestic and international markets. This involves two principal elements, first, encouraging the pastoralists to sell more of their animals, and second, improving livestock quality, and creating a modern efficient marketing system. The former aim is being achieved by a programme of education and the development of services in the rural areas. As more children go to school and families are able to buy more consumer goods they require money and pastoralists are therefore selling more animals. The Livestock and Meat Marketing Corporation (LMMC) is also reducing the number of middlemen in the trade and thereby increasing the prices for producers. One of the problems of developing the livestock marketing system is the lack of an adequate transportation system. The pastoral heartlands of Sudan are in the west of the country which is up to 500 miles from the central market in Omdurman and 1,000 miles from the country's only port on the Red Sea. With the financial assistance of the World Bank and bilateral aid organizations, double-decker cattle trains are being imported and reconditioned to transport animals to market. This reduces the substantial weight loss incurred in long-distance livestock drives. At the same time, primary, modern secondary and terminal markets, as well as railway, holding, transit, and terminal yards are being constructed. In the pasturelands mobile veterinary services, fodder collection points, and radio communications are being developed along the existing stock route from Nyala, in the west, to Khartoum. It is hoped that this assistance to the producers will also discourage the substantial smuggling of Sudanese animals to neighbouring countries. It is estimated that 600,000 sheep are smuggled abroad each year, primarily because livestock taxes are considered too high. The LMMC recognized these resentments and is now trying to persuade the regional authorities, which rely heavily on livestock taxes, to reduce them and thereby increase the legal trade. So far, the 20 per cent export tax on sheep and goats has been lifted and on cattle reduced to 10 per cent, although the 5 per cent development tax has remained in force. One of the other problems for the authorities is that Saudi Arabia, which is the largest market for Sudanese animals, has done little to stamp out smuggling from Sudan by middlemen (Taban, 1984).

Once the animals reach the central markets there have been serious health, hygiene, and theft problems as animals are held in residential areas of Khartoum and Omdurman. This is now being allieviated by the development of a 6,800 feddan holding ground which will be able to accommodate 30,000 cattle or 100,000 sheep. Seven wells supply the animals with water at a normal price and veterinary services, green fodder, dry fodder, and concentrates are also available. Besides this fattening centre, modern slaughter yards are being established near the central market and greater emphasis is being placed on meeting the qualitative requirements of Sudan's export markets. It is hoped that these developments will substantially increase the country's livestock industry which is almost exclusively based on the pastoral system of production. There is scope for such an expansion since it is estimated that, although only 12,101 cattle were exported in 1982/83, there were 1,200,000 which were identified as surplus for export (Taban, 1984).

Turkey

The livestock sector in Turkey is the largest in the Middle East region, whether valued in total production or exports. It is the biggest single holder of sheep and goats in the area and has the second largest number of cattle and poultry. Production of mutton, beef, eggs, and milk is the highest, and of white meat the second highest in the Middle East. Together with Sudan, Turkey is the only sizeable exporter of livestock products, but is alone in its importance as an exporter of grains, fruit, and vegetables. Like Sudan, its production is dominated by the traditional sector although there is increased commercial ranching particularly in the dairy herds. The pastoralists are concentrated in the less fertile eastern regions of Turkey, which are adjacent to the principal export markets of Iran and Iraq. Official exports have not risen at the rate envisaged in the development plans, primarily because of the widespread smuggling of sheep and goats across the eastern frontiers. There are also indications that dairy cattle are being slaughtered in some regions for meat exports to Turkey's neighbours. The increasing domestic demand for meat has led to shortages and speculative pricing and the government has therefore placed periodic bans on the export of all livestock and meat. The current depletion of the herds in the east and the expanding urban demand for meat could soon result in Turkey becoming a net importer (Business International, 1983). The government is therefore trying to regulate and educate the pastoralists who provide the majority of the livestock and meat so that reliable and augmenting production can be achieved.

THE DEVELOPMENT OF THE DAIRY AND POULTRY SECTORS

The most important developments throughout the Middle East have been in the dairy and poultry sectors. As urban centres have expanded, the demand for the fresh milk, cheese, butter, eggs, and chicken has increased. Both the public and private sectors have invested heavily in modern capital intensive schemes which have increased milk and egg production by up to a factor of ten in some of the smaller Gulf countries. While the expansion does not appear to be so dramatic in some of the larger countries, their production is on a much larger scale (Figure 5.1).

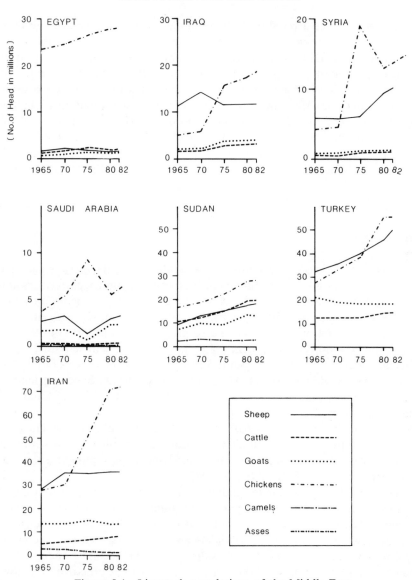

Figure 5.1   Livestock populations of the Middle East

Bahrain

Bahrain is a good example of a small country building up a poultry industry almost from scratch to become virtually self-sufficient. The wholly government-owned General Poultry Company (GPC) and the majority privately owned Delmon Poultry Company (DPC) are the two production companies in Bahrain.

The industry is based around the GPC-built and -owned egg hatchery and feed mill project which produces 24 million eggs a year and can mill eight tonnes of grain per hour. DPC is more involved in poultry processing and it is in the middle of an amibitious programme to rear a total of nine million birds for Bahraini tables. This is likely to have considerable implications for the 30 existing chicken farmers on the island who currently produce about three million birds a year. The new DPC project is both capital intensive and will introduce European standards of hygiene to chicken rearing and slaughtering. It is typical of the new projects in the Gulf area which are also being gradually introduced in some of the larger and poorer countries of the region (MEED, 1982).

Saudi Arabia

Saudi Arabia is another example of a Middle East country which has concentrated on large-scale, capital intensive dairy and poultry projects, mainly at the expense of the traditional sector. The government provides attractive subsidies on animal feed (50 per cent), poultry equipment (30 per cent), and dairy equipment (30 per cent), while also giving free transport for 50 cows. Despite this, the third development plan's target of 'prudent self-sufficiency' (Whelan, 1982) has led to the encouragement of large-scale projects rather than expansion by individual farmers. The largest dairy scheme, Saudi Agriculture & Dairy Company (Saadco), is financed directly by the Ministry of Finance & National Economy because its requirements are more than the state Saudi Arabian Agricultural Bank (SAAB) can lend to individual projects. The bank lends up to 80 per cent of its funds to the dairy and poultry sector with fruit and vegetables, wheat, livestock fattening, and fisheries accounting for less than 4 per cent each in 1980/81 (Whelan, 1982).

The new trend in agriculture in Saudi Arabia is joint stock companies involving local businessmen of which the largest is the National Agricultural Development Company (Nadec). Although the government owns 20 per cent of the stock, there are now about 127,000 shareholders and Nadec shares are so successful they are virtually unobtainable. Besides its 2,300 hectares under wheat production and its flock of 1,000 sheep, Nadec also has a herd of 700 cows which produce an average of 6,500 litres of milk a day. The company plans to increase the herd to 2,000 and later 4,000 animals and also to establish a chicken farm producing eight million eggs per year. One of the principal problems in the dairy sector, however, is that imported Friesian and other breeds are not suited to the climate in the region. It would make far more sense to use the indigenous small black cows which are the naturally selected variety and which give more milk per bodyweight than Friesians (Whelan, 1982).

Turkey

Turkey, which is the largest milk producer in the Middle East, is expanding its dairy industry very rapidly. The national herd is estimated at about five million animals, of which about 1.5 million are probably improved strains including Jersey, Brown Swiss and Holstein-Friesian cows. Several firms are now producing and marketing sterilized milk, butter, cheese, yogurt, and milk powder for the domestic market. The industry's development includes investments in dairy farming, veterinary services, refrigeration, packaging, storage, and distribution. At the same time, there has been a considerable expansion of the poultry industry in the past decade and exports to Iraq and the Gulf countries have increased. The protection of poultry from diseases is a major problem at the moment because there is no legal importation of vaccines against most major poultry diseases (Business International, 1983). Poultry slaughterhouses in Ankara and Mani have a joint capacity for 27,000 birds per day but elsewhere the slaughtering is done by very small units.

Sudan

Despite its relatively under-developed livestock sector, Sudan is also making substantial efforts to build up its dairy and poultry industries. In addition to the traditional sector, there are about 100 poultry farms with 150,000 hens producing approximately 30 million eggs annually, with 250,000 birds for meat each year (Sudanow, 1983). Each province has poultry production units which are run by the government Animal Resources Department and which produce 125,000 chicks per year. There are, however, major problems because of the lack of spare parts for incubators, and inadequate electricity supplies, animal feed, and vitamins. There are now a number of modern poultry projects which have been partially financed by the Gulf countries with FAO participation. They include the Sudan-Kuwaiti Chicken Scheme and the Jebal Aulia Chick Production Project. Similarly, the traditional dairy producers are gradually being replaced by new large-scale plants. The Belgravia Dairy, which was started in 1930, can today only supply about 10 per cent of the demand for the capital (Sudanow, 1983). The state Sudan Development Corporation has recently completed the Khartoum Dairy Products Co. Ltd plant which has a design capacity of 18 million litres of milk per year. It is obvious that, like the rest of the Middle East, the Sudanese dairy and poultry sector will probably expand far more rapidly than the production of other livestock.

PATTERNS OF LIVESTOCK PRODUCTION AND TRADE

In the past twenty years there have been a number of interesting new trends in livestock production and trade in the Middle East. Sheep are by far the most important animals in the region and mutton and lamb are the most popular meats. Turkey is the largest producer with over 50 million animals, followed by Iran (35 million) and Sudan (19 million) (Figure 5.1). The increase in numbers since 1965 has not been particularly spectacular, only doubling in most of the countries. The apparent sharp decline in Yemen AR and the equally dramatic rise in Yemen PDR can probably be explained by the inaccuracies in the statistics.

Sheep and Goats

The trend towards the greater importation of live sheep and goats is quite pronounced. Turkey and Sudan, with 2.5 million and 500,000 animals exported in 1982, respectively, are the only real net exporters. Egypt exported 80,000 live animals despite also being a major importer of mutton and lamb. Saudi Arabia is by far the largest net importer, buying more than the rest of the region put together. More than six million animals are imported at a cost of over 50 million dollars (Figure 5.2) a proportion of which are slaughtered by Muslim pilgrims in Mecca. Given its very small population, Kuwait's import of over 1.5 million sheep and goats is very substantial while other Gulf countries' imports have also risen dramatically. It is interesting to see the changes in Iran's production and trade in sheep and goats. The number of animals has risen gradually from 30 to 35 million but at the same time the country has gone from being an exporter of approximately 500,000 animals in 1965 to a net importer of 3,036,000 in 1978. Following the Islamic Revolution, this has since decreased to 700,000 in 1982 (Figure 5.2). It is a good example of the effect of political events on both the production and trade of livestock and products. The effect of the loss of Jordan's West Bank territory is also shown in the changes in the number of animals the country had before and after 1967. There was a rapid rise in the number of workers from North Yemen who went to Saudi Arabia and the Gulf after 1973 oil price rise. The increase in their remittances is shown in the rise in imports of animals by North Yemen from 706 in 1973 to 180,000 in 1982. The country has also changed from being a net exporter of 2,000 cattle to a net importer of 30,000 head in the same period. Increasing animal imports have been principally caused by rapid rises in disposable incomes among the rapidly growing populations throughout the region, occasioned above all by augmenting oil wealth. Profound changes in the value, variety and quality of livestock food consumed in the region have come about since 1945 but mainly since the oil boom of the early 1970s. The internal dynamic of change arising from oil industry activities has also tended to reduce domestic output of animal products in the oil-exporting states. Elsewhere, there has been no compensating increase in output either to make good losses on the regional scale or in many cases to compensate for local growth in demand. In Syria, for example, the number of sheep has doubled over last twenty years but whereas the country exported 130,200 head net at the beginning of the period, it imported 168,711 net by the close.

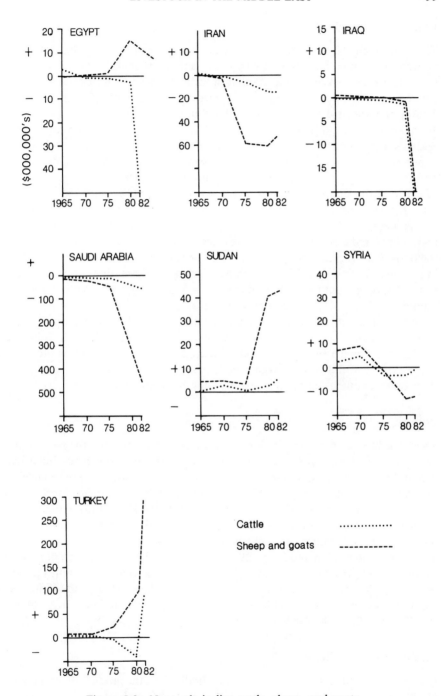

Figure 5.2   Net trade in live cattle, sheep, and goats

Cattle and Draught Animals

The production of, and trade in, cattle is interesting in showing the difference between a commercially orientated country and an underdeveloped and largely pastoralist society. With about 20 million cattle, Sudan has more than any other country in the Middle East and is followed by Turkey with 16 million. While Turkey exported 101,937 animals in 1982, Sudan's exports have never exceeded 26,000 in one year. There is certainly no shortage of a market for cattle since Sudan's neighbours, Saudi Arabia and Egypt, import over 250,000 cattle annually between them. Lebanon and Kuwait followed by North Yemen, Iraq, and Iran also offer increasingly important export markets for beef. Imports of cattle in the small Gulf states increased substantially from a very low base after the 1973 oil price rises.

The increase in use of mechanized transport and the level of affluence in the region have reduced the need for pack and draught animals. It is primarily in the poor countries that they are still used for transport or as sources of power in farming. With over 2,500,000 camels, Sudan has three times as many as the rest of the region put together, and Iraq, in second place, has a tenth of Sudan's numbers (Figure 5.1). There are large herds of wild camels in northern and western Sudan but large numbers are also used for long-distance transport in the region. Iran and Egypt lead the number of asses with 1,800,000 each, followed by Turkey, North Yemen, and Sudan (Figure 5.1). Having a colder winter than the rest of the Middle East, Turkey and Iran are more suited to horses and mules than other countries and consequently they have larger numbers.

Poultry

There has been a dramatic increase in the number of chickens in the past twenty years as poultry production has struggled to keep pace with the increasing urban demand for meat. It would seem, however, that the statistics tend to exaggerate increases achieved. As an example, the number of chickens in Jordan appears to have risen from 5,846,000 in 1979 to 26,000,000 in 1980. Taking the figures at face value, it is the countries with large overall and urban populations which have the largest number of chickens. Iran has about 75 million, followed by Turkey with 56 million and Egypt 29 million (Figure 5.1). The growth in poultry production has been fastest in the oil-rich Gulf states which have built up their output levels almost from scratch. For example, the UAE's production rose from 216,000 in 1977 to 3,000,000 in 1982, while Qatar's increased from 29,000 in 1971 to 380,000 in 1982.

Fresh and Frozen Meat

Given the vast numbers of livestock in the Middle East, the production of mutton, lamb, beef, veal, and goat meat is comparatively small. The preference for

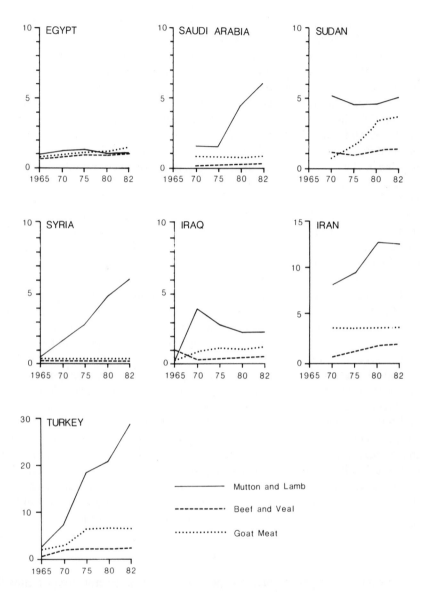

Figure 5.3   Meat production in the Middle East

slaughtering live animals on feast days, unhygienic methods of slaughter and packaging, and a hot climate in which meat does not keep for long have all contributed to the low levels of meat production. Once again, it is the large countries of Turkey and Iran which lead the league table for mutton and lamb, while Saudi Arabia, which is third, is able to afford more modern refrigeration equipment than most. Because of the preference for mutton and the predominant use of cattle for milk rather than meat, beef and veal production is about a tenth that of mutton in most countries in terms of the number of head which are slaughtered. As already noted, however, sheep are much smaller in size and are therefore more suited to a single family unit than a cow.

The import of fresh and frozen meat in the Middle East has increased substantially in the past decade. With the exception of Turkey, which is a major exporter, and Sudan, which now has no trade in meat, all the countries have become net importers. The growth has been most rapid in the larger oil-rich nations of the Gulf where urban demand is increasing very rapidly. Iran's beef imports rose from 1,201 to 65,000 carcasses between 1965 and 1982. In the same period, Saudi Arabia's mutton imports rose from 939 to 32,000 carcasses and beef imports from 619 to 49,400. The effect of the Gulf War on Iraq's meat imports is evident from the statistics; beef imports rose from 10,000 to 80,000 carcasses in the first two years of the war and are probably considerably greater now (Figure 5.3). It is known that the armies of the region are amongst the principal purchasers of meat.

Recognizing that the Middle East is one of the fastest-growing markets for meat exports, Australia, New Zealand, and the countries of the EEC are making further efforts to win a share in the trade (Robie, 1983 and Whelan, 1983). In the case of Iran and Iraq and other centrally planned economies, this has recently meant accepting a barter system in which oil is exchanged for meat and live animals. It has also necessitated the *halal* method of slaughter, bilingual labelling and special packaging being introduced in the supplier countries. The growth of chicken meat imports to the Middle East, now estimated at between 300,000 and 400,000 tonnes per year, is an important and possibly worrying development for the region's traditional meat suppliers (Robie, 1983). At first, it was thought that it was a temporary change, which was related to the rapid influx of migrant workers, but it now appears to be a long-term change in consumer preferences. Together with increasingly competitive exports of both animals and meat from the developing countries such as Turkey and Sudan, these changes will have to be met by the major red meat suppliers to the Middle East.

CONCLUSIONS

While each country is unique, it is possible to detect several broad trends in livestock production and trade in the Middle East. Sheep and goats are both the most numerous animals and lamb/mutton meat is the most favoured in the region. Their numbers have risen gradually during the past twenty years but herd growth has not kept pace with demand. For this reason, the import of live animals as well as mutton and lamb has increased, particularly in the past decade. Only in Sudan do cattle outnumber sheep; but the country has failed to capitalize on the considerable demand from its Arab neighbours, and most notably from Saudi Arabia. With the exception of Turkey and Sudan, the rest of the countries of the region are all net importers of both live cattle and beef. While the numbers of sheep, goats, and cattle have generally risen very slowly, the poultry and dairy sectors have been the fastest-growing parts of the livestock industry. The introduction of modern capital intensive production units, in the Gulf region in particular, has led to a substantial increase in the output of milk, eggs, and chickens. It has not, however, managed to keep pace with the very rapid urban-led growth in demand for these products. Finally it has been seen that, in general, the number of pack animals has fallen in the past twenty years, although some of the poorer countries maintain large numbers for transport and agricultural work. In the oil-rich countries they have been replaced mainly by mechanization, and horses and camels are only kept for social prestige and racing purposes.

The Food and Agricultural Organization has produced indices of livestock production for the period from 1971 to 1982 using 1974–76 as the base years (FAO, 1982). They generally show that, while production has increased, when it is measured on a per capita basis the rise is less impressive. They also illustrate some very interesting points of analysis. It appears that Syrian production has been the most impressive both in total and per capita terms. Taking 1974–76 as 100, it rose to 218.09 in total and 167.66 on a per capita basis (Figure 5.4). While Lebanon appears to have equally impressive figures, it can be seen that production fell during the mid-1970s, during the civil war, and that it has only recently achieved the same levels of production as it enjoyed in 1971. The increase in the number of migrant workers since 1973 could account for Saudi Arabia's poor per capita indices in relation to its overall production, which has doubled in a decade. The effect of the Islamic Revolution in Iran in 1979 is reflected in the decline and then stagnation of production, while the recent very high birth rates will soon begin to effect per capita figures. Turkey, Sudan, and Egypt all show very similar trends, with total production rising by 25–30 per cent since the base year, but population increases reduce this to little more than 5 per cent. It is in the Yemens that there has been the greatest lack of progress in livestock production. It rose by only 5.89 per cent from the base period in Yemen PDR and by 3.79 per cent in Yemen AR. Because of the increase of population, this has led to per capita falls of over 10 per cent in the same period (Figure 5.4). The poor showing in South Yemen can probably be attributed to the rigid government control on farmers in the country. In North Yemen the same trend is more likely to be the result of so many men leaving the land for better paid jobs in Saudi Arabia and the Gulf.

In conclusion, it can be seen that livestock production in the Middle East during the past twenty years has been very disappointing. Given the long-established tradition of nomadic pastoralism and herding and the potential of the physical environment, it seems that the general emphasis on irrigated cereal production is misplaced. The majority of the governments of the region have

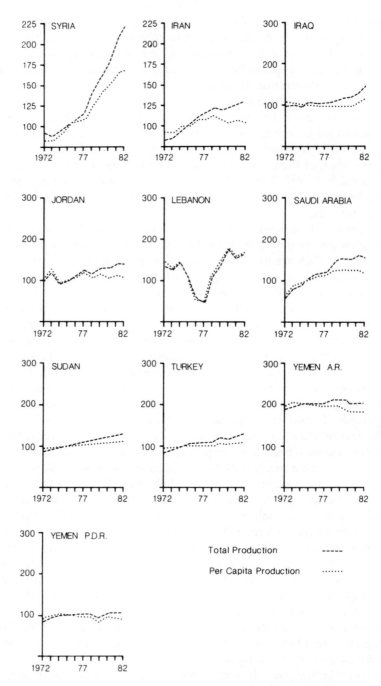

Figure 5.4   Total and per capita livestock production indices

ignored the importance of the livestock sector and its potential for increased production. Instead they have been prepared to increase their imports of both live animals and meat. The only sub-sectors which have received the investment and attention that they deserve are the dairy and poultry industries. Livestock represent one of the region's most important agricultural resources, particularly for those countries which do not have the financial cushion of oil revenues. It is essential that the livestock industry is encouraged to achieve its full potential and thereby reduce the increasingly expensive dependence on animal and meat imports.

REFERENCES

Allan, J. A., McLachlan, K. S., & Penrose, E. T. (1973) *Libya: Agriculture and Economic Development*, Cass, London, p. 184.
Bank of Sudan (1983) *Foreign Trade Statistical Digest*, July-September 1983, Statistical Department, Bank of Sudan, Khartoum, p. 3.
Barker, P. (1982) *Saudi Arabia: The Development Dilemma*, Economist Intelligence Unit, London, pp. 51–55.
Beaumont, P., Blake, G. H., & Wagstaffe, J. M. (1976) *The Middle East: A Geographical Study*, Wiley, London, p. 187.
Birks, J. S. (1981) 'The impact of economic development on pastoral nomadism in the Middle East: an inevitable eclipse?' in Clarke, J. I., & Bowen-Jones, H. (Eds), *Change and Development in the Middle East*, Methuen, London, pp. 82–94.
Braudel, F. (1972) *The Mediterranean and the Meditteranean World in the Age of Philip II*, London, Fontana. Braudel outlines some of the analytical approaches that have postulated inherent conflict between the steppe and the sown, pp. 85–102.
Business International (1983) *Turkey: Opening to the World Economy*, Research Report, Geneva, June 1983, pp. 30–32.
Clarke, J. I. (1981) 'Contemporary urban growth in the Middle East', in Clarke, J. I., & Bowen-Jones, H. (Eds), *Change and Development in the Middle East*, Methuen, London, pp. 154–170.
Ercon (1974) *Hydrogeological Investigation of the Hari Rud Basin between Marwa and Ghuryan*, Phase 2 Report, Volume 3, *Agriculture*, Afghanistan, pp. 127–134.
Evans-Pritchard, E. E. (1940) *The Nuer*, Oxford University Press, London.
Fabietti, U. (1982) 'Sedentarisation as a means of detribalisation: some policies of the Saudi Arabian Government towards the nomads', in Niblock, T. (Ed.), *State, Society and Economy in Saudi Arabia*, Croom Helm, London, pp. 186–197.
FAO (1982) *Production Yearbook*, FAO, Rome.
FAO & Government of Libya (1967) *Land Policy in the Near East*, FAO, Rome, pp. 38–86.
Gurdon, C. (1984) *Sudan at the Crossroads*, Menas Press, London, p. 7.
Haseeb, K. (1964) *The National Income of Iraq 1953–1961*, RIIA/OUP, Oxford, p. 34 and pp. 61–71.
Hill, A. G. (1981) 'Population growth in the Middle East since 1945 with special reference to the Arab countries of West Asia', in Clarke, J. I., & Bowen-Jones, H. (Eds), *Change and Development in the Middle East*, Methuen, London, pp. 130–153.
Lebon, J. H. G. (1965) *Land Use in the Sudan*, The World Land Use Survey, Occasional Papers No. 9, Geographical Publications, London.
MEED (1982) *Bahrain, Middle East Economic Digest* Special Report, September 1982, London, p. 8.

Moliver, D. M., & Abbondante, P. J. (1980) *The Economy of Saudi Arabia*, Praeger, New York, p. 56.

Robie, D. (1983) 'Middle East recovers its taste for lamb', *Middle East Economic Digest*, 11 February 1983, London, pp. 59–60.

Sudanow (1983) *Sudan Yearbook*, Sudanow, Khartoum, pp. 164–172.

Taban, A. L. (1984) 'Developing an industry', *Sudanow*, 9, 2, 22–23.

Tapper, R. T. (1983) *The Conflict of Tribe and State in Iran and Afghanistan*, Croom Helm, London.

Thesiger, W. (1965) *Arabian Sands*, Penguin, London. Thesiger reviews the attitudes of the true bedouin nomad to the external world and the settled population (pp. 85–102).

Watson, R. M. (1975) *Livestock Census of the Democratic Republic of Sudan*, Resource Management and Research, Nairobi.

Whelan, J. E. (Ed). (1982) *Saudi Arabia, Middle East Economic Digest* Special Report, August 1982, London, pp. 39–45.

Whelan, J. (1983) 'Lean years ahead for meat producers', *Middle East Economic Digest*, 25 February 1983, London, pp. 39–45.

Agricultural Development in the Middle East
Edited by P. Beaumont and K. McLachlan
© 1985 John Wiley & Sons Ltd

Chapter 6

The Agricultural Sector in Development Policy

RICHARD I. LAWLESS

INTRODUCTION

Today the countries of the Middle East represent the world's most rapidly growing food deficit area (Paul, 1981). In 1960 the region was a net food exporter, but the 1970s saw rising food imports at a time when world food prices were climbing rapidly. The value of agricultural imports rose from about $4 billion in 1973 to more than $20 billion in 1980. Nearly 60 per cent of the region's wheat requirements are now imported, as are between 15 and 20 per cent of meat supplies. By 1980 Weinbaum (1982) has calculated that Iran was importing between 30 and 40 per cent of its food requirements, Egypt 40–45 per cent, Jordan more than 50 per cent, Libya 60 per cent, and Saudi Arabia as much as 75 per cent. The burden on foreign currency reserves is enormous and is growing. Iran spent over $2 billion on food imports in 1978, compared with $330 million in 1973–74; Saudi Arabia's import bill for 1980 was projected at $4.5 billion, a 50 per cent increase in a single year. In addition there was the very high cost of creating an enlarged system of warehousing, wholesaling, and distribution for the growing volume of foodstuffs now reaching the cities of the Middle East not from their rural hinterland but from the coastal ports.

Increases in domestic food production during the 1970s only just kept pace with population growth in most Middle Eastern countries and fell far short of the demand for better and more nutritious food — a demand stimulated by larger disposable incomes and public policies to raise nutritional levels. The performance of the region as a whole has been unimpressive in terms of both yields attained and comparative rates of growth.

Although agriculture continues to support a high proportion of the labour force (Table 6.1), its share of the employed population has fallen, in some cases dramatically, during the last 20 years. Similarly agriculture's contribution to GDP has contracted relative to other economic sectors, and food and related commodities as a proportion of merchandise exports have also diminished in almost all countries of the region (Tables 6.2 and 6.3). The figures in Table 6.4 confirm that performance in the agricultural sector has been poor in

Table 6.1   Percentage of labour force in agriculture, 1960 and 1978

Country	Percentage in agriculture 1960	1978
Egypt	58	51
Iran	54	40
Iraq	53	42
Israel	14	7
Jordan	44	27
Lebanon	38	12
Libya	53	21
Saudi Arabia	71	62
Sudan	86	79
Syria	54	49

Source: Weinbaum, 1982.

Table 6.2   Distribution of GDP by major sources

Country	Percentage in agriculture 1960	1978	Percentage in industry 1960	1978	Percentage in services 1960	1978
Egypt	30	29	24	30	46	41
Iran	29	9	33	54	38	37
Iraq	17	8	52	—	31	—
Israel	11	7	32	37	57	56
Jordan	—	11	—	29	—	60
Lebanon	12	—	20	—	68	—
Libya	—	2	—	72	—	27
Saudi Arabia	—	1	—	76	—	23
Sudan	58	41	—	—	27	—
Syria	—	20	—	28	—	52

Source: Weinbaum, 1982.

many countries of the region. In Egypt agricultural growth remained stagnant in the 1970s in contrast with higher rates for other economic activities, notably the industrial and service sector; Iraq registered a negative growth in agricultural production in the 1970s at a time when industry and services were expanding at more than 12 and 13 per cent. Measured by the ratio of the sector's contribution to GDP to its share of the agricultural workforce, the region has also experienced a declining efficiency of its agriculture with oil-exporting states such as Iran and Iraq experiencing a particularly notable decline, but also countries such as Sudan where agriculture remains the dominant element in the economy (Weinbaum, 1982). With justification agriculture has been described as the Achilles' heel of Middle Eastern development efforts (Owen, 1981).

Table 6.3:   Food and related commodities as percentage share of merchandise
exports

Country	1960	1977
Egypt	84	49
Iran	9	1
Iraq	3	1
Israel	35	19
Jordan	96	38
Lebanon	—	—
Libya	84	0
Saudi Arabia	0	0
Sudan	100	95
Syria	81	28

Source: Weinbaum, 1982.

Table 6.4   Average annual growth rates by sector, 1960–70, 1970–78

Country	Agriculture 1960–70	1970–78	Percentage growth industry 1960–70	1970–78	Services 1960–70	1970–78
Egypt	2.9	3.1	5.4	7.2	5.1	12.0
Iran	4.4	5.2	13.4	4.0	10.0	16.1
Iraq	5.7	− 1.5	4.7	12.2	8.3	13.5
Israel	5.0	6.6	15.6	5.3	1.5	5.4
Jordan	5.0	2.6	9.9	—	6.4	—
Lebanon	6.3	—	4.5	12.0	4.8	—
Libya	2.2	12.7	—	− 2.7	—	16.7
Saudi Arabia	—	4.0	—	—	—	11.6
Sudan	3.3	8.8	1.7	2.8	—	—
Syria	4.4	7.2	6.3	11.6	6.2	9.5

Source: Weinbaum, 1982.

THE AGRICULTURAL SECTOR IN NATIONAL DEVELOPMENT STRATEGIES

Achieving the goals of economic and social development has become the most important challenge confronting the countries of the Middle East, individually and collectively. Although they differ about the meaning and implementation of the private-public balance issue, in all of them the state plays the key role in economic and social development (Table 6.5).

Development, Khuri (1981) reminds us, implies two corollary processes, growth and distribution; the way one intervenes, overlaps, or intersects with the other shapes the directions of development. Growth refers to gross, general

Table 6.5   Public and private shares of planned investments in some development plans of Arab countries

Country	Plan period	Public share (%)	Private share (%)
Jordan	1964–70	53.0	47.0
	1973–75	56.0	44.0
	1976–80	50.0	50.0
Syria	1960–65	63.0	37.0
	1966–70	70.0	30.0
	1971–75	81.0	19.0
	1976–80	83.0	17.0
Sudan	1961/62–70/71	60.0	40.0
	1970/71–74/75	56.0	44.0
	1977/78–82/83	59.0	41.0
Iraq	1965/66–69/70	78.0	22.0
	1970–74	82.0	18.0
	1976–80	91.0	9.0
Libya	1963–68	52.0	48.0
	1973–75	87.0	17.0
	1976-80	87.0	13.0
Egypt	1960/61–64/65	62.0	38.0
	78–82	89.0	11.0
Saudi Arabia	1965/66–68/69	85.0	15.0

Source: Sadek, 1981.

increase in the capacity, efficiency, and volume of economic activity; distribution on the other hand refers to the mechanisms of allocation and those that interfere in the process of allocation. Who in society gets how much of this growth and why? Throughout the region in designing policies or strategies of development, the agricultural sector has been accorded a low priority by planners and governing elites. By the early 1970s Weinbaum (1982) estimates that no more than 15 per cent of public investment in the region was going to the agricultural sector. These sectoral priorities, he argues, must be interpreted as part of a political process. Basically, public funds have gone disproportionately towards underwriting industry and subsidizing food for city dwellers because influential elites are almost exclusively urban-based.

As such they have had a strong stake in an overall growth strategy based on the pre-eminence of urban-based industry and commerce. Established economic interests welcome the stimulus of public expenditure in the large cities and value the role of low agricultural prices in holding down the cost of raw materials. They have usually opposed policies that might attract larger private investment in agriculture through restricting opportunities for capital speculation in the urban sector. (Weinbaum, 1982)

Moreover, government policy-makers cannot afford to ignore the demands of the often highly politicized urban middle and working classes. Investment programmes carried out in urban areas can bring immediate and highly visible results in terms of increased employment and better services; it is a price many regimes are willing to pay for urban peace. In contrast, investment in rural areas involves long lead times, while traditionally the rural population have been able to exert little or no bargaining power to attract public funds. Investment in rural areas offers only limited political returns to regimes.

To virtually all political decision-makers, modernization has meant, and still means, Westernization. The advanced Western industrial nations became the models for governments who sought to throw off their economic backwardness by means of rapid urban/industrial expansion (Sayigh, 1982). Public investment in heavy industry, transportation, and communications, it was believed, would build strong national economies; dependence on agriculture would keep them weak and poor. Almost unanimously the region's policy-makers accepted the vision of an urban/industrial society as the key to national prosperity and as a means of diversifying and strengthening economies in which either agriculture or oil was the major source of national income. The agricultural sector was too weak to bid successfully for investment funds. Instead the dominant economic policies extracted capital and labour from the rural areas to sustain an expanding urban/industrial sector.

Some examples will help to illustrate the indifference and calculated neglect of agricultural development in more than two decades of public policy in the Middle East. In Egypt the agricultural sector provides some 45–47 per cent of total employment, 30 per cent of GDP, and some 50 per cent of exports. Equally important, it has been estimated that over half of the country's manufacturing sector consists of agriculturally based industries such as textiles and food processing. Under Nasser the share of agricultural investment in total public capital expenditure rose from 11.6 per cent in 1952–3 to 16.8 per cent in 1967–68. Some sources estimate that agriculture's share in public sector investment rose to 25 per cent in the mid-1960s when the Aswan High Dam is included (Richards, 1980). However, most capital formation was directed towards the hydraulic system, with the High Dam accounting for roughly one-third of all capital formation during this period. By 1970–71 investment in agriculture had fallen to 12 per cent of the annual state budget and it continued to fall during the Sadat era when the emphasis was firmly on industry and related economic activities as part of the president's 'open-door' policy; during the planning period 1978–82 investment in agriculture had fallen to a mere 8 per cent (Tubiana, 1981).

In Iran, as government revenues from oil rose from $817 million in 1968, to $2.25 billion in 1972–73 and $19.16 billion in 1975–76, investment in government development plans rose dramatically: the First Plan (1948–56) had a projected expenditure of $350 million, the Fourth (1968-72) $8,284 million and the Fifth (1972–78) was revised after the oil price rises to $69 billion (Halliday, 1979).

In the Shah's efforts to transform Iran into a 'new Japan', rapid industrialization was promoted as the most effective way of diversifying the economy to offset the eventual decline in income from oil exports. Agriculture was given low priority in government economic planning and this sector was deprived of funds throughout the 1960s and 1970s. Under the Third plan (1962–67) agriculture received only 11.7 per cent of allocations if investment in dams and other capital intensive schemes are excluded, and under the Fourth Plan only 8 per cent of the development budget was allocated to agriculture. The declared aim of the Fifth Plan (1972–78) was to redress the sectoral imbalances resulting from the country's rapid economic expansion and allocations to agriculture were increased. Nevertheless, actual spending fell far short of planned allocations. As in other Middle Eastern countries, the agricultural sector proved incapable of utilizing productively even these meagre resources allocated to it. By the end of the Fifth Plan the actual budget for agriculture was no more than 12 per cent; agriculture's contribution to GDP had fallen to a mere 9 per cent compared with 29 per cent in 1960.

The Sixth Plan was overtaken by the Revolution. The new revolutionary government reassessed nearly all Iran's economic and social strategies and announced that priority would be given to low growth rates, a concentration on small-scale projects in industry, emphasis on traditional agriculture and stringent control of oil exports. Having made such an issue of agriculture in the political battle against the Shah, the victorious revolutionaries made its revival one of their priorities. Self-sufficiency in foodstuffs above all else became central to the economic philosophy of the new regime, but, although there was some increase in the resources allocated to the agricultural sector, little improvement in production was recorded. Indeed food imports rose from $2,800 million in 1980, to $3,000 million in 1981 and an estimated $4,500–5,000 million in 1982.

In his study of urban bias in development Lipton (1977) has demonstrated that throughout the Third World it is above all by keeping farm output prices down that both private and public powers transfer savings capacity from agriculture to the rest of the economy. The urban concentration of governmental power and support and the industrializing preferences which underlie planners' ideology lead them to transfer resources out of agriculture by means of 'price-fixing policies'. The Middle East is no exception. By means of a system of depressed crop prices, the agricultural sector over much of the region has been forced to bear much of the burden of capital formation for the urban/industrial sector. Crop price increases, governments argue, would result in rising food costs in urban areas, leading to higher urban wage demands. Higher urban wages would increase the costs of industrial products, make investment in industry less attractive and locally manufactured goods less competitive in overseas markets. In practice the surplus extracted from agriculture has not always been used to finance industrial investment and new jobs in efficient manufacturing enterprises but has been transferred to urban consumers by way of subsidies. In many parts of the region the heavy costs of food subsidies for urban consumers are only possible because of the low prices paid to local farmers for their crops and livestock products. Low procurement prices for agricultural products also ensure higher profits on the sale of export crops, profits which often accrue to governments where marketing is through state-controlled organizations. Taxing the agricultural surplus explicitly and implicitly has left too little for reinvestment in agriculture to increase production.

A recent World Bank report (1980) analyses in detail agricultural price management in Egypt. It points out that Egypt follows an industry-led policy of economic development based on the theory that since productivity is generally higher in manufacturing than in agriculture, priority should be given to industrialization. The policy instruments for achieving this strategy have included taxation of the agricultural surplus to provide industrial investment funds, and maintenance of relatively low industrial costs through cheap food and low wages. Following the land reform measures introduced by Nasser, the state greatly extended its control over the agricultural sector, and a system of state-run cooperatives displaced private entrepreneurs in supplying inputs and became the major marketing channels for the most important crops. Government price and purchasing policies became the principal means of rural tax collection. In theory certain crops are sold exclusively to the government at only a fraction of the international price. Since farmers are assigned quotas of these crops, either the government or urban consumers get the difference.

In 1975 the Ministry of Planning conducted a series of studies into price related issues in agriculture. Their report estimates that in 1974 farmers had received 15 per cent, 16 per cent, and 30 per cent respectively of the actual export prices the state had received for their onions, rice, and cotton. The differentials between foreign and domestic prices, the report argues, brought windfall gains to the state. The weighted average export price of cotton lint rose by 118 per cent from LE 550/ton f.o.b. in 1972 to LE 1201/ton in 1974. World prices for rice climbed from US$ 147/ton in 1972 to US$ 542/ton in 1974, a 268 per cent increase. Meanwhile, farm gate prices for rice in 1974 had risen only by 3 per cent in total over the whole period since 1968. Seed cotton prices rose 21 per cent between 1972 and 1974 after a long period of slow increases. Furthermore transfers to Treasury from profits by the Cotton Organization totalled LE 348 million between 1973 and 1976, whereas the direct producer subsidies paid by the Agricultural Prices Stabilization Fund on all crops for all purposes amounted to only LE 187 million over the same period. If the exchange rate gains are added in, the transfers out on cotton alone more than pay for all the direct producer subsidies, all public sector investment in agriculture and all the current expenditures of both the Ministries of Agriculture and of Irrigation. The report concludes that the taxes on cotton cover all direct and indirect transfers into agriculture and still leave a transfer out. To this must be added the explicit output price flows gained from rice, oranges, and onion exports plus the implicit outflows on domestically consumed crops such as wheat, maize, and refined sugar.

Richards (1980), however, also points to the detrimental effect on the *national* economy of these price policies and resulting crop allocations. The cultivation of cotton, the principal export crop, has been reduced, with the loss of scarce foreign exchange. The reduction in wheat production and the necessity to keep wheat consumption more or less stable, contributed to the increasing burden of wheat imports. Some economists have estimated the losses of the contradictory system of controlling some but not all crops to be roughly the equivalent of the balance of payments deficit in the 1960s. Moreover, the negative effects of these policies have fallen unequally on the different social classes in the countryside. Because not all agricultural outputs have controlled prices, Richards argues that the richer farmers have been able to evade the regulations and to shift to more profitable uncontrolled crops such as vegetables, fruits, full-season clover, and milk and meat products.

Government price-support policies in Iran have also had a negative effect on the agricultural sector, acting more as a disincentive than as a support. Wheat is the most important crop grown in Iran, accounting for over 60 per cent of the land cultivated annually. Since the 1960s the main aim of government policy has been to stabilize the price of wheat which constitutes the basic element in the diet of the vast majority of Iranians. Every year the Cereals Organization set a floor price at which it would buy wheat from farmers. The effectiveness of the wheat price support policy was, however, limited. The level of the announced floor prices remained constant from the early 1960s to 1974 when the government announced new guaranteed minimum prices for wheat. Because the government did not announce its floor price until the wheat harvest was nearly completed, most peasant farmers had already been forced to sell to private wholesalers. The amount of wheat the Cereals Organization purchased in any one year was invariably only a fraction of the amount offered for sale, compelling farmers to sell to wholesalers for less than the government's floor price (Hooglund, 1981). In 1967 and 1968 wheat farmers in some parts of Iran were receiving prices of only 3,500–4,500 rials per ton from private wholesalers at a time when the government floor price was fixed at 6,000 rials per ton. Furthermore, the number of government collecting points was insufficient and no transport facilities to them were provided by the Cereals Organization. Finally the floor prices bore little relationship to the cost of production. By the mid-1970s the cost of producing one ton of wheat exceeded the floor price the government offered for one ton. However, at the same time, in order to stabilize the price of bread for urban consumers, wheat and wheat flour were distributed at subsidized rates to bakers. In January 1976 delivery price to licensed bakers was 7,510 rials per ton, about 75 per cent of the government buying price of 10,000 rials per ton (Aresvik, 1976). Thus by fixing the terms of domestic trade against the farmers, capital was effectively transferred from rural to urban areas, and there was an increasing reluctance to invest in agricultural production leading to increased requirements for agricultural imports.

Subsidized food prices for urban consumers have often been justified because they are redistributive, benefiting the lower income groups in the cities. In fact there is evidence that many less needy groups benefit, and some critics argue that the subsidized foodstuffs regularly enjoyed by the urban middle classes are an economic waste and a misuse of resources (Weinbaum, 1982). Policy makers are unwilling to try to discourage the demand for high-value foodstuffs such as meat and rice through high prices because such a policy is seen as politically dangerous. The price governments pay is gross distortion in their economies and serious balance of payments problems.

Low agricultural prices and subsidized food in the cities have acted as push and pull factors respectively contributing to the movement of peasants from the countryside to urban areas, and to labour markets in other countries, notably in the oil-rich economies of the Gulf and Libya. Low profits from farming and the possibilities of more attractive employment in urban areas have created acute shortages of skilled and even unskilled farm workers in some parts of the region.

In Iran the percentage employed in agriculture has been falling from 56 per cent in 1956 to 33 per cent in 1976, although this relative decline was accompanied by a small, absolute increase in the agricultural population from 3,326,000 in 1956 to 3,445,000 in 1976. Probably the most dramatic decline in the rural population has occurred in Iraq where shortages of agricultural labour have become so acute that efforts have been made to encourage Egyptian and more recently Moroccan farmers to settle in rural areas. In Egypt incomes in agriculture have been depressed by policy instruments. Real incomes in agriculture are falling and are still only one-third of those in industry. Even low-paid jobs in the private sector are preferred to farm employment since urban areas offer more accessibility to consumer subsidies and other income transfers not related to employment. 'Income differentials have been translated into stigmas against working in the agrarian sector' (World Bank, 1980). The government's 'full employment policy' has guaranteed a public-sector job to all university graduates and to those discharged from military service. As most public-sector jobs are in the cities, and yet the bulk of the population lives in rural areas, this policy has resulted in the large-scale movement of young people out of agriculture, adversely affecting the agricultural labour supply and the age distribution of farm workers. Compulsory military service has had a similar effect. Compulsory schooling has also reduced the numbers available to work on the land. Not surprisingly it is the students with the lowest scores in university entrance examinations that are assigned to the agriculture faculties, a policy which adversely affects the quality of professionals employed in the agrarian sector. Agriculture's share in employment has fallen from 54 per cent in 1959 to 42 per cent in 1977 and, although absolute numbers have increased from 4,057,000 workers in 1970 to 4,486,000 in 1977, there is evidence of increasingly tight rural labour markets.

The construction boom in the urban economies of the oil-rich states has acted as a magnet attracting labourers from the agricultural sectors of the capital-poor countries of the region. Rural shortages of labour in Oman, a result of manpower exports to the Gulf states and Saudi Arabia, have led to the neglect of the irrigation systems on which Omani agriculture depends, and have reduced seasonal crops in particular and agricultural output in general. Yemeni migrants for employment have also come mainly from rural areas and so represent a direct withdrawal of labour inputs from the agricultural system. In particular the loss of agricultural infrastructures with the collapse of farm terraces represents a permanent reduction in the potential of Yemeni agriculture. Shortages of agricultural labour in Jordan as a result of international migration have become so critical that replacement migration, mainly workers from Egypt and Pakistan, are now employed extensively on farms (Birks & Sinclair, 1980).

GOVERNMENT PLANNING POLICIES FOR
THE AGRICULTURAL SECTOR

As Weinbaum (1982) has perceptively noted 'agrarian policies are shaped to further a regime's efforts to manage conflict and secure authority. Rural development may help to realise promised revolutionary reforms as well as stave off forces of radical change. Programmes are often geared to augment the status and power of groups with which a government shares values and common political objectives.' In nearly all countries of the Middle East, governments have sought to reorganize the structure of land ownership by introducing programmes of agrarian reform. These reformative strategies have been used by regimes with very different political and economic systems and to achieve a variety of political, economic, and social aims. Invariably political considerations have taken precedence over economic factors. In Egypt and Iraq, land reform programmes were introduced following military coups and were intended to destroy the political power base of those large land owners who had dominated the deposed monarchical regimes. In Syria it was unification with Egypt which prompted reform legislation. The reform in Iran was one of many land reforms carried out by conservative Third World governments to eliminate a real or possible revolutionary threat from a discontented peasantry and to create a new social grouping in the rural areas that would support government policies. Land expropriated from rural elites was distributed to peasant farmers and the prevailing mode of land redistribution has been to extend small-scale private ownership. Some observers have emphasized that in many cases the plots distributed to reform beneficiaries were too small to be economically viable. Land was often of poor quality and those farmers who were obliged to buy their land with the help of long-term government loans found themselves in debt to distant and unsympathetic state banks. In Iran peasants paid at least the equivalent of ten years' rent for their land, and the majority paid more. A peasant who defaulted on the payment of instalments due for three consecutive years or more was liable to be dispossessed by the government and his holding assigned to someone else.

Others stress that a recurring feature of these reforms has been that they were unequal reforms which concentrated on encouraging a class of rich peasants. They argue that the decision to distribute land in this way is often defended on the grounds that as there is too little land to go round, the best way to boost the rural economy is to give land to those who have some experience of farming techniques, normally those who were tilling the land as tenants before the reform. Large sections of the rural population have certainly been bypassed by government reform programmes. Indeed the break-up of the large estates often succeeded in reducing employment opportunities for the large landless rural population, a substantial proportion of rural families throughout the Middle East — some 50 per cent in Iran and 40–45 per cent in Egypt. The beneficiaries of land distributed under the reform were encouraged to farm as family units, and thus had little or no need to hire labour. Such policies have aggravated rather than ameliorated class differences in the countryside.

Although the traditional landlord class has been considerably reduced, not surprisingly some large land owners have been able to evade expropriation, often by the subdivision of lands among relatives prior to the reforms, and to avoid the imposition of legal restrictions on tenancies. Significantly, no regime in the region has tried to abolish tenancy; in Egypt almost one half of the land is still under tenancy agreements (Weinbaum, 1982).

Land reform programmes have frequently been accompanied by the introduction of new institutions, and the reorganization of production relations. A widespread mode of reorganization has been the state-managed farm cooperative created to provide the beneficiaries of reform land with credits and farm inputs and to provide for the marketing of their crops. Membership is often compulsory for beneficiaries of reform land. Farmers retain legal title to land and the family remains the most common labour unit. The cooperative provides certain economies of scale to offset the land fragmentation created by the reforms. The state-appointed administrative and technical personnel are supposed to ensure that through crop planning and the enforcement of quotas the farmers' basic needs are met and the government's agricultural policies at national level are implemented.

In spite of some early successes, cooperatives have experienced many operational problems. Government provision of vital credits for the purchase of seed, fertilizers, and irrigation equipment has been insufficient and often delayed. Farm equipment has been in short supply and the maintenance of agricultural machinery inadequate. Many farmers came to resent their loss of control over decision-making, which increasingly passed to state-appointed bureaucrats, and peasant individualism remained strongly entrenched. One survey of cooperative members in Egypt revealed that the majority of members preferred freedom to deal directly with the private sector. Cooperative officials have been accused of identifying with the richer members and acting against the interests of the small farmer. A common criticism voiced is that cooperatives are dominated by the well-off *Kulak* farmers who manipulate the new institutions to serve their interests.

An alternative to individually oriented farm cooperatives favoured by many development planners in the Middle East is the centrally managed collective or state farm in which members are paid either according to the number of hours worked or by a fixed wage. They are seen as more satisfactory structures to achieve rational production and land use, to bring about vertical integration of production with marketing and food-processing activities, and to absorb the landless peasants. For many planners, cooperatives perpetuate peasant mentalities and attitudes to land and property, and inhibit acceptance of more advanced farming techniques. Some regimes have acknowledged that large, state-owned and -managed farm units may be the only means to increase marketable surpluses, provide more employment, and extract funds needed for industrial development. Nevertheless governments have moved very cautiously in introducing the collective or agri-industrial options which in many cases challenge prevailing social and economic institutions. Farmers accustomed to owning and working their own land have generally resisted incorporation into large-scale enterprises. Experience, moreover, has proved that such units have been too big for effective management; most have suffered from under-financing and low productivity of land and capital. In Egypt, the collective management of the Tahrir Province Reclamation Project experienced serious technological and administrative difficulties and low productivity, and this approach was abandoned by the Nasser regime. Sadat resisted pressures to extend collectivization to other land reclamation schemes, favouring the exploitation of new lands by large commercially oriented private firms.

Attempts by the region's planners to reorganize agricultural production have met with only limited success. No single mode of production, whether family farms, cooperatives, or large state enterprises, has demonstrated any long-term success in satisfying a country's economic, political, and social objectives. All too often new institutions have been given little time to develop fully and to resolve their problems before being replaced by new modes of production. Frequently such abrupt policy changes have been dictated by political expediency rather than by sound economic judgement. 'Repeated political elite turnover results in a discontinuity of development ideas and personnel and a failure to accomplish many realizable objectives. Worthy projects are likely to be abandoned simply because of their identification with the discredited officials and deposed cliques' (Weinbaum, 1982).

Iran provides a classic example of the failure of the region's planners to establish the basis for a modern agricultural system and to improve the economic well-being of the majority of the rural population. The Shah's regime chose to make land reform the symbol of its 'White Revolution' with its promise of 'land to the tiller'. The regime needed an ally to offset the growing influence of the disenfranchised educated middle class and believed that such a force could be found among the upper strata of the peasantry. At the same time, the Shah's foreign advisers were insisting that the country's prospects for economic development were being undermined by the continued lack of a peasant market to act as an incentive to industrial investment. Under the initial land reform decree promulgated in 1962, ownership was limited to one village or to six *dangs* in separate villages. Orchards, tea plantations, and mechanized areas were exempt. Land was to be distributed to those who were already farming land with priority going to those who provided more than just labour. All those receiving land had to become members of cooperatives.

A considerable number of land owners were able to evade redistribution and a decade after its implementation only some 30 per cent of villages had been affected, and of them less than 10 per cent had been wholly redistributed. It has been estimated that some 700,000 families, under a fifth of all peasants, benefited from this phase of reform. The second stage of the reform was designed to cover land which had not been affected during the first stage. A total of about 210,000 families acquired land under phase two, most landlords deciding to retain their lands and to reach tenancy agreements with the peasants. With little improvement in agricultural productivity and peasants reportedly resisting tenancies, a third phase of the reform was introduced in 1969 designed to convert the tenancy arrangements of phase two into ownership. In all about 1.6 million families were given land, under half the total number of rural families in Iran (Richards, 1975).

Membership of a village cooperative was obligatory for beneficiaries of reform land. By the early 1970s there were some 6,700 cooperatives covering about two-thirds of all villages, but many existed in name only. Initially the cooperatives were starved of government funds because the state feared the consequence of strong peasant-run organizations. Even when control of the cooperatives passed to a government body, the Central Organization of Rural Cooperatives, the total sums allocated to those new institutions were only a fraction of the real credit needs of the peasantry. Marketing operations were slow to develop and most cooperatives were confined to providing credit and in some cases consumer goods. Because of the limited flow of government funds to cooperatives the credit needs of many farmers continued to be met by traditional money-lenders. A survey of credit sources for relatively prosperous peasant families in several districts carried out in 1968 revealed that money-lenders still provided 38.6 per cent to 58.5 per cent of the loans received, cooperatives only 19–45 per cent. A member of a cooperative was able to borrow money according to his share in the cooperative which depended on the amount of land owned, a system which favoured the richer peasants (Richards, 1975).

From the mid-1960s dissatisfaction with the performance of the cooperatives prompted the Shah to encourage the creation of farm corporations, professionally managed, collectively owned farm units modelled on the Israeli *moshavim*. Farm corporations were formed provided that 51 per cent of the owners considered eligible gave their consent. Members received shares in proportion to the amount of land and property contributed, but use of their land passed to the corporation; they were paid partly as a wage for their labour input and partly as dividends from their shares. The purpose was to introduce modern mechanized farming techniques and to consolidate uneconomic holdings into units of at least 20 hectares. Despite widespread resistance, by 1971 43 corporations had been established, ranging from the small Aryamehr Farm Corporation with 80 shareholders and 2,000 shares to the huge Shahbad enterprise with 1,246 shareholders and 67,012 shares. By the mid-1970s farm corporations still accounted for only 5 per cent of total agricultural output but they received heavy government investment — an average of 32.5 million rials per corporation in 1970 compared with 2.9 million rials per cooperative.

The Shah's impatience with slow production increases from the peasants brought a further change in agricultural policy. In addition to the farm corporations the regime turned to foreign and domestic agri-business companies to increase production through capital intensive mechanization. Foreign business interests had become involved in Iranian agriculture in the late 1950s when David Lilienthal, former head of the American Tennessee Valley Authority, promoted a similar operation in the province of Khuzistan utilizing water from a dam on the Dez River. When production increases in the Dez Irrigation Project proved slow to develop, mainly because the farmers were not provided with sufficient credit and extension services, the government decided to divide the 250,000-acre area into large mechanized family units. Farmers in some 58 villages within the project area were bought out by the government, their villages destroyed, and the land handed over to six private agri-business companies. Many of the dispossessed peasants were reduced to working as seasonal labourers on land they had once owned. Dramatic increases in production were not realized, and the agri-business ventures were soon submerged by a host of problems. It had been assumed that with economies of scale, crops could be grown more efficiently. Yet experience proved that beyond a certain point economies of scale did not hold and costs began to mount as extra layers of management were needed. A comparison of field crops in California and Khuzistan found that, whereas the Californian farmers obtained lowest minimum costs at somewhat less than 300 hectares, the Iranian units covered thousands of hectares. By the mid-1970s enthusiasm for agri-business began to wane and a new policy of encouraging medium-sized farmers was introduced (Halliday, 1979).

CONCLUSION

In shaping agricultural policies, regimes in the Middle East have been too preoccupied with placating dominant urban interests to commit adequate resources to the rural sector. Rapid urbanization, population growth and rising disposable incomes will continue to raise food consumption far beyond the capacity of domestic production during the next decade. There is little cause for optimism and agriculture seems destined to remain the Achilles' heel of development efforts throughout the 1980s. With a growing urban population increasingly dependent on and accustomed to subsidized food prices, governments may be even more unwilling to increase the prices paid to farmers for their crops. In effect the surplus being extracted from the agriculural sector will probably continue to help maintain peace in the cities rather than be channelled into productive investment. The food riots in Egypt in 1977 and Sudan in 1979 demonstrated the effectiveness of the food issue in mobilizing the disaffected urban masses. Food and agricultural issues are almost certain to become more acute and politically explosive during the 1980s (Weinbaum, 1982), severely restricting government choices in the formulation of policies for the agricultural sector.

In many parts of the countryside the position of the richer peasants has been strengthened and consolidated by state intervention. As governments tend to limit their efforts to 'technical' solutions and to rely on market forces to strengthen output, these wealthier farmers are being given more freedom to choose the types of crops they are allowed to grow and the commercial agreements they make. They have acquired a stake in the maintenance of the established economic and social order and oppose the extension of land ownership to less privileged groups. This dominant class may well succeed in blocking government attempts to increase employment in rural areas and improve the position of the landless peasants. Implementation of ambitious plans to increase non-agricultural employment in rural ares in order to raise rural incomes has been slow and many peasants see migration to the cities as the only path to an improvement in their income and standard of living.

In the privileged urban economy there are serious doubts that the emerging industrial sector, in spite of heavy investment, will succeed in acting as the 'motor' for future development. Industrial capacity is frequently underutilized, foreign outlets for industrial products have proved elusive, and the local market is restricted because of low rural incomes. Significantly as this modern sector is capital intensive it has created only limited employment opportunities and has proved incapable of absorbing more than a fraction of the number of people coming on to the urban labour market each year. At the same time urban-based bureaucracies are already overstaffed and essentially parasitic. It seems unlikely that the urban/industrial development option will be able to satisfy the growing demand for urban jobs that the shift of labour from agricultural to non-agricultural activities has created. Severe and mounting unemployment and underemployment among young and increasingly educated urban dwellers may be one of the most serious and politically dangerous consequences of recent agrarian policies. Furthermore the price paid for some of the more disruptive interventions in the structure of rural society may be the loss forever of alternative approaches to the urban-industrial model of development (Siddle, 1978).

REFERENCES

Aresvik, O. (1976) *The Agricultural Development of Iran*, Praeger, New York.
Birks, J. S., & Sinclair, C. A. (1980) *International Migration and Development in the Arab Region*, International Labour Office, London.
Halliday, F. (1979) *Iran: Dictatorship and Development*, Penguin, Harmondsworth.
Hooglund, E. (1981) 'Iran's rural inheritance', *MERIP Reports*, no. 99, pp. 15–19.
Khuri, F. I. (1981) *Leadership and Development in Arab Society*, American University of Beirut, pp. 1–13.
Lipton, M. (1977) *Why Poor People Stay Poor: Urban Bias in World Development*, Temple Smith, London.
Owen, R. (1981) 'The Arab economies in the 1970's', *MERIP Reports*, no. 100–101, pp. 3–13.
Paul, J. (1981) 'Perspective on the land crisis', *MERIP Reports*, no. 99, pp. 3–6.
Richards, A. (1980) 'Egypt's agriculture in trouble', *MERIP Reports*, no. 84, pp. 3–13.
Richards, H. (1975) 'Land reform and agribusiness in Iran', *MERIP Reports*, no. 43, pp. 3–18 & 24.
Sadek, M. (1981) 'The role of the public sector in development in the Arab world', in Fuad I. Khuri (Ed.), *Leadership and Development in Arab Society*, American University of Beirut, pp. 168–185.
Sayigh, Y. A. (1982) *The Arab Economy: Past Performance and Future Prospects*, Oxford Universtiy Press, Oxford.
Siddle, D. J. (1978) 'Rural development and rural change', in Alan B. Mountjoy (Ed.), *The Third World: Problems and Perspectives*, Macmillan Press, Basingstoke, pp. 112–120.
Tubiana, L. (1981) 'L'Egypte: agriculture, alimentation et géopolitique des échanges', *Maghreb/Machrek*, no. 91, pp. 24–42.
Weinbaum, M. G. (1982) *Food, Development and Politics in the Middle East*, West View Press, Croom Helm, London.
World Bank (1980) *Agricultural Price Management in Egypt*, World Bank Staff Working Paper No. 388.

Part II
Country Studies in Agricultural Development

Agricultural Development in the Middle East
Edited by P. Beaumont and K. McLachlan
© 1985 John Wiley & Sons Ltd

Chapter 7

Nile Water for Whom? Emerging Conflicts in Water Allocation for Agricultural Expansion in Egypt and Sudan

DALE WHITTINGTON
and
KINGSLEY E. HAYNES

INTRODUCTION

The Nile has historically provided Egypt and the Sudan with a water supply which has been the envy of the rest of the Middle East and North Africa. Since the completion of the Aswan High Dam in 1970, water has been in such plentiful supply in Egypt that, before his assassination, Anwar Sadat publicly entertained the idea of selling water to Israel. Economic and population growth are gradually bringing this period of water abundance to an end.

This paper examines some of the issues involving water allocation and agricultural expansion for the downstream states—Egypt and Sudan. In the first section we present a brief introduction to the agricultural sectors in Egypt and Sudan. The second section reviews the existing allocation of Nile waters between Egypt and Sudan as specified in the 1959 Nile Waters Agreement. In the third section we examine the projected water supply and demand balance and the value of water in Egypt and Sudan. The fourth section addresses the question of how water can be reallocated in the long term for more efficient agricultural expansion in Egypt and Sudan.

THE AGRICULTURAL SECTORS IN EGYPT AND SUDAN: AN OVERVIEW

Agriculture in Egypt is almost completely confined to the Nile Valley and Delta, a handful of oases, and some arable land in Sinai. About 2.6 million hectares are cultivated annually, approximately 3 per cent of the total area in the country. About 2.4 million hectares are Old Lands in the Nile Valley and 0.2 million hectares are recently reclaimed areas. Although the agricultural sector currently accounts for roughly 40 per cent of the total employment in Egypt, agriculture's contribution to the GDP is only 24 per cent.

With the completion of the Aswan High Dam, perennial irrigation and multiple cropping have been extended throughout Egypt; an average of about two crops are grown annually (Fitch & Abdel Aziz, 1980). As shown in Figure 7.1, yields for many crops are high by world standards despite the low level of capital investment, often primitive technology, and fragmentation of land holdings. Such comparisons can be deceptive, however, because Egypt's natural factor endowments are high. The Old Lands of the Nile Valley are among the richest agricultural areas in the world, and climatic conditions in terms of temperature and duration of sunshine are extremely favourable. There is still undoubtedly considerable room for improvements in crop yields. A recent United States Presidential Mission on Agricultural Development in Egypt concluded that long-range potentials for increasing output are of the order of 200 per cent and even higher for many crops (USAID, 1982).

Egyptian agriculture operates under a complex 2–3 year system of crop rotations. As illustrated in Figure 7.2, the major winter field crops are clover and wheat; the major summer crops are cotton, maize, and rice. The crop mix is in large part determined by government pricing policy. The productivity of Egyptian agriculture has lured central governments to extract surplus production from farmers throughout history, and this has been the explicit government policy since the 1952 Revolution. The government has set prices substantially below international levels for cotton, basic food grains, and many other

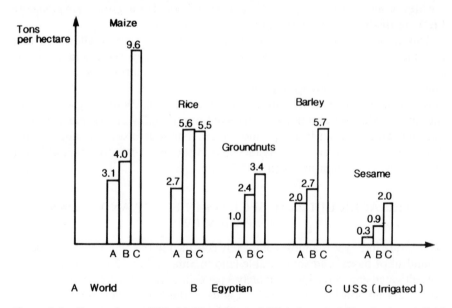

Figure 7.1   Comparison of World, Egyptian and US irrigated yields of selected field crops (1978–80 averages)

commodities, as part of a system of *de facto* taxation. Low prices have been coupled with mandatory delivery quotas for grains and government control of all marketing channels for cotton, plus a system of obligatory acreage requirements. These policies have resulted in an average of 30 per cent net tax on agricultural output, which has contributed to the maintenance of low food prices for consumers, and has enabled the government to obtain foreign exchange to finance industrialization efforts and defence purchases. In one sense this policy has been successful. Real economic growth since 1978 has averaged 8–10 per cent, and the services and petroleum sectors have grown rapidly. The cost has, however, been a low rate of growth in the agricultural sector, about 2.5 per cent annually.

To avoid quota and acreage restrictions, some farmers have shifted to the production of orchard and truck crops. Because the government has not had effective means of controlling meat and fodder prices, there has been a shift to the growing of clover, which at almost 1.3 million hectares is the country's

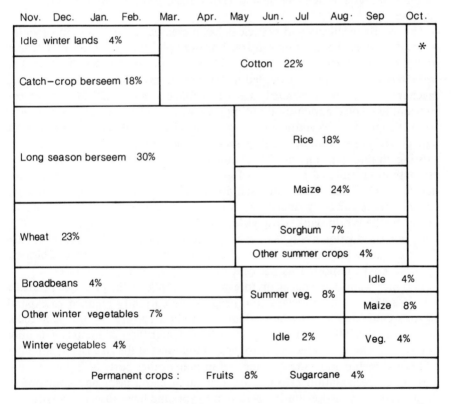

* Represents lands which are temporarily idle between summer crops, e.g. cotton and rice, and winter crops, e.g. berseem.

Figure 7.2 Egyptian cropping pattern in 1977–79

largest single crop. In spite of mandatory acreage rules, land planted in cotton and food grain has declined over the past 20 years. Cotton is currently planted on slightly over 20 per cent of the cultivated lands from March to October; wheat is planted on the same percentage from November to May.

The Ministry of Irrigation delivers water for irrigation to a schedule of rotations. Four- to seven-day rotations are typical, with water available for one-third to one-half of the time. Vegetables and rice may receive water more frequently than other crops. The extremely small size of the landholdings and operational units often makes it extremely difficult, if not impossible, to deliver a specified quantity of water to a particular crop. There is no metering of this water, and currently no charge for its use. With the abundance of summer water provided by the High Dam, the path of least resistance for the Ministry of Irrigation is often to supply excess water in order that the tails of the canals receive ample supplies, regardless of the amount abstracted by other users. Capacity constraints in some of the major canals, however, currently limit the amount of water which can be delivered during peak demand periods in the summer.

The use of fertilizers and pesticides has increased dramatically in Egyptian agriculture over the last thirty years. For example, the quantities of nitrogen and phosphate fertilizers applied in 1975 were over four times those applied in 1950 (El Tobgy, 1976). Coupled with new seed varieties and increased farm mechanization, the expanded use of fertilizers and pesticides produced substantial results; agricultural output grew rapidly (3.5–4.0 per cent per year) from 1955 to 1965. Nevertheless, a lack of fertilizers is still a constraint in some agricultural areas. Moreover, the increased availability of water has exacerbated existing drainage problems. The United States Department of Agriculture estimated that fully 80 per cent of Egypt's agricultural area suffered to some degree from waterlogging and salinity in 1976 (USDA, 1976). A massive programme is currently under way for the installation of tile drains, funded in part by the World Bank. Crop yields levelled off over the period 1965—1975, and grain production has fallen far behind consumption needs. In fact, in 1974 Egypt became a net importer of agricultural products for the first time in history. By 1981 Egypt imported 48 per cent of its staple food commodities.

This increasing dependency on foreign food supplies is the principal rationale for an ambitious land reclamation programme currently planned to expand Egypt's cultivated area by more than one million hectares by the year 2000, more than 300,000 of which have water lifts greater than 20 metres. The bulk of this area is located immediately to the east and west of the Nile Delta, although additional reclamation sites are located in many other parts of Egypt (Figure 7.3). Land reclamation efforts have generated enthusiasm and controversy throughout the post-Revolution period in Egypt, and have absorbed 40 per cent of the investment in the agricultural sector. The recent history is difficult to evaluate (Hunting Technical Services, 1979–80; Voll, 1978), but the gist of the

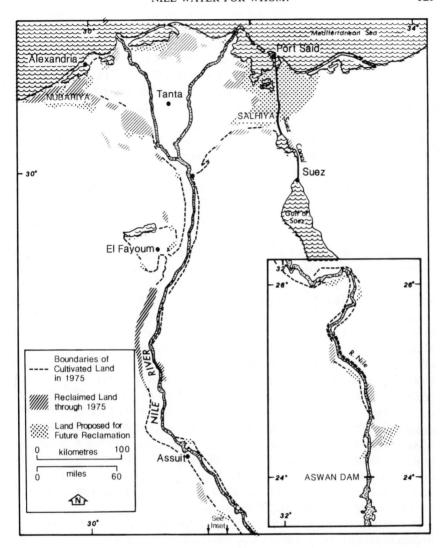

Figure 7.3 The River Nile below Aswan

story is that desert reclamation has proved an expensive and to date a largely unrewarding enterprise. Desert soil properties are generally very poor, and land preparation has taken much longer and has been more expensive than anticipated. The maintenance and operation of the irrigation and drainage systems has proved more difficult, both technically and organizationally, than in the Old Lands of the Nile Valley. Energy costs associated with pumping are high in many areas. The 200,000 hectares of reclaimed lands actually under cultivation account for only 2 per cent of agricultural production. There is little doubt that investment in vertical expansion and intensification in the old lands of the Nile Valley offers greater rates of return than desert reclamation efforts (Pacific Consultants, 1980; Hunting Technical Services, 1979–80; USAID, 1982).

The recent US Presidential Mission to Egypt (USAID, 1982) has recommended a series of changes in Egyptian agricultural policy in order to stimulate agricultural development, including:

(1) eliminating current government commodity pricing policies;
(2) upgrading agricultural research and extension services;
(3) consolidation of existing ministerial responsibilities;
(4) increasing public sector investment for physical infrastructure such as roads, marketing facilities, household electricity, and water supplies;
(5) permitting more private marketing of agricultural produce.

A movement toward a more market-oriented agricultural sector would presumably increase efficiency in a variety of ways, such as increased yields, more rational crop mix, better marketing practices, less spoilage, greater attention to quality differentials, and wiser use of inputs such as water and fertilizer. Increased yields coupled with higher cropping intensity would result in increased crop water requirements, perhaps of the order of 10–20 per cent. More rational commodity pricing policies would also discourage the cultivation of clover — currently Egypt's largest single crop — and promote the planting of crops for which Egypt has a comparative advantage, such as cotton and orchards. However, this shift in crop mix is unlikely to have a significant impact on total crop water requirements. Rice and sugarcane are the only crops with dramatically higher water requirements. (Whittington & Haynes, 1980).

Agriculture plays a proportionately greater role in the economy of Sudan than in Egypt, accounting for over a third of Sudan's GDP and over 90 per cent of exports. Eighty per cent of the Sudanese labour force is engaged in agriculturally related activities, and agriculturally based industries represent 80 per cent of the total manufacturing sector.

In contrast to Egypt, the majority of cultivated land in Sudan is rain-fed. Sudanese agriculture uses about 11.2 million hectares, of which 9.5 million are rain-fed. About a quarter of the rain-fed area is used in mechanized agriculture; the remainder is cultivated with traditional techniques. As much as 35 million hectares in Sudan may be potentially arable, and another 100 million hectares are suitable for grazing land. The most fertile areas are the clay plain in the centre of the country on either side of the Blue Nile. The western provinces of Kordofan and Darfur have light, sandy soils suitable for rain-fed cultivation of sesame, groundnuts, and watermelons. The soils in the Equatorial provinces in the southern Sudan are for the most part poorly suited for cultivation. Along the Uganda border, however, there is an area where coffee, cocoa, and other tropical crops could be grown.

Sudan has developed about 1.7 million hectares of irrigated land, over 95 per cent of which is irrigated with Nile waters. As illustrated in Table 7.1, the total irrigated area in the Sudan has risen continually since the turn of the century. Table 7.2 details the size of the existing and proposed irrigation schemes in Sudan (see also Figure 7.4).

In contrast to Egypt, the irrigation and mechanized rain-fed schemes in Sudan are government-owned and have varying degrees of government involvement

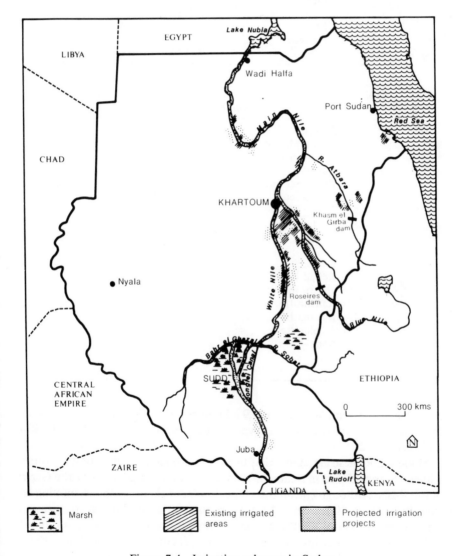

| | Marsh | | Existing irrigated areas | | Projected irrigation projects |

Figure 7.4   Irrigation schemes in Sudan

Table 7.1   Growth of irrigated area in Sudan

Year	Irrigated command area (10³ha)
1920	17
1925	44
1935	126
1955	588
1965	1,176
1975	1,575
1977	1,701
Planned	
1990	2,394

Table 7.2   Existing and proposed irrigation schemes in Sudan (million ha)

Existing	
Gezira-Managil	0.87
Khashem El Girba	0.17
Rahad	0.13
Pump Schemes South of Khartoum	0.59
Pump Schemes North of Khartoum	0.03
Sub-total	1.70
Planned	
Rahad II	0.13
Gezira Intensification	0.11
Kenana II	0.13
Kenana III	0.13
Jebelin–Renk	0.08
Renk–Gelhak	0.10
Sub-total	0.68

in their management. With about 0.9 million hectares, Gezira–Managil is today the largest irrigation scheme in the world under one management (Barnett, 1974; Gaitskell, 1952).

Modern Sudanese agriculture has been built around the cultivation of cotton, which was introduced in Sudan in 1904, and into irrigated areas of the Gezira in the 1920s. Although cotton is still the dominant export crop, accounting for slightly less than 50 per cent of total exports, cotton production constitutes a smaller population of GDP than cereal crops (principally dura or sorghum). Other major export crops include gum arabic, sesame, and groundnuts. Yields of most crops are considerably lower in Sudan than in Europe, but so are

Table 7.3   Trends in production in irrigated agriculture, in Sudan 1970–80

	1970–71	1971–72	1972–73	1973–74	1974–75	1975–76	1976–77	1977–78	1978–79	1980–81	Average 1970–84	Average 1975–80
Cotton												
Area (000 fd)	1,015	1,018	1,037	1,042	1,078	802	851	937	884	836	1,038	869
Yield (tons/fd)	0.70	0.64	0.51	0.63	0.58	0.36	0.51	0.57	0.42	0.31	0.61	0.43
Production (000 tons)	712	660	532	656	631	294	435	534	375	259	638	371
Groundnut												
Area (000 fd)	186	167	270	296	384	553	346	380	311	232	261	372
Yield (tons/fd)	0.68	0.67	1.03	0.89	1.19	0.74	0.74	1.12	0.79	0.69	0.94	0.85
Production (000 tons)	126	111	278	262	456	411	255	426	246	159	247	317
Sorghum												
Area (000 fd)	629	511	406	545	402	628	471	539	591	503	499	529
Yield (tons/fd)	0.42	0.42	0.75	0.54	0.42	0.46	0.45	0.50	0.44	0.42	0.52	0.44
Production (000 tons)	263	215	363	292	171	288	212	268	259	209	261	234
Wheat												
Area (000 fd)	289	288	248	419	591	690	639	601	566	436	367	565
Yield (tons/fd)	0.56	0.43	0.61	0.56	0.46	0.33	0.46	0.53	0.30	0.33	0.51	0.41
Production (000 ton)	162	124	152	235	269	225	294	317	168	145	188	230
Total irrigated area	2,119	1,984	1,961	2,302	2,455	2,673	2,307	2,357	2,352	2,007	2,165	2,335

Source: Dept. of Agricultural Production and Statistics, Sudan.
Note: Excludes sugar production.

One feddan = 0.42 hectare.

production costs. Cultivation is less intensive, and the use of fertilizer and pesticides is lower.

Table 7.3 presents the area in irrigated cultivation and the average yield for the major crops in Sudanese agriculture from 1970 to 1980. As illustrated, cotton yields declined dramatically over the decade, largely because of poor government management and pricing policies in the Gezira–Managil. Under a system of joint accounts for charging tenants for inputs, the management discouraged the use of inputs for the cotton crop. Moreover, tenants actually received only about 20 per cent of the export price for their cotton. In addition, irrigation supplies were often insufficient and unreliable owing to capacity constraints in the irrigation distribution system and a shortage of summer water. Moreover, the amount of water delivered is set at 950 m^3 per hectare regardless of the crop or its stage of growth.

This decline in cotton yields wreaked havoc with Sudan's economic development plans, despite rising cotton prices over the period. In 1981, the World Bank made funds available to Sudan for an emergency rehabilitation effort of the Gezira scheme on the condition that Government pricing and management policies be altered so that farmers responded more closely to real resource costs. As a result, in 1981 and 1982 cotton yields were back to levels at the beginning of the 1970s. However, the Geriza–Managil scheme, as well as other irrigation schemes, are still operating at far below potential. The cropping intensity of the Gezira is about 0.7, and for Rahad and Managil 1.0 (even at this low level of intensity the Gezira–Managil scheme consumes one third of Sudan's Nile waters allocation). Given past investments in infrastructure, rehabilitation of existing irrigation schemes is the top investment priority in the agricultural sector, and perhaps the economy.

Traditional agriculture in Sudan is characterized by shifting cultivation, the use of hand tools, and a lack of pesticides, fertilizers, and high-yielding crop varieties. The majority of Sudan's population is engaged in traditional cultivation, including almost the entire population of southern Sudan. Crop yields are highly variable, owing to large fluctations in annual rainfall, but are in general low. Soil fertility is declining in many areas owing in part to reduced fallows resulting from population growth and encroaching desertification. Millet and sorghum are the main subsistence crops, but the traditional sector also produces significant amounts of groundnuts, livestock, and gum arabic for export markets. Land is legally owned by the government, but is commonly allocated by community traditions. The traditional sector is the focus of USAID development efforts in accordance with its stated policy of helping the poorest of the poor, but research is needed on farm level economics and management to identify appropriate investment strategies.

Although large-scale, rain-fed, mechanized agriculture is only 40 years old in Sudan, it is making a substantial contribution to the economy, contributing about 50 per cent of the country's sorghum production and 25 per cent of its sesame. About 75 per cent of the mechanized farming is done on the heavy clay plains in the Kassala and Blue Nile provinces in eastern Sudan (south-east of Khartoum), within the 400–800 mm rainfall zone. There was previously little traditional agriculture in this region, and population densities were correspondingly low. Approximately 80 per cent of the cropped area is in sorghum, with smaller amounts of sesame and cotton. Soil fertility is declining in many areas owing to the often exclusive cultivation of sorghum and the absence of fertilizer use. About 20 per cent of the land currently used for mechanized cultivation is left fallow every year.

There are two institutional arrangements under which mechanized farming is carried out. Most mechanized cultivation is done by private entrepreneurs who lease 420 hectare blocks of land from the Mechanized Farm Corporation, a parastatal organization within the Ministry of Agriculture. A much smaller portion of mechanized cultivation is carried out on state-owned estates actually operated by the government. In practice, large entrepreneurs with significant capital resources who operate out of larger cities as absentee landlords have controlled most of the mechanized operations. Current government policy calls for a large expansion of the mechanized agriculture sector, better management and agricultural practices in the existing areas, and increased opportunities for participation of cooperative societies and of individuals with limited access to private capital markets.

PRESENT ALLOCATION OF NILE WATERS BETWEEN EGYPT AND SUDAN: 1959 NILE WATERS AGREEMENT

At the turn of the century it was calculated that an additional 3.5 billion m³ of summer water would supply all possible requirements in Egypt, which would be met by annual storage at Aswan and Wadi Halfa. Cognizant of the long-term development potential in the Nile basin, Garstin (1904) noted that such works 'would leave untouched the countries bordering the Nile to the south. Their interests must be safeguarded by such a scheme as will insure them a proportional share in the prospective benefits'. The Nile Waters Agreement of 1929 was the first step in this direction. Prior to the creation of over-year storage on the Nile, Egypt's principal concern was an 'adequate' supply of Nile water during the low flow period between January and July, which was required for the irrigation of its summer cotton crop. The 1929 Agreement allocated only 4 billion m³ to Sudan compared to 48 billion for Egypt, and reserved the entire 'timely' flow for Egypt, obliging Sudan to irrigate the Gezira scheme only during the winter months.

The construction of the Aswan High Dam fundamentally changed the basis of water allocation between Egypt and Sudan. The 1959 Nile Waters Agreement established a more equitable distribution of water between Egypt and Sudan (Table 7.4). It raised Egypt's water allocation to 55.5 billion m³ and Sudan's

Table 7.4   A comparison of the Nile Waters Agreement of 1929
and 1959

	1929 Nile Waters Agreement[a] $(m^3 \times 10^9)$	1959 Nile Waters Agreement[b] $(m^3 \times 10^9)$
Egypt's share	48.0	55.5
Sudan's share	4.0	18.5
Unallocated	32.0	0.0
Storage losses	0.0	10.0
	84.0	84.0

[a]Other provisions included:
  (1) Established Egypt's right to monitor flows in Sudan, to undertake projects without the consent of upper riparian states, and to veto any construction projects which would affect her interests adversely.
  (2) Reserved the entire timely water supply (January 20–July 15) for Egypt.
[b]Other provisions included:
  (1) Established a Permanent Joint Technical Commission to oversee the Agreement, coordinate a unified Egyptian–Sudanese position for negotiation with other riparian states, and undertake construction of water conservation projects.
  (2) Equal sharing of any natural increase in the yield of Nile, of water losses arising from allowed claims by other riparian states, and of investment costs and yield increases from engineering works on the Upper Nile.

to 18.5 billion m³. The agreement also eliminated the timely water clause. This allocation was based on an assumed annual flow at Aswan of 84 billion m³. After Sudan withdraws 18.5 billion m³ as measured at Aswan (20.35 upstream at Sennar), the remaining 65.5 billion m³ enters the Aswan High Dam reservoir. Evaporation and seepage losses are assumed to average 10 billion m³ and Egypt is left with its annual allocation of 55.5 billion m³.

Actually, the mean flow at Aswan over the last century is over 90 billion m³, not the 84 assumed in the 1959 Agreement. The lower estimate is clearly favourable to Egypt because the difference flows into the Aswan High Dam reservoir for her sole use. On the other hand, annual evaporation and seepage losses are closer to 14 billion m³ rather than the 10 billion m³ originally anticipated.

The 1959 Nile Waters Agreement established the Permanent Joint Technical Committee (PJTC) to implement its provisions. The work of the PJTC included planning for the development of the Nile basin, collection of data, and supervision of any construction projects. Sudan and Egypt agreed on a 50–50 sharing of investment costs and of increases in yield from engineering works on the Upper Nile. The 1959 Agreement also called upon Egypt and Sudan jointly to review any claims on Nile waters by other riparian states, and if such claims were accepted, Egypt and Sudan would reduce their allocation equally.

The result of the 1959 Nile Waters Agreement was to sanction Egypt's historic claim to the majority of the Nile waters based upon the principle of prior appropriation—'first in time, first in right'. Sudan, with close to half the population of Egypt and twice the amount of high quality land suitable for irrigation, received 20 per cent of the total water supply. For her part, Egypt feels she has dealt with the issue of Nile water allocation honourably since, until the twentieth century, she was the only state in the basin to rely upon irrigated agriculture.

WATER SUPPLY AND DEMAND

How do these Nile water allocations compare with the existing and forecast water use by Egypt and Sudan? There have been several recent attempts to compare annual water supplies and demands over the next couple of decades all of which focus on average conditions and ignore the risks associated with annual fluctuations in river flow (Egyptian Water Master Plan, 1980; Waterbury, 1979, 1982; Haynes & Whittington, 1981). The results of these calculations depend principally on four assumptions:

(1) crop water requirements per hectare in the Old and New Lands (including conveyance losses in the irrigation distribution system);
(2) extent of agricultural expansion through desert reclamation programme;
(3) completion schedule of upper Nile water conservation projects;
(4) extent of drainage water re-use in Egypt.

Table 7.5 presents a comparison of supply and demand estimates for Egypt in 1980 and 1990 prepared by the Egyptian Water Master Plan (1980) and John Waterbury (1982).

It is important to note that the differences in water use forecasts arise largely from uncertainties in policy variables, rather than natural phenomena or unforeseen technological changes. The expansion of new lands, the crops grown, and irrigation technologies utilized all depend on government decisions. Even the differences in the estimates of crop water requirements are largely a function of assumptions about the ability of the government both to improve field water management and drainage, and to increase crop yields by removing other input constraints besides water.

Similarly, on the supply side it is technologically feasible to complete the upper Nile projects by the end of the century, which would increase the annual supply at Aswan by about 9 billion m^3. Whether this will actually happen is another question. Waterbury (1982) has argued convincingly that given its current economic morass and other pressing investment priorities, Sudan will not be able to undertake any additional upper Nile projects after Jonglei 1 for the foreseeable future. The baseline engineering and environmental impact studies are still lacking. Indeed, at the time of this writing (November, 1983) construction on the Jonglei canal itself has been indefinitely halted by civil unrest in Southern Sudan.

The fact that the uncertainties in forecasts are due to policy variables has two important implications. First, if supply and demand are not in balance, it means that there are conflicting policies which must be reconciled. For example, the implication of the calculations presented by Waterbury in Table 7.5, is that Egypt cannot expand its agricultural sector with present agricultural and water use practices at the scale currently contemplated.

Second, if supply and demand forecasts are in balance, the policies on which the forecasts are based must turn out to be successful. For example, the Egyptian Water Master Plan concluded that Egypt has ample water supplies, to proceed with its desert reclamation plans. This assumes that (1) substantial improvements in field level water management will take place, (2) crop water requirements in both old and new lands will then be reduced significantly from current levels, and (3) the Upper Nile projects will be contracted on schedule.

Table 7.5   Estimates of water demand and supply in Egypt: 1980 and 1990 billion m³

	1980		1990	
	Egyptian Water Master Plan	Waterbury	Egyptian Water Master Plan	Waterbury
Demand				
Old Lands	29.4	32.4	29.4	33.0
New Lands	—	4.8	8.5	11.2
Munic. net loss	1.8	3.0	2.2	4.0
End net loss	0.3	1.0	0.8	2.0
Navigation	3.8	2.5	1.6	1.6
Unaccountable & evaporation	2.7	6.7	2.2	7.0
Drainage	16.0	15.0	14.2	14.2
Demand total	54.0	65.4	58.9	73.0
Supply				
Aswan	57.5	60.0	61.7	58.9
Drainage re-use	—	2.5	5.4	6.0
Drainage return flow	—	4.0	—	4.0
Supply total	57.5	66.5	67.1	68.9
Balance	+ 3.5	+ 1.1	+ 8.2	− 4.1

Source: Waterbury, 1982.

The chief conclusion to be drawn from the available water supply and demand calculations is that if Egypt is to expand its cultivated area through ambitious land reclamation schemes, it has three options:

(1) to increase the efficiency of the irrigation system and improve farm water management practices;
(2) to utilize more efficient irrigation and drainage technologies;
(3) to increase the reuse of drainage water.

The increased supplies from the upper Nile projects alone would not be sufficient, even if they were completed on schedule. Unless some combination of these alternatives is achieved, reclamation efforts will have to be reduced.

Sudan is currently using 15–16 billion m³ of its 18.5 billion m³ allocation. Almost all of this is for irrigation. There is little return to the river. Irrigation schemes are planned which would use an additional 10 billion m³ — all of the new supplies which would be available from the upper Nile projects. Both the upper Nile projects and the irrigation schemes await financing.

Forecasting water supplies and demands is thus essentially an exercise in forecasting Sudan's macroeconomic (and political) conditions, and the priority these water development projects will receive in Sudan's extremely limited capital investment programme. It seems unlikely that Sudan will use its full allocation of 18.5 billion m³ plus its share of Jonglei I before 1990. The intensification of the Gezira scheme is likely to be the only major irrigation project which will come on line during the next decade which will require significant amounts of water.

The water demand forecasts which have been prepared for Egypt and Sudan treat water as a requirement which must be supplied irrespective of its cost. It is commonly assumed that any gap between future demand and existing supplies should be filled by the construction of new water supply projects. These kind of calculations do not address the question of whether the future water uses — specifically for agricultural expansion — are worth the expense of the additional water supplies or who should receive priority in water allocation (see Kindler and Russell, 1983, for approaches for dealing with this problem). To begin to examine this question, in Figure 7.5 we have incorporated our best estimates of the value of water in different uses in Egypt with the data in Table 7.5, and presented the results in the form of a supply and demand diagram.

As is true in other river systems throughout the world, municipalities, and industries are the highest value users of water. We do not know the economic value of water to municipalities and industires, but it is certainly much greater than the amount farmers would be willing to pay for water used in agriculture. Since the total municipal and industrial consumption in 1980 was low, there was plenty of water available to meet these needs. Water releases for navigation purposes are probably not as valuable, but we have included them in the

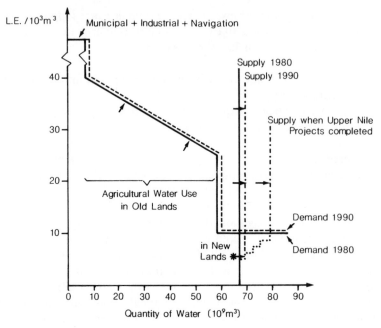

Figure 7.5   Water supply and demand in Egypt: 1980 and 1990

municipal and industrial category because such releases are small and likely to decrease in the future.

The next highest value use of water is for irrigation in the old lands of the Nile Valley. The estimates available for the shadow value of water in the old lands are at best only approximations, but when the current water allocation is fully utilized, a range of 25–40 LE/1000 m³ seems reasonable for planning purposes as the opportunity cost of transferring water from the Old Lands to other uses (Egyptian Water Master Plan, 1980; Bowen and Young, 1982; Whittington and Guariso, 1983). As with municipal and industrial uses, Egypt should almost always have 'adequate' supplies for the irrigation of the highly productive Old Lands of the Nile Valley.

Our hypothetical demand curve becomes horizontal at 10 LE/1000 m³, indicating that more water can be used for hydroelectric power generation up to the capacity of the Aswan High Dam power station and for desert reclamation (see Whittington and Guariso, 1983 for a discussion of the complementarity between hydroelectric generation and irrigation uses). The value of water used for both hydropower and the reclamation of new lands appears to be low.

The supply curves sketched in Figure 7.5 for 1980 and 1990 indicate that the marginal water uses in the system will be for irrigation of new lands and hydropower generation.

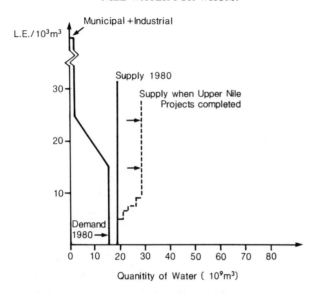

Figure 7.6   Water supply and demand in Sudan: 1980

Figure 7.6 illustrates the comparable supply and demand curves for Sudan. There have been few estimates of the shadow value of water in Sudan (Coyne and Bellier, 1978), but it seems that the value of water in existing Sudanese irrigation schemes must be between the value of water in Egypt's old and new lands.

From an efficiency point of view, there are three factors which support this conclusion. First, water used in Egypt should generally be able to generate additional hydropower benefits passing through the Aswan High Dam turbines. Second, cultivation in Egypt is more intensive and yields are higher. Third, transportation costs of getting agricultural products to both export and domestic markets are lower in Egypt. Partially counteracting these advantages is the fact that some water is lost to evaporation and seepage as the Nile flows from Sudan to Egypt. Nevertheless, we believe that water in existing Sudanese irrigation schemes will be worth 15–25 LE/1000 m³ when there are system-wide water quantity constraints. At present Sudan is not using its full allocation so the marginal value of water is zero.

In Figure 7.7 we aggregate the individual supply and demand curves for Egypt and Sudan. In the long run we assume Sudan's demand for water for new irrigation schemes should be valued higher than Egypt's use of water for desert reclamation. It is clear that adequate supplies are available to meet the forecast municipal and industrial consumption in Egypt and Sudan, the 'requirements' of the Old Lands of the Nile Valley in Egypt, and the needs of the existing irrigation schemes in Sudan. There is, however, insufficient water both to pursue Egypt's desert reclamation efforts, and to irrigate all the potential sites in Sudan,

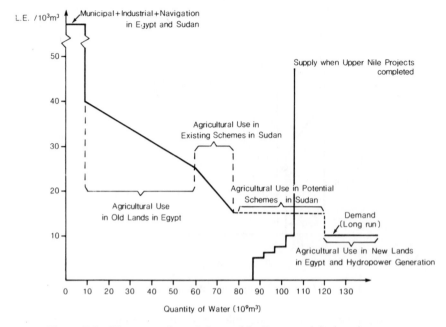

Figure 7.7   Water supply and demand in Egypt and Sudan: long run

particularly given that the claims of the upstream riparian states remain unresolved.

This analysis shows that if Egypt and Sudan were one country, no one would seriously consider Egypt's current plans for massive reclamation of desert lands while deep clay soils of the Blue Nile plain lay idle. The dilemma the existing national boundaries pose for the efficient allocation of Nile waters has long been recognized. Although the 1920 Nile Projects Commission reaffirmed Egypt's historic claims to Nile waters, one dissenting member, H. T. Cory, argued that the allocation of water supplies for future development should be determined by the availability of good lands on which to put the water. In fact, some knowledgeable observers do not take seriously Egypt's announced plans to reclaim an additional one million hectares by the year 2000, but see it as an attempt to lay claim to additional water supplies which can be relinquished in future negotiations at relatively little cost.

EXCHANGE OF WATER AND LAND: THE TERMS OF TRADE

Given Egypt's strong legal claim to the majority of the Nile water and Sudan's excess supply of high quality land for irrigation, what arrangements would permit a more efficient allocation of water for agricultural expansion in Egypt and Sudan? What are the terms on which Egypt and Sudan could exchange water and land? Such a trade has profound political implications. John Waterbury (1982) reported that rioting broke out in Juba when rumours circulated that Egyptian peasants would settle in southern Sudan to take advantage of the alleged benefits of the Jonglei Canal. In the long run, however, the economic benefits of optimal utilization of the Nile water will be enormous and provide compelling arguments to consider seriously how such transactions could be arranged. It is one of the ironies of the situation that although Egypt has long been the chief proponent of a basin-wide accord, a more efficient distribution of Nile water would involve a reallocation of supplies away from Egypt and to upstream states.

The efficient use of land and water in Egypt and Sudan requires that scarce factor inputs cannot be reallocated to increase irrigated agricultural production in Sudan without decreasing agricultural production in Egypt. To state the conditions for efficient allocation slightly more formally, let

Q_E = agricultural production in Egypt
Q_S = agricultural production in Sudan
L_E = land resources suitable for irrigation in Egypt
L_S = land resources suitable for irrigation in Sudan
W_E = Nile waters allocated to Egypt
W_S = Nile waters allocated to Sudan
$Q_E = f(L_E, W_E)$; $Q_S = g(L_S, W_S)$ denoting the production functions which describe how the inputs are converted to outputs in Egypt and Sudan, respectively.

Efficient production then requires that the marginal rate of substitution between water and land (MRS_{WL}) be equal in Egypt and Sudan.

$$MRS_{WL}^{Egypt} = MRS_{WL}^{Sudan} \qquad (1)$$

Since the marginal rate of substitution between two factor inputs is equal to the ratio of their marginal products, i.e.

$$MRS_{WL} = \frac{\text{Marginal Product of Land}}{\text{Marginal Product of Water}} = \frac{\dfrac{\partial Q}{\partial L}}{\dfrac{\partial Q}{\partial W}} = \frac{MP_L}{MP_W}$$

the condition for efficient agricultural production in the two countries is that the ratio of the marginal products of land and water be equated

$$\frac{MP_L^{Egypt}}{MP_W^{Egypt}} = \frac{MP_L^{Sudan}}{MP_W^{Sudan}} \qquad (2)$$

or alternatively that the ratio of the marginal products of land in the two countries equals the ratio of the marginal products of water:

$$\frac{MP_L^{Egypt}}{MP_L^{Sudan}} = \frac{MP_W^{Egypt}}{MP_W^{Sudan}} \tag{3}$$

The logic of this efficiency condition can be shown by considering the Edgeworth box diagram presented in Figure 7.8. The horizontal dimension is the total quantity of high quality potential land suitable for irrigation in Egypt and Sudan; we assume for purposes of illustration that Egypt has three million hectares and Sudan five million hectares. The vertical dimension is the total of Nile water to be allocated after evaporation and seepage losses are deducted, which we assume to be 74 billion m^3 at Aswan. Each point inside the diagram represents an allocation of the total land and water available between Egypt and Sudan. The initial endowments are depicted by point A, at which Egypt has the majority of the water and Sudan has the majority of the land. The curves labelled Q_E^1, Q_E^2, Q_S^1, Q_S^2 represent an isoquant mapping of the agricultural production relationships where Q_E^1 is a given fixed level of agricultural production in Egypt which can be obtained from the different combinations of inputs shown by the curve.

As the isoquants are drawn, the point A is inefficient because by moving from A to, for example, B, agricultural production in Egypt can be increased without decreasing production in Sudan. Alternatively, a movement from A to C also results in an efficient outcome by increasing production in Sudan without decreasing production in Egypt. In both the move from A to B and A to C,

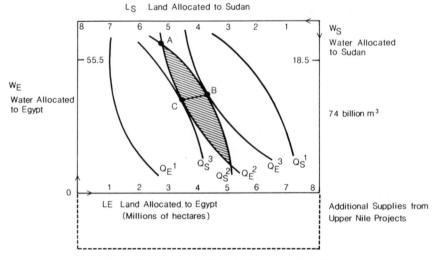

Figure 7.8   Edgeworth box diagram: the Nile

Egypt trades water for land, but the relative value of water is higher over the move from A to B; in this case the terms of trade are advantageous to Egypt, the party with the water.

As Egypt trades water for land, the less valuable land becomes in terms of water, i.e. the shadow price of water to Egypt increases. Conversely, as Sudan trades land for water, the less valuable water becomes in terms of land for Sudan. At both points B and C, the isoquants are tangent, and Egypt and Sudan place the same relative value on land and water (i.e., $MRS_{WL}^{Egypt} = MRS_{WL}^{Sudan}$). By exchanging land and water, both Egypt and Sudan would be better off at any point in the shaded area than at the initial endowment A. However, only those points on the 'contract curve' between C and B are efficient, and trade should continue until the 'efficiency locus' is reached. Where the final transaction will fall on the contract curve is indeterminant, and depends upon the relative bargaining strength of the two countries.

Where then has this exercise in standard microeconomic theory left us? Sovereign nations never exchange territory for water rights. Still this framework is useful for thinking about the magnitude of reallocation necessary for efficient water utilization and the bargaining positions of the two parties.

The standard water planning practice is to assume that crops have fixed water requirements for a given irrigation technology, which implies fixed factor production relationships and that the isoquants in Figure 7.8 have right angles (i.e. L shaped). If both Egypt and Sudan are assumed to have fixed factor production relationships, then the shaded area in Figure 7.8 which depicts the area of potentially profitable exchange will be a rectangle with the vertical side representing the maximum amount of water Egypt could trade and the horizontal side the maximum amount of land Sudan would theoretically trade for water. To the extent that water is substitutable for other factor inputs, this rectangular area will be an overestimate.

Both of these values are straightforward to estimate in a rough sense. As we have seen, when the Upper Nile projects are finished, Egypt will have of the order of 10–15 billion m^3 of water for New Lands irrigation, in excess of that required for municipal and industrial growth. This amount of water would be sufficient to irrigate an additional 1 million hectares in Sudan. However, the quantity of water potentially available from Egypt, and Sudan's share of the Upper Nile is not enough to irrigate all the high quality land available in the Blue Nile plain in Sudan. So water, not land, will ultimately be the limiting factor on the expansion of irrigated agriculture. Still, Egypt's excess supply and Sudan's share of the Upper Nile projects would be sufficient roughly to double Sudan's irrigated acreage.

A next obvious question to ask is whether the exchange could take place in terms of money rather than land. At what price would Sudan be willing to buy water from Egypt? In the long run Sudan should be willing to pay an amount for water slightly less than the value of water on Egypt's old lands — say

25 LE per 1000 m^3. Thus, the annual value of Egypt's excess 10–15 billion m^3 would potentially be several hundred million Egyptian pounds. In the distant future when Nile waters are fully utilized in modern agricultural schemes throughout the basin, the value of this water should be even higher, as evidenced by the high value of water which already exists in Israel and the Jordan Valley. For purposes of comparison, the value of Egypt's total Gross Domestic Product from the agricultural sector in 1980 was 1.6 billion LE, of which only about 50 million LE was from new lands. The total value of agricultural exports and of potential 'water exports' are then of comparable order of magnitude — and water exports would not entail the production costs associated with agriculture. The Water Master Plan (1980) has estimated that water from the upper Nile projects will on average cost 3 LE per 1000 m^3. Even granting that this may be three or four times too low (Waterbury, 1982), the water is still very cheap for modern irrigated agriculture. Environmental criticisms may result in modification of the projects, but will ultimately receive little weight in the face of the pressures from population growth and agricultural expansion. But Sudan will not be in a position to buy Egypt's share of the upper Nile projects for decades to come.

In the medium term, given Sudan's severely limited financial resources, what other type of arrangements of agreements between Egypt and Sudan could promote a more efficient allocation of Nile water? One possibility would be joint Egyptian — Sudanese ventures for the development of large scale irrigation schemes in which Sudan supplied the land and Egypt supplied the water. In fact, the Gerzira scheme was originally organized as a private venture with the British stockholders assuming a long-term lease. The tough questions in any such venture would involve the distribution of land ownership (with water rights) and tenancy arrangements, and the access of these to Egyptian as well as Sudanese farmers. There are a variety of alternative arrangements which might possibly be negotiated without unduly threatening the Sudanese — such as the issuance of long-term work permits for Egyptian agricultural labour, limited immigration of Egyptians to Sudan, Egyptian participation in the scheme's management. Such a joint venture might be more politically acceptable if it was set up as a 99-year lease by the management of both Egypt's water and Sudan's land. The management could thus 'rent' land and water from both countries. Both the land and water have limited value without the other, although Egypt could use its share of water for hydropower production and Sudan could possibly use its share of land for rainfed agriculture. Assuming these are comparable on a per hectare basis, negotiations would probably lead to roughly equivalent allocations of profit, or rents, to water and land.

CONCLUSIONS

The geopolitical implications of the Nile's hydrology and Egypt's total dependence on the Nile for its agriculture — or the 'hydropolitics' of the Nile

Valley as John Waterbury (1979) has aptly described it — have intrigued explorers, poets, and politicians, alike but many of the conclusions drawn from common place hydrological observations have been self-servicing. In 1904 Sidney Peel, a fellow of Trinity College Oxford, wrote —

'The story of the binding of the Nile makes one thing perfectly clear, that all the Nile land is one country. No divided sovereignty is possible; there must be one firm hand over all.'

It was also clear to Peel whose hand that should be —

'It seems a far cry from . . . the Lakes and swamps of Equatorial Africa or the rain-swept hills of Ethiopia to the cotton mills of Lancashire, but the Egyptian peasant lifting water on to the fields of the Delta knows that the connection is close enough. The English have a direct commercial interest in holding the Valley of the Nile . . .'

This vision of the unity of the Nile basin is now promoted by Egypt, and just as during the colonial period this theme served British economic interests, today it serves those of Egypt, which lays claim to the lion's share of Nile waters and is actively seeking more. For the Egyptians this 'firm hand' could be a Nile Waters Commission formed by the representatives of the riparian countries. While in both principle and practice this Egyptian proposal is thoroughly sound, it is not clear that the upstream states are yet in a position to negotiate with Egypt as equals or that they have the technical expertise to know their own interests (Waterbury, 1982). Ethiopia has, in fact, agreed to negotiate with Egypt on the allocation of Nile waters, but only if Egypt renounced the 1959 Nile Waters Agreement. Egypt not only refused to do this, but tried strenuously to block World Bank financing of an irrigation project in Ethiopia (Finchaa) which would use water from a tributary of the Blue Nile. Before the World Bank acted, however, Colonel Qaddafi agreed to finance the project — which only serves to reinforce Egypt's ancient fears that Ethiopia would block the Nile flow for political reasons.

The theme of the unity of the Nile basin is in one sense obviously true; the actions by one country affect the quantity and quality of the water resource available to downstream riparians. But the idea often espoused by England and subsequently Egypt that it is in the interests of all Nile riparians to cooperate on the development of the river is only partially accurate. More water for one party means that less is available for another, and the future appears to hold as much promise for conflict as cooperation.

If Egypt continues to conceive of economic development primarily in terms of the water supply available — as a Malthusian race in which it must expand its water supplies to meet the needs of its rapidly expanding population—it is not only doomed to failure but will also retard economic development throughout the Nile basin. Different strategies for economic development are needed which do not focus to such an extent on access to one resource. Egypt must strike a new ecological balance with its use of the Nile waters, land resources, and population—a task in which she was notably successful for centuries.

REFERENCES AND BIBLIOGRAPHY

Barnett, T. (1974) *The Gezira Scheme: An Illusion of Development*, Cass, London.
Beshai, A. A. (1976) *Export Performance and Economic Development in Sudan*, Middle East Centre, St. Anthony's College, Oxford, Ithaca Press, London.
Bowen, R., & Young, R. A. (1982) 'Allocation efficiency and equity at alternative methods of charging for irrigation water: a case study in Egypt', in *International Water Policy Working Paper 82-2*, Department of Economics, Colorado State University.
Bulletin of the African Institute of South Africa (1975) *The Sudan: Potential Agricultural Giant*.
Coyne & Bellier (1978) *Blue Nile Waters Study*, Vols. I–VII, prepared for the Ministry of Irrigation and Hydro-electric Energy, Sudan, by Sir Alexander Gibb & Partners, Hunting Technical Services and Sir Malcolm MacDonald & Partners: Blue Nile Consultants. (1978) *Nile Waters Study: Main Report* and *Supporting Reports*.
Egyptian Water Master Plan (1980) 'Executive Summary of Main Report', UNDP/EGY-73/024, Egyptian Ministry of Irrigation, Cairo.
El Tobgy, H. A. (1976) *Contemporary Egyptian Agriculture*, Ford Foundation, Cairo.
Fitch, J., & Abdel Aziz, A. (1980) *Multiple Cropping Intensity in Egyptian Agriculture*, Project Paper No. 5, Micro-economic Study of the Egyptian Farm System, Ministry of Agriculture, Cairo.
Fitch, J., Abdel Aziz, A., Khedr, H., & Whittington, D. (1979) 'The economic efficiency of water use in Egyptian agriculture: opening round of a debate', paper presented at the *Seventeenth International Conference of Agricultural Economists* in Alberta, published as a Working Paper by the Ford Foundation, Cairo.
Gaitskell, A. (1952) *Gezira: A Story of Development in Sudan*, Faber & Faber, London.
Garstin, Sir W. (1904) *Report on the Upper Nile Projects*, Ministry of Public Works, Cairo.
Gritzinger, D. (1981) *Water Costs and Returns in North Sinai Agriculture*, Draft Working Paper, Damer & Moore Consulting Engineers, Cairo.
Habashy, N. & Fitch, J. B. (no date) *Egypt's Agricultural Cropping Pattern*, Micro-economic Study Unit, Ministry of Agriculture, Cairo.
Haynes, K. E., & Whittington, D. (1981) 'International management of the Nile: stage three?', in *Geographical Review*, **71**, 1, 17–32.
Hunting Technical Services (1979–80) *Suez Canal Regional Integrated Agricultural Development Study*, EGY/76/001/6, Feasibility Report, Vols. 1–7, Arab Republic of Egypt Ministry of Development and New Communities, UNDP, Cairo.
IBRD (International Bank for Reconstruction and Development) (1979) *Sudan: Agricultural Sector Survey*, Washington DC.
Khiden, M., & Simpson, M. C. (1968) 'Cooperative and agricultural development in the Sudan', in *The Journal of Modern African Studies*, **6**, 4, 509–518.
Kindler, J., & Russell, C. (Eds) (1983) *Modelling Water Demands*, Academic Press, London, p. 240.

Pacific Consultants (1980) *New Lands Productivity in Egypt — Technical and Economic Feasibility*, AID Contract No. AID/NE-C-1645.

Peel, S. (1904) *The Binding of the Nile*, Edward Arnold, London.

Riordan, R. W. *et al.* (1982). *Strategic Implications to the United States of Potential Conflict over Shared Water Resources in the Nile River Basin*, Report No. 82–39, National Defense University, The National War College, Washington DC.

USAID (United States Agency for International Development & Ministry of Agriculture ARE) (1982) *Strategies for Accelerating Agricultural Development: A Report of the Presidential Mission on Agricultural Development in Egypt*, Washington DC.

USDA (United States Department of Agriculture, USAID, & Ministry of Agriculture ARE) (1976). *Egypt — Major Constraints to Increasing Agricultural Productivity*, Foreign Agricultural Economic Report No. 120, Washington DC.

Voll, S. P. (1978) *Egyptian Agriculture since the Revolution*, Ministry of Agriculture, Cairo.

Waterbury, J. (1979) *Hydropolitics of the Nile Valley*, Syracuse University Press, New York.

Waterbury, J. (1982) *Riverains and Lacustriner: Towards International Cooperation in the Nile Basin*, Discussion Paper No. 107, Research Program in Development, Princeton University.

Whittington, D., & Guariso, G. (1983) *Water Management Models in Practice: A Case Study of the Aswan High Dam*, Developments in Environmental Modelling, Elsevier, Amsterdam.

Whittington, D., & Haynes, K. E. (1980) 'Valuing water in the agricultural environment of Egypt: some estimations and policy considerations', in *Regional Science Perspectives*, **10**, 109–126.

Wynn, R. F. (1966) 'Water resource planning in the Sudan: an economic problem', in *Agricultural Development in the Sudan*, Proceedings of the 15th Annual Conference of the Philosophical Society of the Sudan, Khartoum.

Agricultural Development in the Middle East
Edited by P. Beaumont and K. McLachlan
© 1985 John Wiley & Sons Ltd

Chapter 8

The Iranian Village: A Socio-economic Microcosm

ECKART EHLERS

INTRODUCTION

Today, approximately 40 per cent of Iran's population can be termed rural. Agriculture and animal husbandry still are the main source of rural income. So, peasants and pastoralists represent the biggest single socio-economic group in the country.

In spite of tremendous differences in ecology, location, ethnicity, and many other factors, one may nevertheless say that there is something like the 'typical Iranian village'. Its prototype is not particularly marked by formal unity: sizes and forms of villages show as great a variety as the architecture and building materials of the houses. Rather, its unity is based on more or less identical socio-economic structures, characterized by specific relationships between landlords and peasants.

Although there is also a vast range and a remarkable regional differentiation of landlord–peasant relationships in Iran (Lambton, 1953), one may nevertheless say that, basically, there have traditionally always been two socio-economic types of rural settlement:

(1) villages belonging to rural peasant proprietors, *khurdeh malik*, who mostly work their land themselves;
(2) villages belonging to urban-based landlords, many of them large landed proprietors, *umdeh malik*, who had their land worked on a share-cropping basis by peasants, *ra'iyat*, or tenants.

While it is difficult to quantify the exact numbers of either type of village, there are some trustworthy estimates as to the importance of different types of land-management in different parts of the country in pre-land reform Iran (Table 8.1).

On the basis of statistical analyses and air-photo interpretations Bobek (1975/6) has concluded that the free peasant system, characterized by irregular field block systems, is found 'mainly in peripheral and agriculturally marginal areas, and the landlords system prevailed in a quasi-continuous zone running from eástern Azarbaijan, through Qazvin and then towards a centre between

Table 8.1   Socio-economic structure of rural land use in Iran, 1960

Province	Percentage of land use by		
	Proprietors	Tenants	Share-croppers
Tehran	20	2	78
Gilan	10	31	59
Mazandaran	27	45	28
East Azarbaijan	16	1	83
West Azarbaijan	7	2	91
Kurdestan	12	1	87
Kermanshah	10	1	89
Khuzestan	20	18	62
Fars	30	21	49
Kerman	42	1	57
Baluchestan/Sistan	56	1	43
Khorasan	47	11	42
Esfahan	28	13	59
Total Iran	23	13	64

Source: Planck, 1963.

Kermanshah and Arak. It continued beyond Esfahan towards the central plains of Fars. Isolated areas also existed in Khuzestan and Khorasan. Taken as a whole, it included well populated and the economically and politically most important areas of western and eastern Iran'. The formal characteristic of the latter type is the 'open field village with strip division'.

In view of the fact that the overwhelming majority of all villages seems to have belonged to the share-cropping type of agriculture, the following examples will focus on this specific type. Their essentials in regard to land use (Bowen-Jones, 1968), village layout and housing (Planhol, 1968; Rainer, 1977), and agricultural tools and implements (Watson, 1983; Wulff, 1966) have remained more or less unchanged. The continued existence of material culture is in strong contrast to the almost revolutionary changes that have taken place in social and economic organization of agriculture. As a result of the so-called 'White Revolution' of 1962 and of the political changes in Iran since 1979, it seems appropriate to look even at the present-day Iranian village in an historical perspective:

— the traditional village of the era before 1962;
— the land-reform villages and the changes they have undergone;
— the developments since the Islamic revolution of 1979.

THE TRADITIONAL IRANIAN VILLAGE

The main characteristics of the traditional Iranian village, based on a share-cropping agriculture, are a more or less complicated social and economic interrelationship between landlords, a series of different middlemen, share-cropping peasants and tenants, and a great number of various beneficiaries who—although not involved in the process of agricultural production—participate in the harvest. Main partners, however, are landlords and share-croppers. Based on traditional Islamic right, according to which land (soil), water, seed, and draught animals, with tools and human labour, are equally responsible for the success or failure of agricultural production, landlords and share-croppers divided the harvest according to their contribution of production factors, i.e. a 20 per cent share of the harvest for each factor.

Table 8.2 shows, however, that there were many exceptions to this general rule. Based on different specific arrangements between landlords and peasants even within one region or village (Gharatchedaghi, 1967), there were sometimes different shares of the harvest for the same duties performed. Utilizing the fundamental work by Lambton (1953), Planck (1962a) has summarized a few of the more common share-cropping practices in different parts of the country.

As mentioned before, besides these basic divisions between the two partners, there were many other persons participating in and benefiting from the agricultural production process. Normally both landlord and peasant divided

Table 8.2   Shares and duties of share-croppers in different parts of Iran

Region	Shares of the agricultural production (%)	Duties
Arak/Esfahan	67	Labour, Plough[a], Seed
Burujird	50	Labour, Plough, Seed
Fars:		
Mamasani	75	Labour, Plough, Seed
Firuzabad	33	Labour
Fassa	20	Labour
Jahrum	50	Labour, Plough
Niriz (well irr.)	90	All duties
Niriz (qanat irr.)	25	Labour, Plough
Kerman:		
Rafsanjan	50	Labour, Plough, Seed
Rafsanjan	25	Labour
Khuzestan:		
Dezful	80–90	Labour, Plough, Seed
Behbahan	80	Labour, Plough, Seed
Shushtar	67–75	Labour, Plough, Seed

Source: Lambton, 1953, compiled by Planck, 1962.
[a] Plough = tools and animals.

these costs and deductions among themselves, as is shown in Table 8.3 a typical example from Goldasteh near Tehran.

The majority of traditional Iranian villages belonged either totally to single, large landed proprietors (1960: 10,384 *shish dang* villages out of a total of 41,458) or at least partly to them (4,019 villages split into one to five *dang* as properties of individuals). 17,145 villages belonged either to groups of different sizes, or to absent landlords, or to farmĕrs with property titles, *khurdeh malik*. In all these cases, the division of the landlords' share in the harvest normally corresponded to the above-mentioned percentage of production factors involved in the agricultural production process.

Table 8.3   Division of harvest in Goldasteh/Tehran

Recipient	Percentage of harvest	%
Total harvest		100
Blacksmith	0.16	
Carpenter	0.16	
Field-guard	1.0	
Village-headman (*kadkhuda*)	1.0	
Mullah	0.2	
Ox-drover	0.08	
Boneh /farm labourers	5.0	
Total deductions	7.6	7.6
Rest		92.4
Landlord: Two shares for soil and water: 2×18.48	= 36.96	
Urban capitalist (*gavband*): Two shares for seed, plough and tools: 2×18.48	= 36.96	
Share-croppers (*raiyat*) and farm labourers: One share for human labour 1×18.48	= 18.48	
Share-cropped harvest	= 92.4	—

Source: Ehlers & Safi-Nejad, 1979.

Share-croppers, on the other hand, mostly organized themselves into working-units, generally of two to six members each (*bonku, boneh, haraseh, pagav, sahra*, etc.). Working the land together, they, too, split their harvest share among themselves. More often, however, different working-units received annually changing field-complexes within the boundaries of the villages. These, then, were allocated to the different members of the group: '. . . open field villages with strip division appear to be dominated by landlords who organized their villages into fixed numbers of ploughlands of equal size. These were laid out in numerous strips, the use of which rotated annually or from time to time among crop-sharing tenants . . .' (Bobek, 1975/6). This practice of annual redistribution of the landlords' land among their dependent share-croppers, is seen as the main reason for the typical strip-pattern of the fields. Free peasantry field systems, on the other hand, are characterized by irregular blocks (see Figure 8.1) as an expression of individual and permanent property titles to specific pieces of land. Probably the best-known example of the extreme annual changes of land use and land tenure is that of the village of Talebabad (Safi-Nejad, 1974), divided among eleven *boneh*.

Table 8.4   Traditional social structure of rural families in Iran

Property	%	%	Social Status
No property	60	6	Landless labourers etc. (*khoshneshin*)
		54	Landless peasants and tenants (share-croppers, etc.)
Less than 1 ha	23		Share-croppers and tenants
1–3 ha	10	35	with small properties of
			their own
3–20 ha	6		
More than 20 ha	1	5	Independent peasants

Source: Planck, 1962b.

It has already been pointed out that any generalization with regard to the forms of social and economic organization of traditional village life is dangerous and should be avoided. Each of the many detailed case studies published (Goodell, 1975; Kielstra, 1975; Planck, 1962b; Salmanzadeh, 1980; Stobbs, 1976; Tavana, 1983; Watson, 1979) reveals local specifics and a wide variety of organizational structures of village life. This diversity becomes even more apparent when we include all those forms not based on traditional land-lord/share-cropper relationships (Lambton, 1953). A rather crude survey of the prevailing social structure of rural Iran (Table 8.4) around 1960 reveals not only the prevalence of landless villagers, *khoshneshin*, (Hooglund, 1973; Khosravi, 1973) and share-croppers on the one hand, but also the almost total absence of a free land-owning peasantry, on the other. The abolition of these obvious discrepancies was one of the proclaimed intentions of the Iranian land reform laws from 1962 onwards.

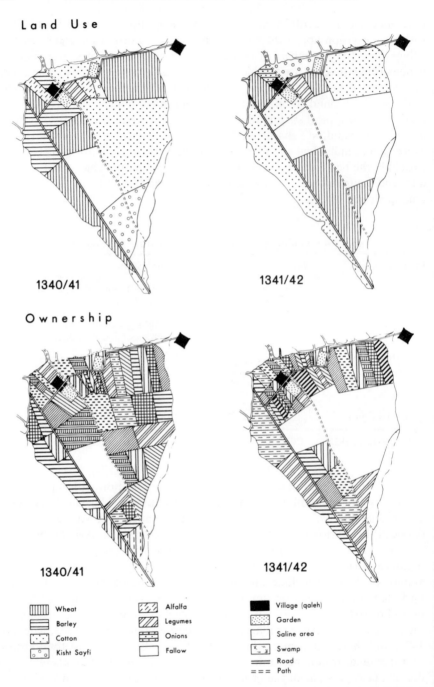

Figure 8.1 Talebabad: land use and ownership 1340/41 and 1341/42 (source Safi-Nejad)

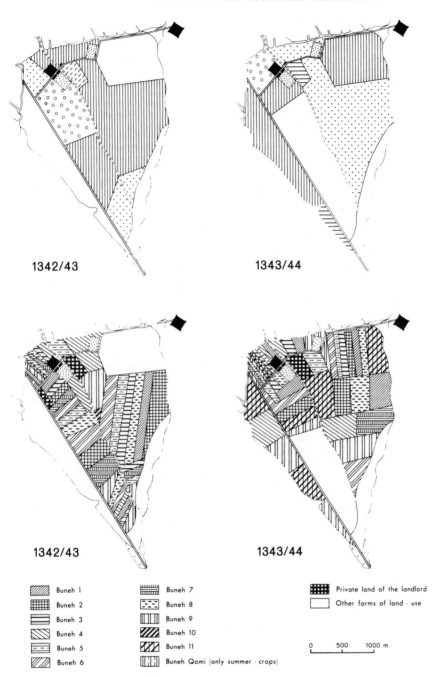

1342/43   1343/44

1342/43   1343/44

Buneh 1	Buneh 7	Private land of the landlord
Buneh 2	Buneh 8	Other forms of land - use
Buneh 3	Buneh 9	
Buneh 4	Buneh 10	
Buneh 5	Buneh 11	0 500 1000 m
Buneh 6	Buneh Qomi (only summer - crops)	

Figure 8.2   Talebabad: land use and ownership 1342/43 and 1343/44 (source Safi-Nejad)

Table 8.5 The development of land reform legislation in Iran

Phase	Main aspects of legislation	Results of its prosecution	
Phase 1: Land Reform Law (1962)	(1) Upper limit of property: 1 village (*shish dang*)	Bought villages Other bought estates	16,333 1,001
	(2) Purchase of land from large-landed proprietors on the basis of 15 annual instalments	Number of recipients of land	777,825
	(3) Sale/distribution of these lands to crop-sharing peasants on the basis of 15 annual instalments		
Phase 2: Additional Articles to the Land Reform Law (1963)	Proprietors of a village can choose from the following:		
	(1) Sale of their lands to those crop-sharing peasants working the land	Seller Purchaser	3,276 57,226
	(2) 30-year leases to those crop-sharing peasants working the land	Lessor Lease holder	223,321 1,232,548

Phase 3: Land Reform Law of 1968 and Additional Articles concerning the Distribution of Lease-holds (1970)	(3) Formation of a joint farming company between landlords and crop-sharers	Landlords 60,126 Crop-sharers 110,126
	(4) Distribution of agricultural land between landlords and former crop-sharers according to local distribution codes	Landlords 18,563 Crop-sharers 156,580
	(5) Purchase of crop-sharing peasants' rights of land use	Peasants 16,485 (Sellers)
	Distribution of leases among lease holders and crop-sharing peasants	281,844 landlords sold their lands to 128,816 crop-sharers; 6,668 landlords distributed their lands among 20,999 crop-sharers according to part (4) of Phase 2
Phase 4: Law for the distribution of public religious endowments (vaqf-e'amm) (1971/72)	Distribution of lands among former lease holders and crop-sharers	Distribution of 1,527 endowments to 47,063 peasants until 1972

Source: Planck, 1975.

THE 'WHITE REVOLUTION' AND IT'S EFFECTS ON
THE RURAL MICROCOSM

The Iranian land reform of 1962, main ingredient of the so-called 'White Revolution', had—in spite of its country-wide uniform provisions and its comparatively simple conception—far-reaching consequences for the traditional village structure of Iran. The first years of the reform programme and their consequences have again been analysed in a profound study by Lambton (1969), the findings of which show the regional variations and problems of this legislation. When the whole land reform programme came to a standstill in the 1970s, not only had millions of hectares of land changed hands, but also the social and economic order of rural Iran had changed. With the removal of the traditional landlord-peasant relationship and the formation of many hundred thousand small-scale family farms, one of the main objectives of the land reform programme—that those who worked the land should also own it— had been reached. On the other hand, there is no doubt that the manifold forms and problems of indebtedness of the new smallholders, partly in connection with the untouched Islamic laws of inheritance, created and extended new forms of dependence between the beneficiaries of land reform and their former landlords, urban businessmen, and/or moneylenders. Thus, in many parts of the country only a decade later land reform legislation and its results were considered a social and economic failure. Some even called it a disaster. While Table 8.5 presents the different stages of land reform legislation as well as their main regulations and effects, two case studies serve to demonstrate the disintegration arising from them.

Case A—Shahquz Koti

Until 1961, the fertile agricultural lands (58 hectares) of Shahquz Koti, a village in the rice- and orange-producing area of the central Caspian lowland near Shahsavar and originally *khaliseh* (i.e. crown land), were cultivated by tenants. Each of them had approximately four hectares of land, some more and some less. In 1961, the lands held under tenancy were distributed among the tenants. Mainly owing to early mortality, Islamic laws of inheritance and comparatively large numbers of legal heirs, the number of farms increased in the ten years 1961–70 from 15 to 55 and the average farm size decreased from an average 3.9 hectares to approximately 1 hectare. A typical example of the catastrophic splitting up of a viable farm unit into a number of small farms is represented in Figure 8.3. It shows that within a period of ten years (1961–70) a farm of initially 5.2 hectares size disintegrated into one unit of 1.7 hectares and two units of 0.7 hectares each. The rest of the land fell to female heirs who either rented it to their brothers or combined it with the farms of their respective husbands.

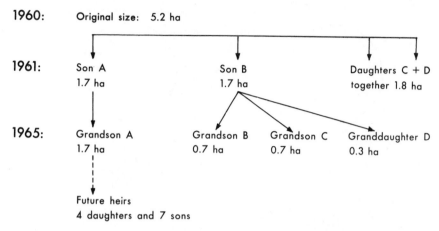

Figure 8.3   Fragmentation of a land reform farm in Shahquz Koti

Because of the individual cultivation of the land—forms of collective land management are more or less unknown in the rice-growing areas of the Caspian coast—inheritance of land was connected with an extreme fragmentation of each field. It is before this background that an official study of the socio-economic situation in Gilan came to the following conclusion: 'The main problem facing Gilan's agriculture is that its patches of lands of insignificant size are sparsely situated. . . The exceptionally small and widely scattered units (on average each holding covers two pieces of land with an area of 0.62 hectare) in Gilan are great impediments to the raising of the farmers' level of income' (Research Group, 1967).

The rapid deterioration of originally viable farm units from 1962 onwards resulted in a continuation or renewal of economic dependencies on mostly urban capitalists. By the mid-1960s many of the former landlords or urban businessmen had regained their old properties: the beneficiaries of the land reform either became wage-labourers on their own allotted farms or migrated to the cities in search of work and new sources of income.

Case B—Zangiabad/Fars

For the past 250 years the village of Zangiabad has been a religious *vaqf*. The village, before land redistribution, consisted of 1,512 hectares of cultivatable land, divided equally into two sections, each containing three *dangs*. Each half of the village had 756 hectares of land and 54 share-croppers; of these, every six made up one agricultural production unit with the agricultural output divided equally between them. Thus, there was a total of 18 productive units in the village, nine in each half. The land cultivated by the peasants in each production unit 'did not remain constant but changed almost annually due to land

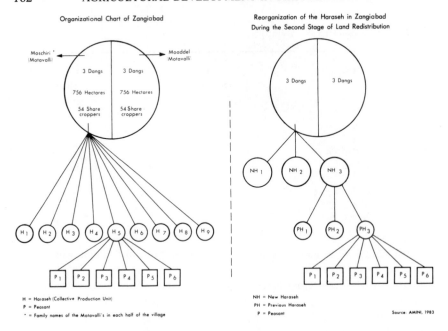

Figure 8.4   Zangiabad: organizational chart and reorganization of the *haraseh* during the second phase of land redistribution

reallocation' (Amini, 1983). Zangiabad thus represents the more or less typical example of collective land use (*haraseh*). Land reform and its first stages hardly affected the *haraseh* structure and its organizational pattern. With the implementation of the third stage, however, farmers could buy the land (*vaqf-e khas*, i.e. personal *vaqf*) and develop personal property rights. Thus within only a couple of years the *haraseh* structure began to dissolve and 'some *harasehs* were completely abolished within two years' (Amini, 1983).

The decline of one specific *haraseh* (Figure 8.5) within a period of five years (1972–76) shows the almost chaotic dissolution of centuries-old and established forms of social organization as a result of the land reform laws. Economically, too, the disintegration of the *haraseh* structure had far-reaching consequences in regard to land and water use:

Members of the old *harasehs* began selling their shares in the wells to those who were financially more stable. The individuals who purchased the wells then began selling excess quantities of water. Those who sold their share in the wells became dependent upon the river for their irrigation needs or had to purchase water from those who owned the wells. Due to the lack of collective care for the wells, there was a rapid increase in the number of wells dug by individuals.

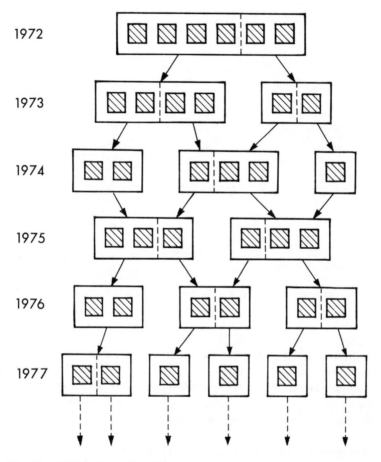

Figure 8.5 The decline of one *haraseh* in Zangiabad between 1972 and 1977 (source: Amini, 1983)

Prior to the third stage of land redistribution there were six semi-deep wells, one for each new *haraseh*. However, in 1977, after the third stage of land redistribution, there were approximately eighteen wells, all owned individually. (Amini, 1983)

The two case studies, typical of the results of a series of similar investigations in different parts of the country, demonstrate the social and economic break-up of the traditional microcosm of the Iranian village. Other factors contributed, too. Urban concentrated industrialization caused many villagers to leave their homes and, very often, their newly acquired agricultural lands. Social and regional mobility were connected with sales and purchases of agricultural lands. Together with permanent fluctuations in land titles as a result of indebtedness,

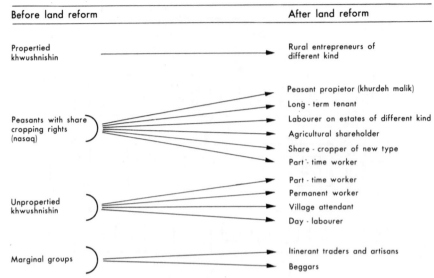

Figure 8.6   Socio-economic differentiation of the rural population of Iran (source: Planck, 1979)

inheritance, and other factors, the 'White Revolution' caused deep changes in the traditional social and economic order of rural Iran.

From a sociological point of view, land reform was connected with a subversion of the traditional rent-capitalist class structure, but it hardly brought about any equality among the villagers. On the contrary, social differentiation within villages became more pronounced than before (Figure 8.6).

> The distances between the classes enlarged inside the villages. People favoured by the land reform climbed the social scale. The majority of the *khoshneshin* had no other choice than to continue its rural life in utter misery or to migrate to the cities. The rural exodus extended enormously. The land reform opened the way for renewing the rural elites. The social positions which had been vacated by the landlords and their local agents became partly occupied by members of the new, rapidly extending bureaucracy. (Planck, 1979)

RE-INTEGRETATION AND CULTURAL CHANGE IN RURAL IRAN

1972 may be considered as the end of land reform activities. In the fall of 1972, in connection with discussion of the objectives of the Fifth National Development Plan, new guidelines for the development of rural Iran became apparent. Their main points were:

(1) With the implementation of the three phases of land reform and the clarification of the legal status of persons engaged in agricultural activities, the way will be paved for all farmers and agricultural workers to become members of the cooperative system or shareholders in farm corporations.
(2) Private agri-businesses and mechanized farming units will receive special encouragement and incentives and investors in such units will receive greater support from the government in the form of technical aid, finance, and credit to facilitate their efforts to establish cooperatives, or agricultural, animal husbandry, or agro-industrial corporations.
(3) Agro-industrial units, private, public, and mixed, will be established to farm 300,000 hectares, using advanced agricultural techniques and massive investment, so as to raise output and improve the quality and marketing of agricultural products, and such units shall receive government support.

The extension of the cooperative movement and the development of farm corporations marked the re-integration stage of land reform (Figure 8.7). Its main purpose seems to have been a re-integration of the numerous small individual farm units, many of which were far below the minimum for economic existence. The re-integrative movement followed different concepts, all of which, however, brought changes to the traditional rural culture as a consequence. In accordance with Planck (1975) four different approaches may be discerned: the *traditional* approach, which tried to revive the old-established forms of collective agriculture. Instead of interpersonal dependencies, however, the community of free peasants should form joint farming companies in order to get the best results from the collective use of land, water, machinery, etc. The *cooperative* approach, sponsored by the government, was based on legislation according to which every beneficiary of land reform had to become a member of a rural cooperative. Their success, although impressive in statistical terms, was, however, greatly diminished by the incompetence, corruption, and mismanagement of the bureaucracy: most cooperatives proved to be social and economic failures. The *revolutionary* approach, enforced in different parts of the country and mostly carried out against the will and interest of the individual farmers, consisted of a collectivization of one or more villages and their consolidation in the form of a capitalist land holders' society or an agricultural joint stock company, both government-controlled. As with the fourth type of re-integration, the *growth-oriented* approach (agri-business or agro-industry), the revolutionary approach was connected with severe social set backs for all beneficiaries of the land reform. In particular, the loss of the individual property rights and compensation in the form of shares caused social and political unrest. It must be seen as one of the main factors for many farmers' readiness to commit themselves in favour of revolutionary political changes (Salmanzadeh, 1980).

In 1978, immediately before the success of the Islamic revolution, the contours of the new socio-economic order began to show up. The overall picture of rural

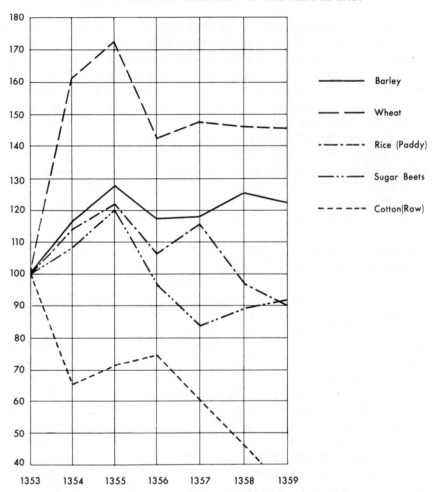

Figure 8.7   Index of major agricultural production 1353 (1974/75)–1359 (1980/81)
(Source: Bank Markazi Iran, Annual reports)

Iran may be described as that of an economy, where farmers' per capita incomes
were well below the national average. Rural Iran's social and economic status
in 1978 was characterized by stark disadvantage in comparison with urban developments. Experiments in developing new forms of agriculture met not only with
great lack of enthusiasm by the farmers but, even in those cases where they had
already been implemented, with open hostility. Ehlers (1977) and Salmanzadeh
(1980) have given specific examples from Khuzistan. The traditional socio-
economic microcosm of the Iranian village was definitely destroyed. New and
viable forms of rural life and promising developments were not visible.

Against this background, the obvious progress of both cooperative and joint
farming is not more than wishful thinking. For sure, while all the institutions

Table 8.6   Rural cooperative societies and unions, joint farm corporations, and production cooperatives

	1969/70	1972/73	1977/78
Rural cooperative societies:			
Number	8,102	8,361	2,925
Membership (thousand persons)	1,400	2,065	2,983
Capital (million rials)	1,984	3,329	8,385
Rural cooperative unions:			
Number	112	127	153
Number of member societies	7,542	7,961	2,907
Capital (million rials)	781	1,580	3,665
Joint farm corporations:			
Number of corporations	—	43	93
Number of shareholders	—	15,250	35,444
Capital (million rials)	—	685	1,515
Rural production cooperatives:			
Number of cooperatives	—	0	39

Source: Bank Markazi Iran (1353–59): Various years.

represented in Table 8.6 may have been in existence, only a vanishing minority of them may be considered as institutions accepted and supported by the majority of their members. A general dissatisfaction and unrest in rural Iran was officially admitted by the end of the Shah's reign:

> The lack of proper attention to the agricultural sector in the previous development plans has resulted in backwardness in this sector. Apart from a shortage of credits and other bottlenecks, the wrong pricing policy for agricultural products produced changes in the terms of trade to the disadvantage of the agricultural sector compared with other sectors. As a consequence not only investment in agriculture was unattractive, but most farmers began to migrate to the cities. (Bank Markazi Iran, 1357)

CONCLUDING REMARKS

The Islamic Revolution of 1979 swept away monarchy and many of its institutions. One of the first activities of the new and, at that time, provisional government of the Islamic Republic of Iran was to formulate a new agricultural policy. Its main essentials were a change to indirect interference in agricultural production, limiting state activities to providing credit and basic input requirements, and a general disbanding of those rural cooperative societies and rural producers' cooperatives where the farmers, as main shareholders, requested it. At the same time, some of the large agro-industrial complexes were turned over to the former owners of the lands from whom these companies had taken it over. In addition, on 25 Shahrivar 1358 (16 September 1979) the Revolutionary Council approved the so called Law for the Transfer of Cultivable Lands. This law conditionally approved the ownership principle of the old land reform act. This first legislation was followed on Farvardin 1359 (April 1980) by an Amendment to the Law for Revitalization and Transfer of Lands. Its main purpose was to foster land distribution among small-holders and owners of unviable farm units. At the same time it was intended to enlarge the existing hectarage of agricultural lands and, thus, to increase productivity.

In spite of all these measures and in spite of different forms of financial support to the agricultural producers (e.g. by increasing the guaranteed minimum price for sale of agricultural produce and considerable reductions in costs of obtaining loans and credits), the overall picture of agricultural production is marked by a distinct downward trend (Figure 8.7). These comparatively drastic decreases in productivity are the more surprising as they are connected with simultaneous extensions of the hectarage of almost all crops.

Reasons for these negative developments in rural Iran are difficult to analyse. While climatic factors, surely, may have influenced harvests here and there, they alone can hardly be held responsible for the set backs. The same is true for more or less official explanations such as: 'However, due to the continued difficulties in the agricultural sector and the rise of new problems during 1359 (1980) such as the outbreak of the Iran/Iraq war fought in certain grain areas, the production of foods did not increase in proportion to the population. Therefore the dependence on import of food items continued' (Bank Markazi Iran, 1359 (1979/80)). As a matter of fact, production of food not only failed to increase in proportion to the population but it decreased absolutely and relatively!

In view of this background, it is probably not wrong to state that the Islamic Republic of Iran has failed so far to establish a convincing and attractive new agricultural policy. The signs of further deterioration in rural Iran are alarming: decrease in production, increase in prices and credits, rural emigration, and ever-increasing imports of basic foods from abroad. While the rural cooperatives are also developed under the new government, there seems to be only one way of successfully promoting rural Iran in the near future: the restoration of the traditional social and economic microcosm on a village level. This is not meant in the sense of the traditional relationships between landlords and peasants, but in that of the Iranian village as a community of free, equal, and independent farmers who work their land individually or collectively. The ancient traditions of collective forms of cultivation and harvesting are a unique precondition for such a successful endeavour. What has been lacking so far does not seem to be the ability of the farmers, but the ability of governments to present a convincing and attractive policy by which agriculture can be acknowledged as an economic activity of importance giving opportunities equal to those of other sectors in the national economy.

REFERENCES

Amini, S. (1983) 'The origin, function and disappearance of collective production units (*Haraseh*) in rural areas of Iran', *Der Tropenlandwirt*, **84**, 47–61.

Aresvik, O. (1976) *The Agricultural Development of Iran*, Praeger, New York.

Bank Markazi Iran (1353–59 (1974/75–1980/81)) *Annual Report and Balance Sheet*, Tehran, pp. 18–27.

Bobek, H. (1975/76) 'Entstehung und Verbreitung der Hauptflursysteme Irans — Grundzüge einer sozialgeographischen Theorie', *Mitteilungen Österr. Geogr. Gesellschaft*, **118**, 274–322.

Bowen-Jones, H. (1968) 'Agriculture', in *Cambridge History of Iran*, Vol. I, Cambridge University Press, London, pp. 565–598.

Ehlers, E. (1977) 'Social and economic consequences of large-scale irrigation developments — the Dez Irrigation Project, Khuzestan, Iran', Worthington, E. B. (Ed.), *Arid Land Irrigation in Developing Countries*, Pergamon Press, Oxford, pp. 85–97.

Ehlers, E. (1983) 'Rent capitalism and unequal development in the Middle East: the case of Iran', in Stewart. F. (Ed.), *Work, Income and Inequality: Payment Systems in the Third World*, Macmillan, London, pp. 32–61.

Ehlers, E. & Safi-Nejad, J. (1979) 'Formen kollektiver Landwirtschaft in Iran: *Boneh*', in Ehlers. E. (Ed.), *Beiträge zur Kulturgeographie des islamischen Orients*, Marburger Geogr. Schriften 78, Marburg, pp. 55–82.

Gharatchedaghi, C. (1967) *Distribution of Land in Varamin: an Opening Phase of the Agrarian reform in Iran*, Leske-Verlag: Schriften des Deutschen Orient-Instituts, Materialien und Dokumente, Opladen.

Goodell, G. (1975) 'Agricultural production in a traditional village of northern Khuzestan', in Ehlers, E., & Goodell, G., *Traditionelle und moderne Formen der Landwirtschaft in Iran*, Marburger Geogr. Schriften 64, Marburg, pp. 243–289.

Hooglund, E. J. (1973) 'The *khwushnishin* population of Iran', in *Iranian Studies*, **6**, 229–245.

Hooglund, E. J. (1981) 'Rural socioeconomic organization in transition: the case of Iran's *bonehs*, in Bonine, M. E., & Keddy, N. (Eds), *Modern Iran*, State University of New York Press, Albany, pp. 191–207.

Khosravi, K. (1973) 'Les paysans sans terre: les *khoshnechin*', in *Sociologia Ruralis*, **13**, 289–293.

Kielstra, N. O. (1975) *Ecology and Community in Iran: A Comparative Study of the Relations between Ecological Conditions, the Economic System, Village Politics and the Moral Value System in Two Iranian Villages*, Academisch Proefschrift Universiteit van Amsterdam, Amsterdam.

Lambton, A. K. S. (1953) *Landlord and Peasant in Persia*, Oxford University Press, London.

Lambton, A. K. S. (1969) *The Persian Land Reform 1962–1966*, Oxford University Press, London.

Planck, U. (1962a) 'Der Teilbau im Iran', *Zeitschrift für Ausl. Landwirtschaft*, **1**, 47–81.

Planck, U. (1962b) *Die sozialen und ökonomischen Verhältnisse in einem iranischen Dorf*, Westdeutscher Verlag: Forschungsberichte des Landes Nordrhein-Westfalen 1021, Köln-Opladen.

Planck, U. (1963) 'Berufs- und Erwerbsstruktur im Iran als Ausdruck eines typischen frühindustriellen Wirtschaftssystems', *Zeitschrift für Ausl. Landwirtschaft*, **2**, 75–96.

Planck, U. (1975) 'Die Reintegrationsphase der iranischen Agrarreform', *Erdkunde*, **29**, 1–9.

Planck, U. (1979) 'Die soziale Differenzierung der Landbevölkerung Iran infolge der Agrarreform', in Schweizer, G. (Ed.), *Interdisziplinäre Iranforschung*, Reichert: Beihefte zum Tübinger Atlas des Vorderen Orients, Reihe B (Geisteswissenschaften), 40, Wiesbaden, pp. 43–58.
Planhol, X. de (1968) 'Geography of settlement in the land of Iran', in *Cambridge History of Iran*, Vol. I, Cambridge University Press, London, pp. 409–467.
Rainer, R. (1977) *Traditional Building in Iran*, Akademische Druck und Verlagsanstalt, Graz.
Research Group (1967) 'A study of rural economic problems of Gilan and Mazandaran', *Tahqiqat-e Eqtesadi*, **4**, 11/12, 135–204.
Safi-Nejad. J. (1974) *Talebabad*, University of Tehran, Tehran.
Safi-Nejad, J. (1978) *Asnad-e Bunehha*, University of Tehran, Tehran.
Salmanzadeh, C. (1980) *Agricultural Change and Rural Society in Southern Iran*, Menas Press, Cambridge.
Stobbs, C. A. (1976) *Agrarian Change in Western Iran. A Case Study of Olya Sub-District*, Ph. D. dissertation, School of Oriental & African Studies, University of London.
Tavana, M. H. Zia (1983) *Die Agrarlandschaft Iranisch-Sistans. Aspekte des Strukturwandels im 20 Jahrhundert*, Marburger Geogr. Schriften, 91, Marburg.
Wulff, H. E. (1966) *The Traditional Crafts of Persia*, MIT Press, Cambridge, Mass.
Watson, A. M. (1983) *Agricultural Innovation in the Early Islamic World*, Cambridge, Cambridge University Press.

Agricultural Development in the Middle East
Edited by P. Beaumont and K. McLachlan
© 1985 John Wiley & Sons Ltd

Chapter 9

Agricultural mechanization in Iran

SHOKO OKAZAKI

INTRODUCTION

In the last century when many countries embarked on development programmes, Iran failed to industrialize. Consequently, it became an essentially agricultural country, selling primary produce to the advanced states of Western Europe. During the rule of Reza Shah, 1925–41, various attempts were made at industrialization. However, they were unsuccessful and the fundamental structure of the country's economy remained unchanged. Agriculture continued to be its most important sector.

As a result of the increased output of oil, the relative importance of agriculture in the national economy decreased. For example, in the first quarter of this century agriculture was estimated to have made up as much as 80 to 90 per cent of GNP, whilst by 1956 its contribution had fallen to 33 per cent. However, agriculture continued to rank first in the sectoral contributions to GNP until 1967, when oil took first place (Bharier, 1971). Except for oil which, for example, accounted for 73 per cent of total exports in 1959, agricultural produce was the most important item, constituting 87 per cent of the total non-oil exports in 1961. In the field of employment the agricultural sector has always been of high importance and has absorbed the largest proportion of the country's labour force. For example, at the beginning of the century, agriculture employed 90 per cent of the total working population and, in 1966, 46 per cent (Bharier, 1971).

One of the characteristics of Iranian agriculture is its unchanging nature. Large areas of the country had belonged to large landowners and owner-farmers' lands had occupied only 10 per cent of the total. Although various types of land ownership—private ownership, *iqta, tuyul, vaqf, khaliseh*—had come into existence in the course of history, the proportion of each varying in different periods, no substantial change in the relationship between landowner and peasant had taken place and the long-standing order continued until the land reforms of 1962. The share-cropping system was predominant and the production team system known as *boneh*[1] amongst other names, had continued to exist for long periods. As far as agricultural methods were concerned, following the technical innovations in irrigation, namely the introducion of *qanat* irrigation, which took place in the eighth century BC, no significant developments occurred before recent modernization. Archaic traditional farming methods were followed for

171

a long time, contributing considerably to the stagnation in Iran. Never in the country's history has agriculture provided any impetus to encourage change. However, after the late 1950s Iranian agriculture began to change. The driving forces behind the change were two-fold: the development of agricultural mechanization and the implementation of land reform. Mechanization advanced from the late 1950s onwards in one part of the country, changing its agricultural environment drastically. Land reform was enforced from 1962 onwards, contributing to the dramatic change in the long-standing land tenure system and political situation of Iran. In addition, land reform facilitated the rapid development of mechanization throughout the country and changed traditional agricultural methods greatly. In this paper, development of agricultural mechanization in Iran will be dealt with as a case study and the motivation, implementation, and economic effects of mechanization will be discussed.

EARLY MECHANIZED FARMING

The first agricultural machinery appeared in Iran during the period 1910–18, and from the mid-1920s onward tractors began to be imported (Bharier, 1971). But this new technology did not spread over the whole country, owing to various obstacles: the lack of service and repair facilities, spare parts, and fuel; the reluctance of tenant-farmers to introduce machinery; and an unfavourable land tenure system (Banani, 1961). Above all, landlords had no incentive to replace draught animals with machinery. Thus the primitive method of ploughing by oxen remained predominant, except on the Crown lands where some tractors and other machinery had been introduced.

After the Second World War, however, a new era in agricultural mechanization began in Gorgan, the north-eastern part of Iran. With the abdication of Reza Shah in 1941, his direct control over the Crown lands, most of which were in the fertile Caspian littoral, came to an end and, in 1942, the lands which he had confiscated or purchased at low prices were restored to their former owners (Lambton, 1953). Some of the land owners whose lands were returned started a new type of farming. Besides managing the cultivated lands on which peasants had worked on the basis of the paternal landlord–peasant relationship, they set out to operate directly-managed farms for profit. They employed hired labourers, reclaimed marsh and permanent pasture lands, and bought tractors which had been used on Crown lands from the Agricultural Office at Gorgan. This was the beginning of mechanized agriculture in this region. These land owners were from the Torkoman tribe in Gorgan, people who never relied solely upon the land for their income, but were also engaged in agricultural trade. Following these wealthy Torkoman merchants, several capitalists from Tehran came to Gorgan in order to operate mechanized farms, and their pioneering efforts resulted in the formation of the Gorgan Dry Farming

Company, established in 1949 by high-ranking officials such as the Agriculture and Labour Ministers of the time and some enterprising land owners. As a result of the impetus given by the establishment of this company, the Wheat Planting Company was formed in the following year by a number of politicians, agricultural engineers, high-ranking officials, and land owners living in Tehran. In addition to these companies, large numbers of mechanized farms were set up one after another in the first half of the 1950s by people from Tehran. Local merchants and land owners soon followed their example. Those entrepreneurs who set up mechanized farms in this period generally had funds enough to carry out large-scale operations. However, in the latter half of the 1950s, especially from 1958 onwards, retired army officers, middle-ranking government officials, and others who did not have the means or the social and political prestige of their earlier counterparts, started operating mechanized farms. It has even been said that anyone who could afford to buy an area of over 50 hectares established a mechanized farm in this period, even if he did not have the necessary experience. Farms set up by such people, though smaller in scale, were large in number, scattered throughout Gorgan, causing the area to become a large-scale mechanized farming area, rather like north Syria (Warriner, 1952).

These farms were completely new in Iranian agriculture as far as agricultural management was concerned. First, instead of share-croppers or fixed-rent tenants, the farms employed hired labourers. Secondly, land owners managed their farms themselves. Thirdly, since farms were large enough in scale to use large machinery, namely 50 to 6,000 hectares, farm operation was based as far as possible upon mechanization, with all of the agricultural work involved in wheat planting—ploughing, disking, sowing and harvesting—being done with big tractors and combines from the earliest stages. Thus this area became a centre of agricultural mechanization in Iran and was in striking contrast to the rest of the country.

The establishment of capitalist farms, which was the main cause for the development of mechanization, was facilitated by the existence of vast, fertile, uncultivated lands. Having inherited extensive lands from his father, the Muhammad Reza Shah decided to dispose of his own lands and in 1951 he issued the Decree for Crown Land Distribution, deeming that division of his lands among the peasants would be a wise policy in order to reduce the political criticism and strengthen his regime.[2] Along with this measure, he decided to lease or sell uncultivated Crown lands. From the end of the 1950s, in particular with the accelerated distribution of the cultivated Crown lands among the peasants, vast uncultivated lands such as marsh, shrub, and pasture land were also disposed of to those wishing to develop them. This policy gave many people an easy opportunity to obtain the land they required to set about mechanized farming.

Gorgan, unlike most parts of the country, is blessed with enough rainfall to allow dry farming to be carried out, so it was not essential to invest in irrigation in order to carry out agriculture, allowing for economic farming. Besides the plentiful supply of land, the market for agricultural products was brisk, and the pioneering entrepreneurs showed how profitable mechanized farm operation could be. This greatly influenced other people who were looking for ways of investing their money. Throughout Iran's history, investment in agriculture by those wishing to amass wealth has been a common phenomenon and, once security had been established, wealthy people used to extend their economic activities to agriculture. Running a mechanized farm was a modern aspect of the traditional pattern of enterprise by the rich.

At first, the farms' sole product was wheat. The natural conditions of Gorgan, however, were favourable for the cultivation of the more profitable cotton. In some districts of Gorgan, such as Kordkui, cotton had already been grown and exported to Russia since the early twentieth century. In the late 1950s, cotton was in great demand in the internal market as a result of the development of the textile industry, as well as in the international market. Consequently, cotton began to be cultivated on the mechanized farms too and land sown with cotton increased rapidly, especially after 1958 (Okazaki, 1968). On almost all farms where cotton was grown, ploughing, fertilizer-spreading, disking, and sowing were carried out with tractors and on some leading farms weeding and pest control were mechanized in later years, this practice expanding to other farms as well.

In the early days, dry farming was predominant. Gorgan has a rainfall of twenty inches or more annually and satisfactory returns were brought about even with dry farming. But, since the best scheme for increasing production was to acquire more water at the right time, the farm owners turned their attention to investment in irrigation. Irrigation, especially on land bearing cotton, was enormously effective. According to the 1960 agricultural census, the cotton yield off irrigated land was 2,329 kilograms per hectare, whereas that off dryland was only 965 kilograms per hectare. Therefore, much of the new investment was applied to digging artesian wells 20 to 30 metres deep, tube wells 20 to 130 metres deep with pumping equipment, and even some reservoirs. After 1959, almost all farms had these new irrigation facilities and some pioneering farms even introduced sprinkler irrigation systems.[3]

The development of mechanized farm operations brought about significant economic changes in this area. First of all, since the farms were set up on virgin land, the area under cultivation increased considerably and by 1962 almost all cultivable lands had been reclaimed and were producing crops. Secondly, employment opportunities increased. Farms needed labourers, but surplus labour was not available in Gorgan. Accordingly, the necessary labour was supplied by other regions and farm owners recruited farm labourers from the neighbouring province, Khorasan. With the intensification of cotton cultivation, farms began to need more labourers, both seasonal and temporary, which were sought from the underdeveloped regions of Khorasan and Sistan. Many of them

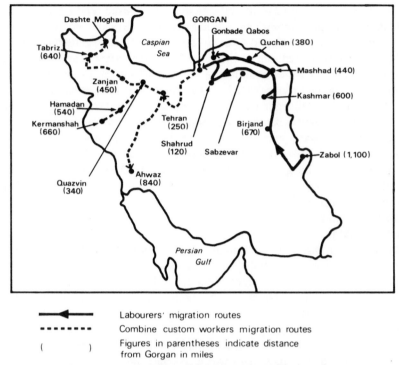

Labourers' migration routes
Combine custom workers migration routes
( ) Figures in parentheses indicate distance
from Gorgan in miles

Figure 9.1   Migration of labour and combine workers

came to Gorgan in April and returned home in December. The most remote
supply of labour was Zabol, 1,100 miles away from Gorgan (see Figure 9.1).
Thirdly, the trade in farm produce such as wheat and cotton, farm machinery, and
fertilizer developed. Consequently, Gorgan became an area of economic growth.

AGRICULTURAL DEVELOPMENTS IN GORGAN

The expansion of farming activities by the entrepreneurs brought about dramatic
changes in the archaic agricultural structure of Gorgan. Another change was
the adoption of mechanized farming by peasants who had previously used
primitive methods. While these peasants had been engaged in difficult and
inefficient cultivation using oxen, the mechanized farms of entrepreneurs had
carried out their activities speedily and effectively with tractors and combines.
The contrast was enough to provide the traditional peasants with the impetus
to modernize. After the mid-1950s, with the conversion of extensive pasture
into cultivated land, grazing for cattle became limited. Furthermore, the land
tenure system of Gorgan was amenable to mechanization. In Gorgan, unlike
the Iranian Plateau regions, the production team system called *boneh*[1] did not

exist and an individual peasant could manage his leased land on his own. He could decide independently whether or not to introduce machinery. Furthermore, in this region peasants enjoyed a kind of tenancy right and, generally speaking, could continue to be tenants of the same lands year after year. In the *boneh* system any new measures require the consent of all the other members. In addition, no tenancy right existed in the plateau regions, which cover a vast area. In this respect, Gorgan was an exceptional case and there was no obstacle to mechanization as far as the land tenure system was concerned.

Even though peasants wanted to introduce tractors and other machinery, they could not obtain them while financial obstacles remained. This major problem was resolved by the government. In August 1952, in order to promote agricultural production the government established the Agricultural Machinery Development Bureau as an external arm of the Plan Organization.[4] Up to 1956 the Bureau imported machinery directly and sold it on the basis of a twelve-monthly instalment plan. However, this proved unsuccessful and the number of tractors sold through the Bureau was no more than 299 in five years. Accordingly, the Bureau abolished the direct import-selling system in 1956, instead granting loans to those who bought a designated brand of machinery from appointed dealers. Thanks to this policy, the purchasers became able to buy tractors and combines with a down-payment of only 20 per cent of the purchase price, the Bureau providing the remainder, which was paid back by the purchasers over four years at an interest rate of 6 per cent. Furthermore, the selling price of the machinery was also controlled by the government and installation of service facilities by the dealers was made obligatory. One of the principal obstacles to mechanization was thus eliminated. An increase in the funds allocated to the Bureau by the Plan Organization made it possible for almost all the applicants to obtain loans to buy machinery and more than 80 per cent of the tractors and combines sold were purchased with the loans from the Bureau (Iran, Ministry of Interior, 1956).

Rich peasants who had managed rather extensive lands now started to buy tractors and combines. However, their lands were not large enough to fully utilize the machinery and their possession of it caused an increase in their expenses so that they began to do custom work. Ploughing and harvesting by machinery, as a result of hiring custom workers, could be done much faster than by using draught animals. Also, harvesting by combines resulted in less waste. With those who hired being saved from onerous labour while they hired custom workers, and with the cost of custom work being less than that done by oxen and labourers, there was an increase in income from custom fees, and the generation of high profits. The success of rich peasants in introducing machinery had a powerful effect on other peasants, the scale of whose holdings was smaller. With the stimulus of the profitability of custom-work, purchasers of farm machinery increased. Only rich peasants purchased tractors before 1960, but after that date, even those who cultivated only a few hectares of land bought them. The same applied to combines. For example, in 1961 a peasant whose holding was only 3 hectares bought a tractor; and a peasant with a 5-hectare holding purchased a combine. These were not exceptional cases. Furthermore, those who owned no land, such as tractor drivers, began to purchase machinery.

Thus, a great number of tractors and combines were introduced into the villages of Gorgan. No accurate data are available concerning the amount of farm machinery owned by peasants. However, the statistics on Crown land distribution published in 1959 give us some idea. As of December 1959 in 58 villages in which Crown land distribution was carried out one tractor had been introduced for every 242 hectares (USOM/Iran, 1960).

No data are available for the same villages in later years; however, in three villages among the 58 in which the writer made surveys, the number of tractors had doubled by 1963. This tendency was not peculiar to the villages of the former Crown lands, but was common to the whole area of Gorgan. As a result of the introduction of tractors and combines by peasants, Gorgan came to be the most advanced area in agricultural mechanization in the country. During the period 1957 to 1962, out of 6,346 tractors and 1,101 combines purchased with the Agricultural Machinery Development Bureau's loans, 1,974 tractors and 609 combines, corresponding to 30 and 55 per cent of the total respectively, were in this limited area, which comprises only 1.5 per cent of the total cultivated land of the country (Table 9.1).

The development of mechanization in Gorgan, which had permeated the stagnant agricultural system in a short space of time, resulted in great alterations in farming methods. Oxen, which had been the main means of ploughing, were replaced by tractors and disappeared from sight, except in mountain areas where

Table 9.1 Number of tractors and combines purchased with the loans of the Agricultural Machinery Development Bureau

Year	Tractors Total	Tractors Gorgan	Combines Total	Combines Gorgan
1952–56	299		7	
1957	725	143	48	38
1958	1,301	401	176	93
1959	1,169	453	305	140
1960	1,283	490	207	150
1961	1,223	371	168	101
1962	645	116	197	87
Total (1957–62)	6,346	1,974	1,101	609

Source: Plan Organization (undated). Unpublished statistics of the Agricultural Machinery Development Bureau, Tehran, Iran.

machinery could not be used. Furthermore, human labour in harvesting was entirely replaced by combines.

As stated above, the new mechanized farms introduced irrigation facilities through investment in wells, thus attaining high levels of production. The great differences in yield between irrigated farming and dry farming were readily visible and induced peasants who had grown up with dry farming to invest in irrigation. Rich peasants raised money and put it into irrigation, thus enhancing productivity in the same way as capitalist farm owners.

The rapid introduction of heavy machinery resulted in a great change in the economic condition of those who used these means in cultivation. By making use of machinery, some peasants succeeded in enlarging their holdings. On the other hand, many proved unsuccessful with the new methods. This was especially true of the small-scale peasant farmers who purchased tractors and combines during the farm machinery boom which began after 1960. As a result of the increase in owners of agricultural machinery doing custom work, fees fell sharply. For example, the ploughing fee in 1963 was only half of what it had been in 1958 and the rates for harvesting showed a similar downward trend. Furthermore, the increase in farm machinery resulted in a decrease in the area on which one tractor or combine was able to work. These factors produced a decrease in the custom fee income of machinery holders. For example, a peasant in the Kordkui district of Gorgan, having sold his land and bought a tractor, had an income of 190,000 rials in 1960, but in the next year he could earn only 130,000 rials, of which he had to pay 100,000 rials as an instalment of the Agricultural Machinery Development Bureau's loan.

Unlike capitalist farms, which had their own service stations, peasants generally speaking could not maintain machinery in good condition for the same length of time as could mechanized farm owners as a result of their inability to hire good drivers and to repair machinery at an early stage. This resulted in the expenditure of large sums of money on repair work in a relatively short time. This factor, with the decrease in custom fee income, was severely damaging to the small-scale machinery owners. Those who could not maintain a level of income sufficient for their expenses tried every means available to cover the deficit. Tractor owners who had done custom work only in the vicinity of their village started to seek jobs in other districts in Gorgan. The same was true of combine owners who, in addition, after harvesting in Gorgan began to seek custom work in remote provinces such as Qazvin (340 miles from Gorgan), Zanjan (450 miles), Tabriz (640 miles), Hamadan (540 miles), and Ahvaz (840 miles). This practice of long-distance custom working, was not peculiar to Gorgan, but was a common phenomenon (Salmanzadeh, 1980) starting in 1960 in Gorgan, and spreading rapidly in 1961 and 1962, owing to the poor harvest in these years. If the income from custom working in other provinces could not meet expenses, it was necessary to find other measures. Some machinery owners raised money through the pre-harvest sale of cotton on disadvantageous terms, bringing about exploitation by merchants and creating a vicious circle. Others sold their land to cover the deficit and the rest were obliged to part with their machinery or to have their machinery confiscated by the Agricultural Machinery Development Bureau for being in arrears of their instalment payments. In other words, mechanization became one of the important factors which fostered class differentiation amongst the peasantry in this region.

IMPORTATION OF MACHINERY AND
ITS CONSEQUENCES

In a sense, the change in Gorgan was exceptional. It was a newly populated area, blessed with naturally advantageous conditions and there were those factors already mentioned which had facilitated the rapid development of farm mechanization. However, there was also a densely populated area where agricultural machinery had appeared at a relatively early stage. This was the Caspian littoral rice-growing district, including Gilan and Mazandaran. In 1958 a few mechanical tillers were introduced on an experimental basis in these provinces. In the following years the number of tillers increased and after 1963 more than 4,000 tillers were introduced each year. In addition to an increase in importation, an assembly plant for tillers began to operate in 1968, and mechanization in paddy-fields advanced rapidly. It is reported that at the beginning of the 1970s tillers had been used in about 70 per cent of paddy-fields in the Caspian littoral (Aresvik, 1976) and according to official statistics for 1974, 9,866 tillers were in use in Gilan, where the total area of paddy-fields was 149,691 hectares, and 13,428 tillers in Mazandaran, where the paddy-fields covered a total of 116,391 hectares. That is to say, one tiller was in use for every 15.2 hectares in Gilan and every 8.67 hectares in Mazandaran.

The main causes of mechanization in Gilan and Mazandaran were quite different from those in Gorgan. Unlike Gorgan, the southern Caspian littoral had neither a shortage of labour force, nor capitalist agriculture, nor newly reclaimed lands. One of the incentives to mechanization in this area was the positive selling activity of foreign tiller makers, namely those of Japan (Table 9.2). Seeing the area as a potential market, Japanese tiller makers, who had faced over-production at home, made advances here and competed with one another for customers. They provided purchasers with favourable terms through their agents and stationed service men in this region. These factors were very helpful in removing the obstacles to mechanization.

Unlike the Iranian plateau region, where group production team systems such as *boneh* were prevalent, in Gilan individual peasants had their own leased land. As they were free to manage their land individually, it was possible for them to buy tillers without the consent of others. Furthermore, they enjoyed relatively stable tenancy rights.

Table 9.2 Imports of tillers and two-wheel
tractors from Japan

Year	Number
1957	1
1958	7
1959	54
1960	117
1961	424
1962	763
1963	4,031
1964	4,369
1965	1,190
1966	2,309
1967	3,913
1968	4,495
1969	2,946
1970	437
1971	2,022
1972	10,050
1973	6,450
1974	3,610
1975	8,304
1976	4,950
1977	8,500
1978	1,710
1979	940
1980	2,876

Source: Ministry of Finance, *Japan Exports and
Imports*, Japan. For years 1957–75 tariff
nos. 712–113 and 712–511; for years 1976–80
tariff nos. 84.24–420 and 87.01–911

Before, as rice growing had depended upon rainfall, the yield had been low and unstable. According to the 1960 statistics, the yield per hectare was around 1,400 kilograms, while the average rent was 627 kilograms, corresponding to as much as 45 per cent of the yield. The peasants' economic situation had therefore been very harsh. However this state of affairs was improved by a new irrigation programme. Under the Second Seven-Year Development Plan of 1955–63 a dam was constructed on the upper reaches of the Safid Rud and irrigation water was channelled to paddy-fields in the region. Owing to the stable supply of water, the crop yield increased greatly, reaching 2,340 kilograms per hectare. As the rent remained the same, peasants were able to enjoy the increased yield fully. Furthermore, thanks to irrigation, better-paying varieties such as *sadri* and *binam* which needed more water could be cultivated in greater quantities. According to the writer's research, the cultivation of these highly-priced varieties, which had formerly been grown on less than 30 per cent of the total land, came to be grown on as much as 80 per cent. This greatly contributed to an increase in income because, while the rent was paid primarily by the cultivation of the low-priced *champa* variety as before, additionally peasants could sell expensive varieties for a favourable price at the market. In the case of Hasanabad, which was surveyed by the author in 1965, the output per hectare in that year, if computed in terms of currency, corresponded to 32,526 rials, whereas the rent was 6,864 rials as before the introduction of better varieties. Therefore, the rent was substantially reduced to 21 per cent of the total yield. In short, after the irrigation programme, the economic situation of peasants was considerably improved, and they had money to spare. This economic surplus allowed them to buy machinery and was, therefore, another important factor contributing to mechanization.

Since ploughing in paddy-fields was very laborious in comparison with the work done in grain fields, the peasants' desire for the introduction of tillers was strong. Furthermore, the price of a tiller was not as high as that of a tractor, making them within reach financially.

As in Gorgan, custom working became popular in this region. For example in 1965 in Hasanabad none of the villagers had tillers, but in 1964 four of the eighteen peasants there hired custom workers and in 1965 twelve did so. At first, they employed custom workers for only one or two days per season to do the ploughing, and used draught animals for the rest of the work, but gradually tillers began to be used for levelling and for other work as well.

Rice growing is very labour-intensive. According to the writer's research, on average every one hectare of paddy-field needs one male and one female labourer. Therefore, peasants with two hectares or more, had in general to hire male labourers called *barzegars* throughout the ploughing season and female labourers called *karachis* throughout the planting and weeding seasons. In the case of Hasanabad, ten peasants out of the total of eighteen had employed fourteen *barzegars* and eleven peasants had hired twenty-two *karachis* in 1965. An increase in the introduction of tillers and in the hiring of custom workers by rich peasants naturally brought about a great loss of employment opportunities for lower class villagers. This was a serious problem for them.

Hitherto, ox manure had played an important part as a fertilizer in paddy-fields. However, the introduction of mechanized farming to paddy-fields inevitably reduced the number of oxen, and therefore the quantity of manure available. This caused the adoption of chemical fertilizers. In other words, mechanization resulted in an increase in expenditure on rice growing. The expense of purchasing and maintaining machinery, hiring custom workers and buying fertilizer became unavoidable. This fostered a commercial approach to rice growing and the use of market-oriented varieties causing its production as a cash crop to increase rapidly. Thus, the peasants in this region became more and more involved in a cash economy.

CHANGES RESULTING FROM LAND REFORM

As in many other countries, the land tenure system in Iran was reformed after the Second World War. From 1952 onwards the lands belonging to the Shah were sold to the peasants, and after 1962 land reform was carried out all over the country. These reforms were enforced under strong American influences in order to survive the contemporary political crisis and were in fact politically motivated. In the short term the reforms achieved their purpose for the Shah and America, bringing about political stability, reinforcement of the Shah's power, and the prerequisites for rapid industrialization.

The land reform of 1962 contributed to the elimination of traditional large land owners and to a change in the long-standing landlord–peasant relationship, although, having many defects, it was far from ideal. Another effect of land reform was the rapid development of agricultural mechanization. The 1963 Additional Articles to the Land Reform Law exempted mechanized lands from the application of the Law. That is to say, so long as the land was worked by mechanized means and provided the land owner had personally, with his own capital or otherwise, mechanized, he would continue to own it (Lambton, 1969). This meant all mechanized lands could be retained by land owners irrespective of the size of the holding. Consequently, large-scale mechanized farms, already mentioned, continued to exist. Furthermore, by law it was made obligatory for land owners to work with agricultural machinery on the lands they were allowed to retain. This facilitated mechanization and modernized agricultural management by land owners.

Changes arising from land reform can be illustrated by reference to one village near Tehran. The land owner of this village had previously managed his lands on the basis of the traditional share-cropping system. But, in 1965, following the Land Reform Law, he was obliged to lease 210 hectares to 60 former share-croppers on the basis of fixed rent, while keeping the remaining 108 hectares for himself. On this land, he set about operating a mechanized farm producing wheat and cotton. Besides introducing agricultural machinery into every stage of the farming process, he employed hired labourers from the village. Furthermore, he dug a deep well with pumping equipment, obtaining enough water to irrigate his land. Previously, the water had been provided from only two *qanats*. The plentiful water supply made it possible to apply more fertilizer. Since he had no need to get any consent from the villagers, as in the case of the traditional share-cropping contract, he could adopt any measures he wished in order to raise productivity. These endeavours resulted in an increase in production, the fruits of which fell entirely into his hands. For example, before the Land Reform Law when the share-cropping system was in operation, the yield of cotton per hectare was 1,627 kilograms. However, later yields reached as much as 2,800 kilograms on the land owner's mechanized farm, while the yield on the villagers' lands was as little as 1,500 kilograms (Safi-Nejad, 1974). Thus, the land owner could gain handsome profits from farm operation. After the reform, his income never decreased, but rather increased.

Thus land reform not only released peasants from the yoke of the old landlord–peasant relationship, but also freed land owners from the old, paternal way of land management. Although their agricultural income did not actually increase following the reforms, peasants obtained the freedom to decide what to cultivate and how to work their land. Land owners also now became free to decide how to manage their lands. Previously, under the paternal landlord–peasant relationship, any changes often used to be met by resistance from the peasants. Land reform made it possible for land owners to become free-minded men of enterprise. Some of them even extended their activities to commerce or industry. Others, seeing the management of mechanized farms to be profitable, became agro-capitalists, operating mechanized farms on the lands which they were allowed to hold, like the land owner of the above-mentioned village. Generally speaking, in the early stages at least, they were able to earn considerable amounts from agriculture, thanks to the favourable prices of farm produce.

This new type of agriculture was not only carried out on cultivated lands, but also extended to unutilized lands. After the Land Reform Law was introduced, many landowners turned their attention to cultivating new land (Lambton, 1969). Employing hired labourers, they managed these newly reclaimed lands by mechanized means, introducing machinery as much as possible. Agricultural machinery gave them a sense of modernity.

The government which had adopted the policies of modernization, deemed agricultural mechanization to be a key measure in its modernization and promoted the importation of tractors and other agricultural machinery.[5] In 1967, a government-sponsored tractor assembly plant was established, and tractors were sold to farmers on favourable credit terms. Between 1969 and 1973, as many as 25,000 tractors were marketed by this plant (Salmanzadeh, 1980). This government measure greatly facilitated the development of farm mechanization. In the pre-land reform period, tractor ploughing was, except in a few advanced areas like Gorgan, not prevalent. Indeed, the number of tractors in operation in 1963 was estimated to be only 6,400 (Okazaki, 1968). However, after the reforms, farming activities by land owners and the modernizing policy of the government brought about the rapid development of agricultural mechanization. As shown in Table 9.3, the importation of tractors increased considerably, and the number of tractors in use augmented year by year. In 1972 the total number of tractors in use was estimated at around 23,000, and in 1977 it reached 50,000. Amongst the various provinces, the more fertile regions show a higher concentration of tractors. Tehran, Mazandaran (including Gorgan), Azarbaijan and Esfahan were the most developed regions in terms of agricultural mechanization. According to the statistics for 1974, one tractor had been introduced for every 115 hectares in Mazandaran, every 124 hectares in Esfahan, every 165 hectares in Tehran, and every 277 hectares in Azarbaijan (Statistical Centre: 1974). It is reported that throughout the 1970s the use of machinery for ploughing and harvesting had become common on holdings of over ten hectares.

Such increasing mechanization of agricultural production inevitably resulted in the reduction of the number of hired workers needed and the total working hours available for those employed. Thus, landless labourers and poorer peasants suffered a great loss in labour opportunities, being compelled to migrate away from the villages towards the urban areas in search of jobs (Hooglund, 1982).

Table 9.3   Imports of tractors

Year	Imports	In use
1952–56	299	
1957	725	
1958	1,301	
1959	1,169	
1960	1,283	
1961	1,223	
1962	645	
1963	n.a.	
1964	n.a.	
1965	n.a.	
1966	3,093	16,000[a]
1967	5,817	17,500[a]
1968	5,294	20,000[a]
1969	4,870[a]	21,000[a]
1970	5,100[a]	n.a.
1971	5,500[a]	21,500[a]
1972	7,000[a]	23,000[a]
1973	9,348	24,200[a]
1974	7,500[a]	27,000[a]
1975	12,702	29,000[a]
1976	9,768	30,000[a]
1977	4,700[a]	50,000[a]
1978	2,000[a]	55,000[a]
1979	2,780[a]	57,000[a]
1980	n.a.	58,000[a]

Sources: 1952–61, Okazaki, S. (1968). 1962–80, Food and Agricultural
Organization, *FAO Trade Year Book, and FAO Production
Year Book.*
[a] FAO estimate.

CONCLUSION

One of the interesting points to be noted in Iranian mechanization lies in the fact that the government had fulfilled a very important role in facilitating it. The government's aim during this period was the dismantling of the old institutions and their replacement by modern ones. In this connection, the creation of up-to-date farming methods was one of the important targets in the modernization of agriculture, which was carried out as a part of the Shah's modernizing policies. The government tended to attribute the low level of production to the backwardness of farming methods and believed that the development of mechanization would bring about an increase in production (Hooglund, 1982). Consequently, in order to finance the purchase of machinery, the government allocated considerable funds, made possible by increased oil income. This greatly contributed to the development of farm mechanization.

Agricultural mechanization brought about various changes in Iranian agriculture. Vast unutilized lands began to be cultivated due to the fact that mechanized farming had also been carried on to virgin lands. With increased investment in irrigation projects, the development of agricultural mechanization was a key factor to increasing the area under cultivation after the Second World War. Furthermore, in some regions, cultivation became more intensive and land came to be used annually without leaving it fallow (Salmanzadeh, 1980). Also, mechanization caused considerable changes in the agricultural labour market. This had two contradictory aspects. On the one hand, in the case of previously uncultivated lands, mechanization created employment opportunities; for example, mechanized farms in Gorgan, set up on uncultivated lands and involved in labour-intensive cotton cultivation, were of relatively high importance in providing employment for poor villagers in remote, underdeveloped areas. On the other hand, in the case of cultivated lands, mechanization brought about unemployment. Thus, mechanization came to be an important factor in emphasizing differences amongst the peasantry. The main promoters of mechanization in Iran were agro-capitalists and land owners. However, in some areas small-scale producers were also involved in farm mechanization. Some people benefited from mechanization, succeeding in enlarging their economic capacity; however, in contrast, many were unsuccessful in introducing machinery, being driven to ruin.

In industrially underdeveloped countries like Iran, generally speaking, the government's role in mechanization is of paramount importance, because the industrial sector can play no useful part. Furthermore, in such countries, land owners had performed an important role in land management. Inevitably, the government's policy was favourably disposed towards land owners. In this connection, Keddie's point of view, which holds that the Iranian Government's policy of facilitating mechanization chiefly benefited landlords (Keddie, 1980) would seem to be correct.

Another interesting aspect of Iranian mechanization lies in the fact that the pioneering and important promoters of mechanization were agro-capitalists. In the countries of semi-arid or arid zones like Iran, which have vast unutilized but cultivable lands, large-scale farm operations for profit are possible, provided that the market for produce, foreign or internal, is favourable to farm owners and security is maintained. Consequently, under such circumstances, agro-capitalists will fulfil a role in the development of agricultural mechanization in those countries which follow a path dedicated to modernization.

NOTES

1. The *boneh* is a system peculiar to Iran. Under this system, share-croppers are grouped into a certain number of units and all of the unit members work together under the supervision of their landowner or his bailiff.
2. It is interesting that religious circles were opposed to this policy. The origin of the mullahs' antagonism towards the Shah's land policy can be traced back to 1951. Another interesting point is that the Shah made this policy public in 1949 in New York when he visited the USA to make a bid for economic assistance. This implies that there was strong pressure from America behind this decision for the implementation of the Crown Land Distribution.
3. The development of irrigation in one of the leading farms was as follows: in 1948 three *qanats* were constructed; in 1959 five wells 100 metres deep were sunk; in 1962 four artesian wells were dug; and in 1963 a sprinkler irrigation system was adopted.
4. The foundation of this Bureau was presumably promoted by the influences of the countries exporting the machinery, such as the USA, UK, and West Germany.
5. One of the policies for this was the 1966 agreement with Romania, which arranged the importation of some 20,000 tractors, over a period of several years. See Aresvik, 1976, p. 162.

REFERENCES

Aresvik, O. (1976) *The Agricultural Development of Iran*, Praeger, New York, pp. 160 & 163.
Banani, A. (1961) *The Modernization of Iran 1921–1941*, Stanford University Press, Stanford, California.
Bharier, J. (1971) *Economic Development in Iran 1900–1970*, Oxford University Press, Oxford, pp. 60, 131, & 141.
Food and Agriculture Organization, *FAO Production Year Book*, and *FAO Trade Year Book*, Food and Agriculture Organization, Rome.
Hooglund, E. (1982) *Land and Revolution in Iran 1960–1980*, University of Texas Press, Austin, pp. 69 & 97–98.
Iran, Ministry of the Interior (1956) *Regulations granting credit for Agricultural Machinery*, Agricultural Machinery Development Bongah, Tehran.
Iran, Ministry of the Interior (October 1960) *First National Census of Agriculture*, Tehran, Iran, pp. 1–2.
Iran, Ministry of the Interior (1971 & 1974) *Statistical Year Book of Iran*, Iranian Statistical Centre, Tehran.
Japan, Ministry of Finance (various dates) *Japan Exports and Imports*.

Keddie, N. (1980) *Iran: Religion, Politics and Society*, Cass, London, p. 174.

Lambton, A. K. S. (1953) *Landlord and Peasant in Persia*, Oxford University Press, London, pp. 256–257.

Lambton, A. K. S. (1969) *The Persian Land Reform 1962–1966*, Oxford University Press, Oxford, pp. 45–46, 168, 193, 195–196 & 261.

Okazaki, S. (1969) 'Shirang-Sofla: the economics of a Northeast Iranian Village', in *The Developing Economies*, 1, No. 3, pp. 261–283.

Okazaki, S. (1968) *The Development of Large-scale Farming in Iran: the case of Gorgan*, The Institute of Asian Economic Affairs, Tokyo, pp. 19 & 38.

Safi-Nejad, J. (1974) *Boneh qabl as Eslahat-e Arzi*, Tus, Tehran, pp. 203–206.

Salmanzadeh, C. (1980) *Agricultural Change and Rural Society in Southern Iran*, Menas, Press, London, pp. 189–191.

USOM/Iran (1960) *The Turkoman and his Land*, Tehran, Tables 2–7.

Warriner, D. (1952) *Land Reform and Development in the Middle East*, Royal Institute of International Affairs, Oxford University Press, London, pp. 71–112.

Agricultural Development in the Middle East
Edited by P. Beaumont and K. McLachlan
© 1985 John Wiley & Sons Ltd

Chapter 10

The Agricultural Development of Iraq

HERBERT W. OCKERMAN
and
SHIMOON G. SAMANO

INTRODUCTION

Iraq, as shown in Figure 10.1, is located in south-west Asia, bordered to the east by Iran, to the north by Turkey, to the west by Syria and Jordan, and to the south by Saudi Arabia and Kuwait (Meliczek, 1973; Government of Iraq, 1976; Ministry of Information, 1980).

The total area of Iraq is 438,446 square kilometres, consisting largely of lowland plains less than 200 metres above sea-level. Indeed, only 13 per cent of the country exceeds an altitude of 500 metres. Iraq has four land regions: the upper plain, the lower plain, the mountains and the desert (*The World Book Encyclopedia*, 1973).

The upper plain is a region of dry, gently rolling grassland lying between the Tigris and Euphrates rivers north of the city of Samarra. The largest hills in this region rise approximately 1,000 feet (305 metres) above sea-level (*The World Book Encyclopedia*, 1973).

Beginning near Samarra and extending to the Gulf is the lower plain, which includes the fertile, well-watered delta between the Tigris and Euphrates where most of Iraq's population live. The southern part of the lower plain has many swamps and two marshy lakes, the Hawr Al-Hammar and Hawr As-Saniyah (*The World Encyclopedia*, 1973).

In northeastern Iraq the mountains form part of the Zagros range, running north into Turkey as the Taurus range. A number of peaks rise more than 11,000 feet (3,353 metres) above sea-level. Kurdish peoples live in the foothills and valleys of the region, concentrated in the districts of Dohuk, Arbil, and Sulaimaniya (Kinnane, 1964; Harris, 1958).

Deserts cover southern and western Iraq, extending into Jordan, Kuwait, Saudi Arabia, and Syria. Much of this region of limestone hills and sand dunes forms part of the Syrian Desert (Fisher, 1978).

The Iraqi Planning and Follow-up Department (1981) suggested a division of Iraq into four zones, comprising 20 per cent in a region of sedimentary deposits as the first, 60 per cent desert highland as the second, and 20 per cent

Figure 10.1   Iraq: location map

mountain zone, within which 75 per cent is undulating, as the third. Overall, some 27 per cent of Iraq is classified by this same source as potentially useful agricultural land, of which 40 per cent is currently cultivated.

AGRICULTURE IN IRAQ

Agriculture is the main source of employment (42 per cent in 1981) and the most important contributor to national income (Fisher, 1979; Government of

Iraq, 1976). The size of agriculture is shown in Table 10.1 together with important dates that have shaped it in Table 10.2. In the decade from 1977, Iraq was 56 per cent self-sufficient in cereals and agricultural imports amounted to $867.4 million, which was 22.2 per cent of total imports. Although the National Economic Plan of 1976–80 gave high priority to industry, agriculture remained important, with the ambitious aim of producing an agricultural surplus for export. The development plan allocated funds for several land reclamation and integrated farming projects, adding nearly one million hectares to the country's arable land. According to the 1976–80 economic plan, the agricultural sector was to receive appreciable financial resources with the objective of achieving food self-sufficiency (Fisher, 1978; Government of Iraq, 1976; Al-Doori, 1979). The plan was never fully implemented, but it is thought that agriculture received approximately $ one billion during its implementation. Major projects under the 1981–85 plan included a 1,927 ha Kirkuk irrigation scheme, a 1,700 ha Abu Ghraib irrigation system, and a 1,285 ha Kifl-Shenafiyah agricultural project on the Euphrates (US Department of Commerce, 1982).

Table 10.1   The size of Iraq's agriculture

Relative share of the socialist sector in agriculture	39 per cent
Number of agricultural cooperatives (1979)	1,987
Numbers of farmers in cooperatives (1979)	355,000
Loans granted by the Cooperative Agricultural Bank (1979)	32 mn dinars
Land reclaimed by irrigation and drainage (1976–80 plan)	3,320 donums
Chemical fertilizers used (1980) of which:	
Ammonium sulphate (21 per cent)	44,000 tons
Urea (46 per cent)	109,000 tons
Superphosphate (27–45 per cent)	20,000 tons
Compound fertilizers	14,000 tons
Potassium sulphate	420 tons
Tractors (1,444 hours/year): harvesters (340 hours/year) (1980)	31,000
Pest control	1.3 mn donums
Seed for field crops (imported 1980)	51,884 tons
Wheat seed	5,344 tons
Potato seed	23,660 tons
Egg production	1,000 mn/year
Poultry meat production	61,000 tons/year
Fish catch at sea	54,700 tons

Source: Ministry of Culture and Information, 1980; Planning and Follow-up Department, 1981.

Table 10.2   Important dates in the development of modern agriculture in Iraq

July 28, 1968	Revolution gave farmers possession of pumps and agricultural implements
August 4, 1971	3,500 farmers' families living in Baghdad request to return to the countryside.
May 23, 1973	Iraq begins to create its first fishing fleet.
September 30, 1975	Issue of law for the participation of state farms in the socialist transformation.
July 18, 1977	Inauguration of the first stage of the Main Drain (Third River) project.
May 10, 1978	Start of resettling 28,000 farmers' families in new homes in modern villages.
October 1, 1978	Start of largest field survey of livestock in the country.
October 11, 1978	Sowing wheat and barley was completed in 2,057,000 donums. Pioneer agriculture for the winter season was completed in 87,000 donums.
March 1, 1979	Allocation of 26,425,000 dinars to subsidize prices of commodities and consumption of agricultural services.

Source: Ministry of Culture and Information (1980).

Iraq is one of the least densely populated countries in the Middle East, but has the greatest agricultural potential. It does not suffer from acute pressure of population on agricultural resources, but in recent years labour shortages at peak seasons and particularly at harvest time have emerged as serious problems (Fisher, 1978, 1979; Baali, 1956, 1966).

A more comprehensive agrarian reform law was enacted in 1970 further reducing the maximum size of individual holdings. No compensation was paid for land expropriated, while land was redistributed to peasants free of charge from the beginning of 1975. Some 10.2 million donums had been expropriated since 1958, 4.2 million donums under the first Agrarian Reform Law and 6.0 million under the second, but, in fact, 5.8 million donums had been redistributed to 157,862 beneficiaries. Most experts agree that, until the 1958 revolution, inequality in Iraq's land tenure relationships contributed largely to the low standard of living of the peasants. In a political system dominated by great land owners, land reform was impossible and it proved difficult to enforce even minimal legislation for controlling rent (Government of Iraq, 1963, 1976; Fisher, 1979).

The current and projected land and water use is demonstrated in Table 10.3.

It will be seen that in Iraq a division into two climatic provinces—a wetter (rain-watered) region and an arid (irrigated) region with the 375 millimetre isohyet as the boundary has great significance, not only in relation to settlement, but also in relation to the crops grown, because it coincides roughly with another

Table 10.3 Agricultural land and water use in Iraq

	Year	Thousand hectares	Million cubic metres	Cubic metres
Cultivated land				
Total	1975	7,000		
Irrigated	1975	2,895		
Cropped land				
Total	1975	3,000		
Irrigated	1975	1,567		
Springs and rivers	1975	1,547		
Underground water	1975	20		
Total water	1975		20,000–27,000	
Water per hectare	1975			15,950
Projected	1985	2,825		
Projected	2000	3,850–4,400	55,000–60,000	

Source: UN Economic Commission for Western Asia: 1980.
(ECWA = Regions or countries lying east of the Suez Canal.)

important line of demarcation, that of the northern limit of the date palm. Thus, to the south, barley, rice, and dates are staple crops all being grown under irrigation; while in the north, barley, wheat, and Mediterranean fruits dominate (Agrawal, 1967; Fisher, 1978, 1979).

In 1977, the agriculture sector received approximately $1.4 billion. The agricultural system is dominated by cooperative communities and state farms. In the year 1977 land reform activities benefited 242,975 persons who were in receipt of land grants. An estimate of communities is as follows: local cooperative communities 1,935; shared cooperative communities 245; gathered cooperative communities 79; and specialized cooperatives 56 (Ministry of Information, 1980; Al-Doori, 1979; Ministry of Culture and Information, 1980).

In 1981, the government appeared to change its policy and reduced direct government involvement in farming. Farmers no longer had to belong to local cooperatives in order to have access to credit, supplies and equipment. Cooperatives and state farms still exist, with the state farms being used to demonstrate new techniques, raise certified seed and handle such crops as cotton that require a large area of land for efficient production. The government is strengthening this policy by making available capital for loans to private farmers (US Department of Commerce, 1982).

SOILS OF IRAQ

Iraq is situated in the belt of unproductive soils in the Middle East. It is typical of arid lands, with high temperatures and very little natural vegetation. As compared with other areas, however, it has the advantage that most of its irrigable soils are alluvial and rich (Agrawal, 1967).

About 60 per cent of the land that is flow-irrigated is affected by high salt levels and 20 to 30 per cent of it was abandoned during the period between 1937 and 1957. This accumulation of salt was caused by using too much water and the lack of a proper drainage system (Al-Samarrae, 1968).

Within the last fifteen years, separate drainage projects have been developed. These involve the construction of large outfalls which lower the water table in the subsoil by leading off groundwater either into rivers or natural depressions. Sometimes, gravity flow is possible, but in most cases pumping is required. Over twelve major dams and irrigation projects are either under way or in preparation, and statistics on two of the larger ones can be found in Table 10.4. These dams should provide an additional four million donums of land in addition to the sixteen million donums that are already irrigated (Al-Samarrae, 1968; Ministry of Culture and Information, 1980).

Geologically, most of Iraq is underlain by rocks of the Cretaceous and Eocene eras, mainly limestone, with bands of salt and gypsum. On the lowland plains of the rivers, the surface is covered by fluvial deposits and alluvium. In these latter areas are found azonal, alluvial, hydromorphic, irrigated soils. In the west and south, zonal desert-steppe soils are found, sometimes stony, sometimes sandy, and often calcareous. The mountain soils of the north are varied — red, brown, and chestnut forest soils — and in places are merely skeletal where rock outcrops occur. Gypsum is often present in the red and brown soils of the plains and foothills (Fisher, 1979; Al-Samarrae, 1968; Meliczek, 1973).

IRRIGATION

Early developments in agriculture were based on irrigation since Iraq has suitable soil and a wealth of water resources. Up-river water storage and irrigation in Syria and Turkey have reduced the flow of the Tigris and Euphrates river system

Table 10.4   Major dam projects in Iraq

Dam	River	Use	Capacity	Date
Mosul	Tigris	Flood control Hydro-electric irrigation (Jezireh Project)	10.7 billion cubic metres	start late 1980
Hadeetha	Euphrates	Hydro-electric irrigation (Euphrates basin)	6.4 billion cubic metres	Started early 1978; completion date 1985

Source: Ministry of Culture and Information, 1980

Table 10.5   Consumption of water by crops in the northern, central
and southern zones of Iraq (in millimetres)

Crops	Northern	Central	Southern
Winter crops			
Wheat and barley	320	430	520
Linseed, sunflower	340	410	—
Legumes (peas, beans, lentils)	220	300	370
Berseem, clover	400	540	590
Sugarbeet	670	840	850
Potatoes (late)	460	490	520
Onions (dry)	740	900	900
Other vegetables	320	370	490
Summer crops			
Cotton	790	1,080	1,170
Sesame	730	790	800
Rice	1,850	1,970	2,145
Tobacco	620	—	—
Legumes (green grams, peas)	940	1,240	1,240
Vegetables (tomatoes, watermelons, etc.)	830	1,130	1,210
Maize, sorghum	780	920	970
Perennials			
Orchards and palm trees	920	1,480	1,730
Alfalfa	1,280	1,820	1,920
Sugarcane	—	—	1,880

Source: Kharrufa *et al.*, 1980.

and much of Iraq's best land has been degraded by salt concentration in the soil. In the 1970s, the government developed additional irrigation systems and used desalination techniques to reclaim land in the Tigris–Euphrates zone (US Department of Commerce, 1982). The water required for various crops in different areas of Iraq is shown in Table 10.5. Major flood control and irrigation projects are shown in Tables 10.6 and 10.7.

Seven per cent of the land is irrigated and major drainage systems can be found in the areas of Hawija, Shanafiya, and Kerbela.

STOCK-REARING

Livestock forms a considerable supplement to plant agriculture in most of the Tigris–Euphrates lowlands and predominates in the Jezireh and parts of Kurdestan so that, on balance, stock-rearing is more widespread than in many other countries of the Middle East. The reason is the difficulty of irrigating some of the terrain and the fact that some areas can only be used for grazing. However, owing to the climate, difficulties in feeding, and the prevalence of disease, livestock yields are low (*The World Book Encyclopedia*, 1973).

Table 10.6   Major flood control and irrigation projects in Iraq

	Lake area (km²)	Discharge capacity (million m³)	Length (m)	Height (m)	Discharge capacity (m³ per second)
			Dam or regulator		
Wadi Tharthar	2,700	3.25	502	7	9,000
Habbaniyah, Lake	430	850			
Majarra, Regulator			67	12	1,400
Darbandikhan, Dam	120	3.0	535	128	11,400
Dokan, Dam	270	6.8	360	116.5	4,600
Ramadi, Dam			209	10	3,600
Hindiya, Dam			249	7.7	2,900
Bakhma, Dam					
Aski Mosul, Dam					
Samarra, Dam			252	12	11,000
Diyalah, Dam			472	12	4,000
Kut, Dam			692	10.5	6,000
Abu Dibbis, Dam	180	25.5	650	14	5,500
Dhebben, Regulator			42	13	400
Warrar, Regulator			196	10	3,000
Gharraf, Regulator			68	9	450
Mishkhab, Regulator					750

Sources: Meliczek, 1973; Ministry of Culture and Information, 1980; Planning and Follow-up Department, 1981.

Table 10.7   Methods of water supply in Iraq

Irrigation method	Area irrigated (1,000 hectares)	%
Rainfall	3,861	51.3
Gravity flow	2,166	28.7
Water pump	1,449	19.2
Water wheels	57	0.8
Other means	3	—

Sources: Meliczek, 1973; Ministry of Culture and Information, 1980.

The main farm animals found in Iraq are:

(1) Sheep (numbering 9.7 million) predominate, being most numerous in the Jezireh and on the right bank of the South Euphrates. Other pastoral regions are the southern margins of the Assyrian plains, between the Jebel Hamrin and the first foothills of the Zagros, and the Gharraf region between Kut and Hilla. Generally speaking, the sheep are bred for wool which is exported in considerable quantities. Some is retained at home for the weaving of carpets and cloaks but most of the remainder goes to eastern

Europe, Hong Kong, and Communist China (Ministry of Information, 1980; Fisher, 1978, 1979).

(2) Goats (2.06 million) tend to be restricted to the north and east and their numbers have declined in recent years. Milk is the most important product, forming a large portion of the diet of the tribesmen. Some meat is eaten and goat hair is used in the making of tents (Fisher, 1978, 1979; Ministry of Information, 1980).

(3) Cattle (2.9 million) are kept mainly by settled and semi-settled cultivators living near the rivers and marshes where green fodder is available throughout the year (Fisher, 1978). Cattle are valuable in two ways: as draft animals for ploughing, transporting, and operating irrigation wheels and as producers of milk and meat. Breeds such as Sharabi are found in the north and cross-breeds with the Zebu (the Janubi breed, and more recently with the imported Sindi, Ayrshire, and Friesian strains) in other areas. Export of live sheep and cattle to Syria, Lebanon, Jordan, and Kuwait takes place by rail and road.

(4) Water-buffalo (240,000) are raised chiefly by the marsh Arabs of the extreme south. Eighty per cent of the total are used in the river areas of the four southern provinces of Amarah, Diwaniyah, Muntafiq, and Basrah (Ministry of Information, 1980; Fisher, 1979).

(5) Camels (52,000) are bred in the southwestern deserts. In recent years they have been bred for meat and very rarely as a means of transport (Fisher, 1978).

(6) Horses (69,000) are bred as a means of transport but their numbers have been declining in recent years. Also, horses are used for ploughing.

(7) Mules (28,000) are also bred for transport and ploughing in the northern part of Iraq.

(8) Donkeys, too, (460,000) are raised for transport and ploughing (Government of Iraq, 1976).

Government funds have been channelled into projects for the production of poultry, eggs, meat, and sugar, as well as the construction of dams and the improvement of agricultural mechanization (see Table 10.6). In April 1978, three new cattle and dairy centres were opened at a total cost of 60 million dollars. The centres were built with assistance from Dutch companies (Government of Iraq, 1976). 707 million eggs, 38,103 tons of red meat and 71,759 tons of poultry meat had needed to be imported in 1980 to meet local requirements (Planning and Follow-up Department, 1981).

MAIN CROPS IN IRAQ

The area and production of principal crops in Iraq can be found in Tables 10.8 and 10.9. Iraqi crops ripen from south to north and harvesting equipment is usually able to follow the harvest.

Table 10.8 Area and production of principal crops

	1974 Area[a]	1974 Production[b]	1975 Area[a]	1975 Production[b]	1976 Area[a]	1976 Production[b]	1977 Area[a]	1977 Production[b]	1978 Area[a]	1978 Production[b]
Winter crops										
Wheat	6,624.0	1,339.0	5,630.6	845.4	6,070.4	1,312.4		696		910
Barley	2,185.0	533.0	2,269.2	437.0	2,399.3	579.3		458		607
Linseed	6.0	0.9	9.2	1.3	4.0	0.7				
Lentils	16.0	2.6	20.6	4.8	22.8	5.1				
Vetch	3.0	0.4	1.7	0.4	2.7	0.3				
Broad beans	65.6	17.9	—	—	—	—				
Summer crops										
Rice	130.0	69.0	119.5	60.5	212.6	163.3		195		172
Sesame	58.0	6.0	46.7	7.6	53.6	7.0		5		5
Green grams	39.8	9.2	52.3	7.0	56.4	—				
Millet	—	—	—	—	—	—				
Giant millet	—	—	—	—	—	—				
Maize	—	37.8	37.8	23.5	81.8	54.9		82		85

Sources: Agricultural Engineers Union, 1980; Fisher, 1979.
[a]Area expressed in 1,000 donums
[b]Production expressed in 1,000 tons

THE AGRICULTURAL DEVELOPMENT OF IRAQ 199

Table 10.9 Crops grown in Iraq, 1979–80

	Area cultivated (in 1,000 donums)	Average yield, kilograms		Yield (in 1,000 tons)
		Traditional cultivation	Pilot cultivation	
Wheat	5,655	179	381	976
Barley	3,659	206	406	682
Rice	239	614	865	167
Sugarbeet	6.5			32
Tomatoes	143			348
Potatoes	18			97
Winter vegetables	130			248
Summer vegetables	561.5			1,150
Maize	131			61
Cotton	64			15

Sources: Fisher, 1978; Government of Iraq, 1976; Planning and Follow-up Department, 1981.

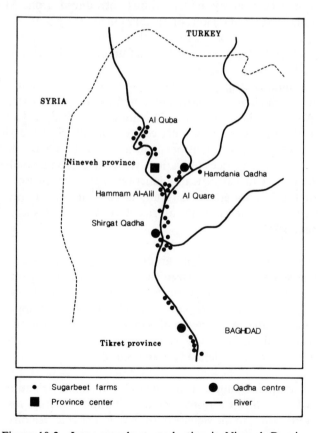

Figure 10.2 Iraq: sugarbeet production in Nineveh Province

(1) Wheat is the most important crop in terms of the area cultivated and the quantities produced, although occasionally in the past, as in 1965–66, the production of barley actually exceeds that of wheat. Wheat is a winter crop and is mostly grown in the wetter areas. Hence, almost 75 per cent of the Iraqi production is from the north and east, chiefly from the *muhafadha* of Nineveh. Other productive areas are along the middle and upper courses of the two rivers, on the Tigris between Baghdad and Amarah, and on the Euphrates in the vicinity of Deir Ez-Zor, Ramadi, and Hillah. The introduction of new wheat species has resulted in improved yields.

(2) Barley is also a winter crop but, unlike wheat, the area under cultivation has declined by about 50 per cent during the last ten years. Yields, however, are higher than wheat. Barley is more tolerant of aridity and soil salinity, and its growing period is shorter, a feature which reduces losses as a result of the sunna insect pest. Formerly the native varieties of barley predominated, but increasingly within the last 20 to 30 years these, too, have been replaced by better strains introduced from Morocco and California. The main regions of production are the plains of Assyria—the Mosul, Arbil, and Kirkuk districts, the middle Tigris valley (*Muhafadhas* of Diyalah and Wasit), and the middle Euphrates (*Muhafadha* of Babylon) (Fisher, 1979; Meliczek, 1973).

(3) Rice is a summer crop grown in the lower valley but is naturally restricted to districts which are regularly inundated by water, or where irrigation is easy. Although the area under cultivation is small (less than 10 per cent of that of barley and about 4 per cent of that of wheat), a yield of up to three or four times as great per land area can be obtained. Since 1971, however, the area of rice cultivation, yields, and production have declined considerably, with a significant reduction in 1975 because of a drop in the flow of the Euphrates. As a result, large quantities of rice have to be imported. The chief regions of rice growing are (Fisher, 1978, 1979; Meliczek, 1973):

(a) The marshes of the Amarah district (Tigris River).
(b) The Shamiyah area between Najaf and Diwaniyah on the Euphrates.
(c) The lower Euphrates from Nassiriyah as far as the head of Hawr al-Hammar. Minor regions of production are the upper part of the Shatt al-Gharraf, the area of Samarra, and the lower Khabur valley. Small-scale paddy-fields are also a feature of the river bank of Assyria, chiefly around Sulaimanyia and Kirkuk.

(4) Millet is another summer crop which also demands irrigation in the later stages of growth. As a result, the distribution is sharply restricted, even in the wetter zone, to the vicinity of the rivers. The chief region of production is obviously the south, where irrigation water is more abundant. Millet is

used for animal fodder and as an ingredient of bread, but the area under cultivation has declined by over 50 per cent in the last ten years and as yields have not improved, production has fallen sharply (Meliczek, 1973).

(5) Maize, also grown in summer, is confined mainly to the middle and upper parts of both river valleys. The area under cultivation is still small but has been extended in recent years and yields have risen accordingly (Mhalhal, 1967; Fisher, 1979).

(6) Sesame is cultivated only on a limited scale.

COMMERCIAL CROPS

(1) Dates. Iraq produces about 80 per cent of the total world supply of dates. For production and export value see Table 10.10. The region of Shatt al-Arab is pre-eminently suited to date production, and both banks are lined with date groves extending inland for as much as two kilometres. Some 30–35 million palms are cultivated throughout the region and, of these, one-half are found close to the Shatt al-Arab, the rest occurring along the river banks as far north as the 33°N latitude line. About 130 varieties of dates are known, but three are most widely grown. The Halawi palm produces the finest fruit, but gives only a moderate yield (20 kilograms per tree). Sayir palm produces slightly less but is more tolerant of inferior growing conditions and is probably the most widely grown. The Zahidi palm does not give a fruit of high quality, but the yield is over 50 kilograms per tree which compensates for the low prices obtained. The date palms can flourish in dry, saline, sandy, loamy, or even waterlogged soils, although the maximum yield of fruit is obtained when abundant water is available. Date picking begins in August. Besides forming an extremely important part of the local diet, dates are used in distilling arak — the chief fermented beverage.

(2) Cotton is the second most important commercial crop. For the area planted in cotton and production figures, see Table 10.11. Large-scale production for export began after 1920. Iraq had no modern textile machinery before 1940. The area under cultivation has changed little during the last ten years, but yields have risen appreciably. In some years there is an export surplus

Table 10.10   Date crop of Iraq showing weight and value

	Year			
	1973	1974	1975	1976
Date crop (tons)	385,000	350,000	400,000	371,000
Export in million dollars	33	30	40	

Sources: Fisher, 1978; Fisher, 1979; Government of Iraq, 1963; Government of Iraq, 1976.

Table 10.11   Cotton production in Iraq, showing area and quantity

	Year			
	1973	1974	1975	1976
Area (donums)	143,270	113,000	105,100	101,320
Production (tons)	45,310	40,000	38,000	33,890

Sources: Fisher, 1978; FAO, 1967.

Table 10.12   Land use and yield for commodities in Iraq

	Total cropped area	Rainfall (1971–75 average)	Irrigated	Per cent change per year 1961–75
Land use patterns (1975)	Percentages			
Commodity				
Cereals	82.0	98.0	61.8	1.6
Pulses	2.0	1.2	3.0	—
Industrial crops	2.6	0.1	5.9	—
Vegetables	6.4	—	14.4	—
Fruits	6.1	0.3	13.3	—
Fodders	0.9	0.4	1.6	—
	Tons per hectare			
	1974–76 average			
Yield				
Wheat	0.43	1.06	1.9	
Barley	0.54	0.99	1.5	
Maize	—	2.90	8.0	
Rice	—	1.91	5.1	
Lentils	0.94	—	—	
Chickpeas	0.65	—	—	
Potatoes	—	15.60	—	
Tomatoes	8.66			
Sesame	0.52	0.64	—	
Seed Cotton	1.90	1.02	6.8	
Sugarbeet	—	24.40	—	

Sources: UN Economic Commission for Western Asia, 1980; Oram, 1980.

and cotton is shipped to Japan, Hong Kong, and central Europe (Fisher, 1978; FAO, 1967).

(3) Minor crops:

(a)   Tobacco is widely grown in the foothills of the north-east section of the country, particularly in Kurdestan, but the product is consumed only within the country.

THE AGRICULTURAL DEVELOPMENT OF IRAQ 203

Table 10.13   Total vegetable and fruit production in Iraq from
1976 to 1978

Year	Vegetable production (tons)	Fruit production (tons)
1976	1,906,000	697,000
1977	1,965,000	906,000
1978	2,067,000	917,000

Table 10.14   Area of vegetable production in Iraq

Kind of vegetable	Area (in donums)[a]	Number of plastic houses	Number of greenhouses
Watermelons	67,850		
Other melons	44,139		
Tomatoes	51,695	256	5
Cucumbers	25,932	469	7
Broad beans	27,054		
Eggplants	14,010		2
Turnips	7,167		
Spinach	3,454		
Lettuces	2,115		
Carrots	2,518		
Cabbages	922		
Potatoes	605		
Cauliflowers	475		
Green peppers	1,516		
Okra	16,391		
Green beans	3,010		
Squashes	9,141		
Onions	18,177		
Others	1,537	131	2
Total	296,908	856	16

Sources: Agrawal, 1967; Agricultural Engineers Union, 1980; Planning
and Follow-up Department, 1981.
[a]One donum = 2,500 square metres.

(b)   Vines are an important feature in the northeastern foothills and in some
districts of Kurdestan grapes may be said to be the principal crop.

Nitrogen-fixing crops are also increasingly grown and vetches, peas, and
beans are especially gaining in importance. Mulberry-growing for silkworm
use is also practised. Cane sugar and sugarbeet are being encouraged and
the area devoted to both crops has more than doubled in the last ten years,
but yields have fallen short of the targets set. Cultivation of both crops

is at present confined to the governorates of Nineveh and Maysan (Hanoody, 1979; Mhalhal, 1967; Rechinger, 1964).

A summary of land use patterns and yield for commodities can be found in Table 10.12.

The total production of vegetables and fruit in Iraq for the years 1976 to 1978 can be found in Table 10.13. The area devoted to the various types of vegetables is shown in Table 10.14. In 1980, 34,305 tons of vegetables and 199,255 tons of fruit were imported to meet local requirements (Planning and Follow-up Department, 1981).

MARINE RESOURCES

Fish technology is in its infancy in Iraq. The General Institute of Fish Technology is the official organization responsible for management and utilization of fish resources and the General Institute of Agricultural Marketing is responsible for the regulations and marketing of all agricultural products including fish (Shamoon, 1980; Arabic Second Conference for Science and Food Technology, 1979). Most of the fishing is done privately and is marketed under the supervision of the official Institute mentioned previously (Shamoon, 1980).

The General Fish Company was established in 1970 and was developed into the Institute in 1979. The General Institute of Marketing was also established in 1979. Fish products consist of the following:

(1) River fish: the quantity of fish harvested between 1970 and 1979 was between ten and fifteen thousand tons annually, and was marketed to consumers by the fishermen themselves. One quarter of the total catch was sold alive and the remainder was sold fresh, stored in ice, without any further technical processing.
(2) Marine fish: salt water fish are marketed fresh by the fisheries, kept in ice, and/or sometimes in salted ice. The annual production averages 1,500–2,000 tons per year.
(3) Shrimp: the supply of these seafood items is limited. They are marketed fresh, and the annual production ranges between 50 and 100 tons per year.
(4) Frozen marine fish: using modern technology some of the marine fish are frozen at $-35°C$ on the hunting ships and at $-18°C$ in storage. This frozen fish is classified according to variety, packed in boxes containing between 27 and 30 kilograms and transported by refrigerated cars. Some of the marine fish are made into fillets and frozen. A small portion is kept as fish pieces and is marketed frozen (Shamoon, 1980).

In 1980 there were 2,465 tons of fish imported to meet local requirements (Planning and Follow-up Department, 1981). This supplemented the 57,176 tons of catches in 1980.

Undesirable types of fish, other marine animals, and fish waste are used to produce ground feed (fishmeal). The production ranges between 500 and 600 tons per year (Shamoon, 1980). Fish storage is provided in refrigerated and frozen warehouses and refrigerated and frozen means of transport are available. In 1980 a total of 73 cold storage units were recorded (Planning and Follow-up Department, 1981). Production of oil from fish is just beginning and fish feed production is being investigated (Arab Second Conference for Science and Food Technology, 1979).

POULTRY PRODUCTION

Because of increased individual incomes in Iraq, the utilization of poultry products is increasing. This caused an increase in imported poultry from abroad, and generated the need for a transport and storage system for perishable products. For these reasons, the government has initiated a project to establish and develop large poultry projects to supply the country with sufficient poultry and poultry products. The capacity of the government production farms is 1,000 million eggs per year in addition to that of the cooperatives and private farms. Specialized poultry farms were also established. Two projects are currently being established to produce 140 million eggs per year and three are being developed with the capacity of 36 thousand tons of poultry meat per year.

Even though tremendous progress has been made in poultry production, the average Iraqi individual still consumes less poultry than the average person in developed countries. Additional efforts are being made to increase the average daily production of poultry so that the utilization of poultry and poultry products in Iraq can equal that of the average person in a developed country (Agricultural Engineers Union, 1980; Arab Second Conference for Science Food Technology, 1979).

CONCLUSION

The Iraqi endowment in renewable natural resources that support agriculture is considerable, especially in water supplies. Successive Iraqi governments have used the country's growing oil wealth to create a sophisticated system of hydraulic controls on the great rivers. Yet, despite the apparent richness in water and soil resources allied to high state expenditures on agricultural development, production from the land has tended to disappoint. One or two areas of great success, such as in poultry output, have been offset by a general deterioration in Iraqi agricultural productivity. In the period since September 1980, the situation has been much impeded by the effects of war and, beginning during 1982, the slow down in government development expenditures.

A return of peacetime conditions might bring a return of interest in agricultural investment, enabling a renewed effort at laying the water control and irrigation base begun earlier. The dilemma will then be how far Iraq can create an efficient farming sector despite a high level of urbanization, the long-term impact of high oil revenues and a diminished service to the rural communities. Past responses to the Iraqi agricultural question do not entirely suggest that serious solutions will be found that improve upon such established prescriptions as increasing food imports and the introduction of foreign farm labour.

REFERENCES

Agrawal, B. L. (1967) *Land Use and Agricultural Production in Iraq*, Baghdad.

Agricultural Engineers Union (1980) Conference on Food Processing for the Nation and the Means of Achieving it (Arabic), Agricultural Engineers Union, Baghdad, p. 167.

The Encyclopaedia of Modern Iraq (1977) A. M. Al-Ani Khaled (Ed.), The Arab Encyclopaedia House, Baghdad, Vol. I, p. 311.

Al-Doori, I. E. (1979) *Fundamentals and Principles of Agricultural Policy in Iraq*, Arabic Government Advertising and Printing Corporation, Baghdad.

Al-Farhan, K. M., Hammadi, I. A., & Al-Sharifi, M. J. (1973) *Agricultural Extension in Iraq* (Arabic), Ministry of Planning, Agriculture Office, Baghdad, p. 259.

Al-Samarrae, H. A. (1968) 'A proposed educational programme for agricultural development in Iraq', Ph. D. dissertation, Ohio State University, Ohio.

Arab Information Centre (1967) *Educational Planning in the Arab World*, New York.

Arab Second Conference for Science and Food Technology (1979) *Food Processing Development in Iraq*, Kingdom of Saudi Arabia, Riyadh.

Baali, F. (1956) *Relation of the People to the Land in Southern Iraq*, University of Florida Monographs, Social Sciences, University of Florida Press, Gainesville.

Baali, F. (1966) 'Social factors in Iraqi rural–urban migration', *American Journal of Economic and Sociological Research*, **XXV**, 359–364.

Encyclopaedia Britannica (1968) 'Iraq', William Benton, Chicago, Vol. 12, p. 587.

FAO (1967) *Production Yearbook*, Rome.

Fisher, W. B. (1978) 'The Tigris and Euphrates lowland' in *The Middle East*, Methuen, London, p. 615.

Fisher, W. B. (1979) 'Iraq physical and social geography', in *The Middle East and North Africa*, Europa, London, p. 951.

Government of Iraq (1963) *The Achievement of the Ministry of Land Reform*, Baghdad.

Government of Iraq, Ministry of Planning (1976). *Statistical Abstracts*, Baghdad.

Hanoody, A. J. (1979) *Factors associated with adoption of improved agricultural practices among the sugarbeet farmers in the province of Nineveh*, Ph. D. thesis, The Ohio State University, Ohio.

Harris, G. L. (1958) *Iraq, Its People, Its Society, Its Culture* (Survey of World Culture Series), Human Relations Area Files Press, Newhaven.

Iraq Tourist Map (1976) Issued by Summer Resorts and Tourism Administration.

Kharrufa, N. S., Al-Kaway, G. M., & Ismail, H. N. (1980) *Studies on Crop Consumptive Use of Water in Iraq*, Pergamon Press, Oxford.

Kinnane, D. (1964) *The Kurds and Kurdistan*, Oxford University Press, London.

Meliczek, H. (1973) 'South and East Asia Oceania' in *World Atlas of Agriculture II*, IGDA, Novara.

Mhalhal, A. (1967) *Agriculture Extension*, Ministry of Agriculture, Baghdad, p. 89.

Ministry of Culture and Information, Ministry of Planning (1980) *Revolution and Development in Iraq*, Baghdad.

Ministry of Industry and Minerals (1979) *State Organization for Food Industries*, Ministry of Industry and Mineral Press, Baghdad.

Ministry of Industry and Minerals (1980) *Organization for Food Industries*, Ministry of Industry and Mineral Press, Baghdad.

Ministry of Information (1980) *Geography of Iraq Agriculture, Rivers, Vocational Education and Higher Education*, Al-Hurriya Press, Baghdad.

Oram, P. (1980) *Development of Appropriate Crop Technology in Irrigated Agriculture in Semi-Arid Regions*, Pergamon Press, Oxford.

Parker, O. D., & Parker G. G., (1965). *Iraq. A Study of the Educational System of Iraq and Guide to the Academic Placement of Students from Iraq in United States Educational Institutions*, American Friends of the Middle East, Washington.

Planning and Follow-up Department (1981). *Republic of Iraq: Agriculture is Everlasting Oil; Accomplishments of the Ministry of Agriculture Agrarian Reforms along the Revolutionary March*, Baghdad.

Rechinger, K. H. (1964) *Flora of Lowland Iraq*, Hafner, New York.

Shamoon, A. R. (1980) 'Fish technology', Fisheries Research Centre, State Fisheries Organization of Iraq, Personal Communication.

Simmons, J. L. (1965) 'Agricultural development in Iraq: planning and management failures', in *Middle East Journal*, **XIX**, 129–141.

The World Book Encyclopaedia (1973) 'Iraq', Field Enterprises Educational Corporation, Chicago, Vol. 10, P. 416.

Union of Iraqi Industries (1976) *Bulletin of Iraqi Industries, Food, Sweets, and Alcoholic, Non-Alcoholic Beverages*, Al-Adeeb Press, Baghdad, p. 753.

UN Economic Commission for Western Asia (1980) *Irrigated Agricultural Development in the ECWA Region*, Pergamon press, Oxford.

US Department of Commerce (1982) *Overseas Business Report*, Agricultural Section, Washington.

Agricultural Development in the Middle East
Edited by P. Beaumont and K. McLachlan
© 1985 John Wiley & Sons Ltd

Chapter 11

Agricultural Development in Saudi Arabia: The Problematic Path to Self-sufficiency

E. G. H. JOFFE

INTRODUCTION

Over the past decade, the agricultural sector of the Saudi Arabian economy has increasingly manifested a dynamic confidence in its ability and imminent success in satisfying much, if not all, of the Kingdom's food needs within the foreseeable future.

Production indices, relative to base of 100 in 1969/70, have risen increasingly rapidly, both in absolute and *per caput* terms, from 90 and 89 in 1971 to 205 and 150 respectively in 1980 (McLachlan, 1984, p. 112–113) (see Table 11.1).

Table 11.1   Production indices, 1971–1980 (base 1969/70 = 100)

	1971	1972	1973	1974	1975	1976	1977	1978	1979	1980
Overall	90	90	120	145	140	174	160	161	190	205
Per caput	89	85	110	128	122	144	129	125	145	150

(Source: McLachlan, 1984, p. 112)

Table 11.2   Cereal production 1971–79 ('000 tonnes)

Year	Wheat	Barley	Sorghum	Millet
1971[a]	38.9	9.3	52.4	17.2
1972[b]	63.7	11.0	27.3	9.9
1973	153.4	15.4	115.9	11.5
1974	132.9	16.7	127.9	10.6
1975	92.5	12.0	153.4	16.6
1976[c]	124.6	13.5	143.0	12.6
1978[d]	147.0	16.0	194.0	13.0
1979[d]	158.0	17.0	215.0	13.0

Sources: [a]Lackner, 1978, p. 183; [b]Moliver & Abbondante, 1980, p. 55, p. 65; [c]Barker, 1982, p. 53; [d]Barker, 1982, p. 53 — estimated.

Table 11.3   Wheat production 1980–84 ('000 tonnes)

1980	1981	1982	1983	1984
135	200	350	750	1,300

Sources: Estimates—*Middle East Economic Digest*, 6.7.1984;
Financial Times, 9.6.1984.

In the past four years the rise must have been even more startling, if output figures for major commodities such as cereals (Tables 11.2 and 11.3), are a reliable guide—even though detailed statistics are not yet available.

Similar patterns have occurred with other commodities where production levels have often risen dramatically, particularly since 1978. Tomato production has increased from 181,000 tonnes in 1972 to 218,000 tonnes in 1979, while citrus fruits rose from 13,900 tonnes to an estimated 34,000 tonnes over the same period. Date and melon production, however, experienced slight falls during the period under examination—thus not fitting into the general picture—mainly because they are traditional agricultural products and there has been a preferential increase in interest in other produce (Moliver & Abbondante, 1980, p. 55). Red meat production also fell between 1970 and 1980, mainly because of overgrazing, although it has now begun to rise again. From 1980 to 1982 output doubled to 28,000 tonnes—equivalent to about 10 per cent of domestic consumption (Prestley, 1984, p. 28).

On the other hand, dairy produce, eggs, and poultry have also shown rapid growth during the period in question. Eggs production has risen from 490 million units in 1978 to 750 million in 1980, 1.23 billion in 1982 and 1.4 billion in 1983. By 1985, Saudi Arabia expects to be a net exporter of eggs (Bowen-Jones & Dutton, 1983, p. 29). When it is borne in mind that egg production in 1973 was only 75.8 million units and had only reached 270 million units in 1976, the twenty-fold increase over the past eleven years is highly impressive. Production of broiler chickens increased five-fold between 1973 and 1980 to 40 million birds and, even though the total only represents one-fifth of overall demand, the rise is still impressive. Output of milk is expected to reach 50,000 tonnes by 1985, once again bringing Saudi Arabia close to self-sufficiency in this commodity also (Bowen-Jones & Dutton, 1983, p. 29; Cottrell, 1980, p. 65).

THE UNDERLYING REALITY

However, the figure that has caught the public imagination, both in Saudi Arabia and abroad, has been the extraordinary ten-fold rise in wheat production over the past decade to an expected 1.3 billion tonnes in 1984. Indeed, production has apparently doubled in one year, from 1983 to 1984 (see Table 11.3). It has led to proud claims that, despite recent indiscretions in official circles in Washington to the effect that the Kingdom's agricultural plans are 'crazy'

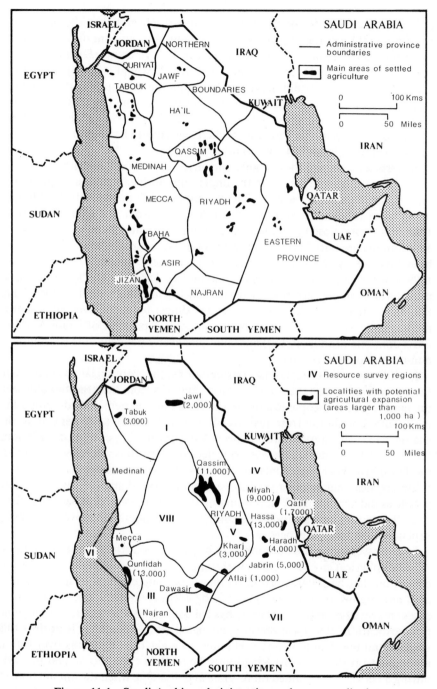

Figure 11.1   Saudi Arabia: administrative and resource districts

(MEED, 6.7.1984), self-sufficiency *is* within reach and that the apparently insurmountable combination of an overwhelmingly dominant oil sector and an extremely harsh ecological environment can be overcome (*Financial Times*, 9.6.1984) (Figure 11.1). Such claims, if justified, would clearly be extremely significant, for they would make Saudi Arabia the only state within the Gulf region to have countered the apparently irresistible trend towards agricultural decline and increased dependence on imports as a direct result of the distortions created by oil in national economies where agriculture has become a marginal activity (McLachlan, 1984, pp. 134–6).

Nonetheless, despite these highly encouraging production trends, there are other factors which suggest that the underlying problems of Saudi agriculture are still far from being resolved. The Kingdom is still the world's third largest importer of foodstuffs, at a cost which has risen from just over $3 billion in 1981 to a figure close to an anticipated figure of nearly $6 billion in 1984. In addition, the dramatic rise in domestic food production in recent years has only been possible as the result of massive investment—some $18 billion since 1979—and a generous subsidy programme (see Table 11.4). Domestic wheat prices, for example, are sustained at the artificially high level of SR3.5 per kilogram—five times the world price and almost twice the average domestic production cost. Yet agriculture represents only 1 per cent of Saudi Arabia's gross domestic product (GDP) and only just over 2 per cent of the Kingdom's non-oil component of GDP.

In addition to these economic considerations, there are social and ecological factors that must be taken into account. Saudi Arabia has traditionally seen the majority of its people located in the rural environment and, despite a fall in the rural population from just over 50 per cent in 1970 to 40 per cent at the end of the decade, agriculture still provided work for 25 per cent of the labour force in 1980, a figure which was projected to fall to 20 per cent by 1985 under the Third Five-Year Plan. Only the service sector was more important in terms of employment and no other productive sector employed anywhere near as many people as did agriculture. Yet, few among the rural labour force have benefited from the spurt in agricultural production, since it is capital intensive and heavily dependent on foreign expertise.

In any case, legitimate questions can be posed about the suitability of Saudi Arabia's physical environment to withstand the pace and extent of agricultural expansion that has characterized recent years. Suitable land is at a premium and, although water supplies appear to be adequate overall, quality and local access may well pose financial and physical constraints that vitiate the sanguine prognostications that have recently emanated from Riyadh. It is against this background that Saudi planners will have to make their decisions over future investment priorities, as the Kingdom's Third Plan comes to an end in 1985, during a period of restrained oil revenue and at a time when the annual budget is running at a substantial deficit for the second year in succession. Against

Table 11.4 Incentives for agricultural production

Type	Amount	Source
Production input:		
Fertilizer	50% of cost	MOAW
Animal feed	50% of cost	SAAB
Potato seed	5 tons free	
	SR1,000/ton thereafter	
	up to 15 tons	MOAW
Machinery and equipment:		
Poultry equipment	30% of cost	SAAB
Dairy equipment	30% of cost	SAAB
Engines and pumps	50% of cost	SAAB
Fish trawlers	Variable	SAAB
Transportation:		
Air transport of cows	100% of cost	SAAB
Output:		
Wheat	SR3.50/kg[a]	GSFMO
Rice	SR0.30/kg	MOAW
Corn	SR0.25/kg	MOAW
Millet/barley	SR0.15kg	MOAW
Dates	SR0.25/kg	MOAW
Date palms planted	SR5.00/tree	MOAW
Agricultural credit:		
All types	Variable conditions	SAAB
Agro-industrial credit:		
All types	Variable conditions	SIDF
Land acquisition:		
Land distribution	Free	MOAW

[a] Purchase price.
Sources: El Mallakh, 1982, p. 96; Prestley, 1984, p. 30.
· MOAW = Ministry of Agriculture and Water.
SAAB = Saudi Arabian Agricultural Bank.
GSFMO = Grain Silos and Flour Milling Organization.
SIDF = Saudi Industrial Development Fund.

such a background, future investment in agriculture on the scale that has been experienced in the past may seem less than attractive, whatever the political imperatives.

THE ENVIRONMENTAL CONSTRAINTS

Saudi Arabia falls squarely within the world's most extensive arid zone. The Arabian peninsula is, furthermore, notable for its lack of any significant permanent river, a distinction which it shares only with Libya within the Middle East and North Africa (Gischler, 1979, pp. 5, 9). It is not, therefore, surprising that agriculture should traditionally have been a marginal activity. Within Saudi Arabia itself, however, agricultural potential and location have been highly differentiated, depending on such obvious factors as topography, soil type, rainfall, and groundwater access, as well as less evident considerations such as social organization and patterns of political control. All these elements have continued to have a role to play in the post-war, oil-rich era and have informed the choices that planners have made in the last decade over the nature and purpose of the recent agricultural revolution.

Given their immutable nature, most important amongst these elements, perhaps, have been the environmental factors. Only 3 per cent of Saudi Arabia's 2.15 million square kilometres (World Bank, 1983, p. 149) receives more than 200 mm of rainfall each year and, of the remainder, 20 per cent of the land area consists of non-arable terrain, such as desert (Bowen-Jones & Dutton, 1983, p. 2).

Basically, Saudi Arabia falls into four distinct geographic regions, each of which is characterized by a different pattern of agricultural use:

(1) Along the western seaboard are the Tihama coastal plain and the western highlands, running from the northern Hijaz to the Tihama around the holy cities of Mecca and Medina, to the southern Asir uplands close to the Yemeni border.
(2) The eastern seaboard, between Kuwait in the north and the United Arab Emirates in the south, is dominated by the critical al-Hassa region with its fertile oases.
(3) The southern interior is dominated by the vast Rub al-Khali desert.
(4) The northern part of the interior contains the socially and economically critical regions of the Nejd and the Nefud, together with the Summam and Dahna plains—traditionally important nomadic grazing areas (Fisher, 1978, p. 464).

The deficiency in rainfall, which is less than 200 mm annually on average in all regions except the Asir and small pockets in western Saudi Arabia, is made up by exploitation of groundwater resources. Indeed, the availability of groundwater for irrigation is the vital element that has traditionally made Saudi agriculture a viable component in supporting the country's dispersed population. It has enabled sedentary agriculture to complement pastoralism in maximizing food production in the past.

Traditionally, the major sedentary agricultural centres have been located in the western and eastern coastal regions, with a scatter of oases running through the centre of the Nejd, in the interior. Nomadic pastoralism, on the other hand, has tended to be concentrated in the northern and northwestern interior— between the deserts of the Nefud and Dahna and towards the Medina-Riyadh axis. In the most intensive sedentary agricultural regions—the Jizan and Asir, in the south of the western seaboard region—nomadic pastoralism was tradition- ally almost non-existent. Elsewhere in the Kingdom, nomads formed a minority of between 10 and 20 per cent of the total population (Hajrah, 1982, p. 37).

The traditional symbiosis between sedentary agriculture and pastoralism underlines the fact that the critical environmental constraint on agricultural activity in Saudi Arabia has not been a shortage of suitable land, but one of access to water (Cottrell, 1980, p. 660). In the 1960s, for example, it was calculated that some 385,000 hectares were actually cultivated, although suitable soil for cultivation totalled 4.2 million hectares. The problem was that, of that total, only 82,550 hectares had suitable access to sufficient water (Bowen-Jones & Dutton, 1983, p. 18).Two decades later, the total cultivated area was 525,000 hectares, 0.3 per cent of the total land area of the Kingdom — although the Ministry of Agriculture and Water calculated that the total usable land area, with potential access to water, could in theory be extended to about one million hectares (Cottrell, 1980, p. 659). In short, despite the massive financial inputs into agriculture in the past 30 years, there has not been a comparable expansion of land under cultivation because of difficulties over water.

The location of areas of potential expansion are limited by similar factors. Agriculture based on rainfall is largely concentrated in the Asir uplands where the monsoon provides 150 to 300 mm of precipitation annually in summer. Cultivation is found mainly on the steep upper slopes of wadi valleys which themselves are subject to flash flooding. On the upper slopes, where altitude creates a cooler microclimate, the land is terraced and is used for cereals, dates, coffee, and bananas in a cultivation regime that is unique to the area (Fisher, 1978, p. 470). Fisher notes that the Asir upland region is unique to the area in another respect, in that it is ecologically similar to neighbouring regions of Africa, such as Ethiopia and Sudan, with a savannah-like landscape and summer rainfall.

The lower areas of the Asir wadis, as they debouch into the drier coastal Tihama plain together with the eastern watershed of the Asir uplands and adjacent areas of the Tihama to the North, are ideal for stream diversion agriculture, although the intense heat of these lowland western coastal regions, together with very high evapotranspiration rates, means that agricultural potential is limited to scattered pockets (Bowen-Jones & Dutton, 1983, p. 18). Traditionally, the areas both of rain-fed cultivation and stream diversion agriculture along the western seaboard have been fully exploited by a sedentary peasantry working on land covering between 350,000 and 400,000 hectares (El Mallakh, 1982, p. 82). Possibilities for extension of dryland agriculture are thus extremely limited.

The sole area in which environmental constraints permit an expansion is that associated with groundwater resources. Here, traditionally, the limiting factor has been accessibility, since all groundwater has had to be lifted by human or animal power and this has generally meant that aquifers more than ten metres deep cannot be exploited. As a result, agricultural activity has been restricted to scattered oasis communities and limited urban market-garden activity in the Hijaz around Mecca and Medinah. In western Saudi Arabia, on the eastern slopes of the Madian mountains and along the watershed of the Asir, the groundwater aquifers have been fed by rainwater runoff in a system typical of the geological structure of the Arabian shield region. To the east of the Arabian shield region, however, the groundwater picture changes radically. Here there are massive resources trapped in porous sedimentary strata laid on top of the shield that form a set of massive aquifers running from west to east across the Kingdom. Although some recharge occurs to these aquifer systems from runoff from the shield mountains and from more eastern outcrops such as the Jebel Tuwaiq escarpment, most of the water in these aquifer systems is believed to be between 15,000 and 30,000 years old. Some 30 aquifer systems have been identified, of which seven are of major importance (COMET, 1983, p. 23).

An idea of the significance of this resource is given by the fact that, in 1975, groundwater use in eastern Saudi Arabia was estimated to be 1.723 km^3 annually, while the Wajid, Minjur, Saq, and Tabouk aquifers alone were believed to have resources equivalent to $1,000 \text{ km}^3$ each (Gischler, 1976, p. 111). One problem is that deeper aquifers contain water that is increasingly saline and, above 1,250 ppm salinity, desalination is necessary. Nonetheless, the aquifer which supplies Riyadh has been producing 236 million litres per day for sixty years, for example, and only 10 per cent of its reserves have been depleted (Moliver & Abbondante, 1980, p. 57). Admittedly, this system does obtain a 15 million m^3 recharge annually from runoff from the Jebel Tuwaiq (which absorbs about 10 per cent of the rainfall it receives), but this only represents about one-sixth of the demands made upon it (Bowen-Jones & Dutton, 1983, p. 20). However, the fact that fossil groundwater is a finite resource must not be overlooked.

Traditionally, the presence of groundwater resources of this type has generated major agricultural activity in the eastern and central oases. In addition to the great eastern coastal oases of al-Hassa and Qatif, there is also a line of oases in the Nejd, running from Ha'il southeastwards via Qassim and al-Kharj towards Jibrin and then southwestwards towards Wadi Dawasir, all of which have traditionally depended on the major aquifer systems for water. The oases, together with associated grazing regions of the Nejd and the eastern coastal zone — estimated at some 210 million hectares (Cottrell, 1980, p. 660) — have traditionally supported the sedentary agriculture–pastoralist symbiosis that characterized the pre-oil era. The oases, too, provided an adequate base for urban development such as Hofuf in the east or Riyadh and Ha'il in the centre which could act as points of economic exchange. In the west, the more shallow aquifers provided a base for oasis cultivation and urban development in a line along the eastern edge of the Madian mountains from Tabouk to Medina and Mecca. This, in turn, allowed a mercantile system linking Yemen to the Levant to develop, which also, in part, depended on nomadic participation in caravan trade. The rain-fed Asir and Tihama agricultural base also entered into this complex tripartite pattern, which depended exclusively on the location and exploitation of water resources (Lackner, 1978, p.3).

Given the overriding importance of groundwater in Saudi Arabian agriculture, its availability must be critical to any potential for expansion. Indeed, all recent expansion has taken this into account and has been based on surveys carried in the late 1960s of the potential cultivatable land available. Potential for expansion was identified in six of the eight areas into which the Kingdom was divided, with only the Rub al-Khali desert in the south and the central Arabian shield area, between Riyadh and the western coastal mountains of the Hijaz, being considered as unsuitable or suitable only for limited grazing. Although, on the basis of soil type, it was clear that the 385,000 hectares under cultivation could be expanded more than ten times to 4.181 million hectares, a realistic evaluation of water potential made it evident that the true potential for expansion, in terms of groundwater access, was really 82,550 hectares over the existing 143,700 hectares of irrigated land already in use (Hajrah, 1982, 85–88). One quarter of this 20 per cent expansion in cultivatable land was located in the south-west, around Qunfidah, just to the north of the coastal town of Jizan. The rest, however, was located in the centre of the Kingdom and along the eastern seaboard — at the oases dependent on the major aquifer systems — with 13,000 hectares at al-Hassa, 10,000 hectares in Greater Qassim, and another 8,000 hectares at Wadi Dawasir alone (Hajrah, 1982, p. 88).

Despite the potential indicated by groundwater access, the environment does exercise other constraints. Chief amongst them is the extreme fragility of much of the soil in Saudi Arabia, given its location in an area of intense aridity. The land is often of very low fertility and, worse, can be extremely difficult to drain. The latter feature is, itself, often the consequence of high water table and shallow soil depths, together with frequent strata of impermeable deposits. Salinity is also a danger and, in the past, these features have had catastrophic consequences. In the al-Hassa oasis complex, for instance, salinity played a major role in cutting cultivatable land area in half, until an ambitious recovery programme was undertaken (Vidal, 1980, p. 91).

SOCIAL LIMITATIONS

Quite apart from the constraints imposed by the physical environment, traditional patterns of social organization in the rural sector and their relationship to ownership have also had a major role to play in the marginalization of agriculture in Saudi Arabia. The problem is not only one of sedentary agriculture, however, for the creation of the Saudi Arabian state in the early part of the twentieth century is intimately linked to the issue of sedentarization and the destruction of traditional patterns of land control has played an important role in determining land settlement patterns since the Second World War. All three considerations have entered into planning decisions in recent years that have led to the current massive expansion of private sector activity in agriculture.

Traditionally, the sedentary agricultural sector has been based on owner-occupation, with a very small amount of share-cropping—typically it is claimed that up to 90 per cent of all plots are owner-occupied (Bowen-Jones & Dutton, 1983, p. 5) and, in one case, the figure is set as high as 96.7 per cent (Hajrah, 1982, p. 82). However, in the east there has grown up an increased practice of share-cropping, mainly because land owners have moved into more renumerative activities connected with the oil sector and, increasingly, the new industrial sector (Lackner, 1978, p. 182). In such circumstances, agricultural holdings in the al-Hassa province may be up to 50 per cent exploited by share-croppers.

One problem of the high degree of owner occupation is that much sedentary agricultural land suffers from severe that much sedentary agricultural land suffers from severe '*morcellement*'—fragmentation resulting from inheritance patterns derived from Islamic and customary legal practice (Hajrah, 1982, p. 82). This has meant that more than 50 per cent of all holdings are less than 1 hectare in size and more than 75 per cent are less than 3 hectares (Bowen-Jones & Dutton, 1983, p. 23). In the north, average holding size is 3 hectares, while in the south-west it is only 1.3 hectares. Even in the eastern oases, average holding size is about eight hectares, but here the situation is worsened by the fact that trees are owned separately from land. In any case, much land and water was controlled traditionally by the *waqf* and thus often did not contribute towards the peasant economy (Lackner, 1978, pp. 3, 182). It is evident that such land ownership patterns would inevitably hinder effective modernization of the agricultural sector and, no doubt, it was for this reason that sedentary agriculture did not play a part in initial attempts at agricultural reform.

The first attempt to create a dramatic change in the patterns of agriculture in Saudi Arabia arose, ironically, for reasons quite unconnected with agricultural production. In the early stages of the creation of modern Saudi Arabia as a unitary state ruled by the Ibn Saud dynasty, Ibn Saud himself evolved the idea that sedentary populations with the same warrior spirit as the nomadic *Ikhwan*—activists within the Wahhabi movement that the Saudis led—would provide a far more reliable support base for his regime than the *Ikhwan* themselves did (Fabietti, 1982, p. 190). The result was the al-Hijar settlement scheme. Here *Ikhwan* groups were settled, usually in the tribal *diras* (grazing grounds) on sites that had arable land, water, and grazing. The sites were designed for between 1,500 and 2,000 persons and provided a mosque around which, in many cases, the *Ikhwan* settled in tents (permanent dwellings were, apparently, rare). The main crops produced were wheat and dates, together with fruit and vegetables, if the water were not too saline. By 1930, 222 *hijras* had been founded, but, after the *Ikhwan* revolt in that year, the experiment was abandoned and the settlements collapsed as the defeated *Ikhwan* went back to pastoralism. Indeed, most had never really abandoned it, for it is estimated that

37 per cent of the *Ikhwan* had continued to be semi-nomads, even after settlement, mainly because of their lack of an agricultural tradition (Moss Helms, 1981, pp. 135–141).

Despite its failure, the experiment had underlined the two critical problems that faced the Saudi dynasty, both in building a nation state in the peninsula and in attempting to reform land ownership patterns. The latter consideration related to the problem of pastureland control — politically the mainstay of tribal autonomy and economically the brake on sedentarization. Pastureland was always tribally controlled — whether it was *dira* land, belonging to pastoralists and typically covering some 55,000 km², or whether it was *hima* land, belonging to a settled community with an average size that depended on the community's potential for defence. The largest *hima* was in Qassim, where it totalled 3,500 km² (Moss Helms, 1981, pp. 47–49). The importance of the *dira* concept cannot be underestimated, for the fact that the *Ikhwan* were settled on their own *diras* was a powerful factor in guaranteeing acceptance of the *hijra* settlement programme. If such tribal practices had not been observed, the programme would have collapsed even more rapidly.

Indeed, even before the *Ikhwan* revolt, Abdulaziz Ibn Saud had realized this and, in the 1925 land distribution law, had determined to break tribal hegemony and tribal structures by removing the *diras* from tribal control (Barker, 1982, p. 51). This law allowed *dira* and *hima* land to be distributed on individual title, thus creating a market commodity out of what had been a vital element in tribal political structures. Since such land grants were in the gift of local *emirs* distribution would create a new class of land owners and would destroy the power of the tribal notability, as the new land owners looked towards the Saudi dynasty as guarantors. In fact, the law had relatively little effect, but was to act as the precursor of a much more effective law in the late 1960s.

Far more critical for the *dira* and *hima* lands was the 1953 decree which effectively destroyed autonomous tribal control over them. By this decree, all grazing lands were thrown open to any pastoralist. The danger of this process soon became evident, when pastoralists acquired transport and began to range more widely in search of pasture — particularly after droughts in the 1960s. Many pasturelands were destroyed by overgrazing and consequent erosion, particularly when nomads were given subsidies of $9 per sheep and $15 per camel to increase livestock, since it was regarded as a national resource. The problem was that pastoralists, seeing herd size as a measure of wealth, did not sell animals for consumption and the purpose of the subsidies was consequently lost (Bowen-Jones & Dutton, 1984, p. 61). Since the mid-1960s, care has been taken to preserve remaining pasture.

Another consequence of the drought in 1960/61 was a further attempt at the sedentarization of the *bedu*. The Ministry of Agriculture and Water (MAW) offered blocks of 1 to 3 hectares, together with free pumps, seeds, fertilizer, and other inputs, together with regular subsidies for produce to 644 individuals on 1,000 hectares in Tabouk, Quriyat, and al-Jawf (Bowen-Jones & Dutton, 1983, p. 23). The project was companion to a much more important project designed to sedentarize the *bedu* in the eastern province at a previously virgin site at al-Haradh. This attempt formed part of the second serious effort to settle the nomads of central and eastern Saudi Arabia, after the failure of the *hijra* attempt in the first thirty years of the century and began in 1958, in al-Haradh, Wadi Sirhan, and Jabrin (Fabietti, 1982, p. 191). The al-Haradh project was particularly instructive. Known as the al-Faisal *bedu* scheme, it was designed to settle members of the al-Murrah tribe, located at the edge of the Rub al-Khali, who had begun to suffer from the consequences of the development of oil and industry in the east of Saudi Arabia. The scheme, designed to provide a variety of farming styles which would fit into the growing urban demand typical of the new oil-rich society developing in the Kingdom, was also designed to persuade the al-Murrah *bedu* of the advantages of settlement (Bowen-Jones, 1980, p. 133).

The project failed for a variety of reasons. Firstly, the al-Murrah *bedu* had no experience of sedentary agriculture, secondly, the drought ended and those involved made a rational decision to return to activities they were more suited to exploit, thirdly, nomadism, the National Guard or urban employment offered a far better economic solution than did sedentary cultivation in the al-Haradh. In short, it became evident from this experience and from those that preceded it that the *bedu* sedentarize only if pastoral production is inadequate. However, sedentarization projects will fail if they do not alter tribal structures, for then the relations of production do not change, even if access to production does (Fabietti, 1982, p. 192).

THE DEVELOPMENT OF THE MODERN SECTOR

By the end of the 1960s, it had become evident that scope for agricultural modernization would be severely limited if it relied on altering the existing relations of production. In the sedentary sphere, there was little scope for any kind of modernization without a throughgoing change in property ownership patterns. Nonetheless, land was available for expansion, but it fell under the rubric of *hima* land and its status was still ambiguous. It was also clear that any attempt at sedentarization of pastoralists, or of integrating them into national markets, was unlikely to be successful while traditional pastoral patterns continued to be viable (either through objective economic factors or as a result of subsidies) or as the alternatives of military service (a traditional outlet for the *bedu*) or of urban work which seemed to offer an even better outlet. In short, Saudi Arabia was confronted with the dilemma typical of many oil economies over the issue of agriculture—whether it was better to attempt to resolve the growing alienation and sense of marginalization of rural populations or whether the growing problem of food insufficiency should take precedence (Bowen-Jones & Dutton, 1984, p. 52).

The government response was to create a totally new situation in which the problems inherent in the existing social and economic agricultural patterns could be circumnavigated — although existing agriculturalists would be able to find opportunities to participate if they wished. The first move was to liberate control of land. In 1968, the Saudi government issued the Public Lands Distribution Ordinance which confirmed the 1925 legal abolition of tribal hegemony on *dira* and *hima* land, transferring such lands to the state for distribution — with the result that collective land could now be turned into private property, while tribal structures based on control of land collapsed. At the same time, the patronage of the emirs — which had preserved other traditional structures — had been ended

Table 11.5 Water resources and utilisation (million cubic metres/year)

	1979/80	1984/85	1999/2000
Resources			
Non-renewable	3,450	3,450	3,450
Renewable	1,145	1,145	1,145
Desalination	63	605	1,198
Reclaimed effluent	—	140	730
Total	4,658	5,340	6,523
Utilisation			
Urban/industrial	502	823	2,279
Agriculture (irrigated)	1,832	1,873	3,220
Rural/livestock	27	28	38
Total	2,361	2,724	5,537
Surplus	2,247	2,616	986

Source: Bowen-Jones & Dutton, 1983, p. 20.

eight years previously, so that private land grants could no longer be used to create patronage links. The last vestige of the system, the right to royal grants of land of up to 500 hectares, was abolished in 1971 (Bowen-Jones & Dutton, 1983, p. 24). The effect of the measures was to destroy permanently *bedu* control of land and thus, eventually, *bedu* society itself. For the *bedu*, the only solutions were to join in the new private ownership system, or to leave the sector altogether (Fabietti, 1982, p. 197).

The next stage was to provide means by which the newly acquired domain land could be distributed and worked. A Public Land Management Department was created in the Ministry of Agriculture and Water in 1967 for this purpose and, in conjunction with the water survey undertaken between 1964 and 1970, it arranged for distribution of appropriate land for private sector development. By 1980, 123,000 hectares of public domain land had been distributed to 19,400 individuals and to 87 companies who wished to institute major agri-business projects. The land included 83,000 hectares of fallow land distributed for livestock and dairy projects and 4,000 small-scale holdings of between 5 and 10 hectares each (Bowen-Jones & Dutton, 1983, p. 24; Barker, 1982, p. 51). Much of the land, of course, was distributed with a specific eye towards groundwater availability, in line with the conclusions of the hydrological survey undertaken in the late 1960s.

Parallel to the land distribution policies, attempts were made to provide an elaborate subsidy system which, by the Third Plan, which started in 1980, had come to cover almost every aspect of private sector agriculture (see Table 11.4). In addition, elaborate extension services were provided, so that agricultural efficiency could also be improved. In large measure, these aims mirrored those of the First and Second Plans, from 1970 to 1975 and from 1975 to 1980, respectively. They are notable for the fact that, although the plans were aimed specifically to improve rural incomes, reduce import dependence, and protect soil and water resources, the very nature of the support was such as to promote large-scale developments and to exclude the traditional agricultural workforce from such benefits (Lackner, 1978, p. 157). The plans have also devoted much attention towards the critical issue of water resources (Table 11.5).

It is evident that agriculture will continue to dominate water use and that non-renewable resources (fossil water) will continue to dominate the supply picture for the foreseeable future. Nonetheless, the pressure of urban and industrial needs is clearly going to be a major factor in the next five years — given the demands of the oil industry which takes four times as much as urban needs and the demands of the new industrial sectors at Jubail and Yanbu. At the same time, the projected expansion of irrigation needs by 78 per cent by the end of the century implies a further potential danger because of the dangers associated with inappropriate water use in Saudi Arabia. The domination of the private sector in development means that there are few means by which water use can be controlled. Yet the dangers of salinity and waterlogging are critical and, if there is not careful management of water, much of the SR13.277 billion to be invested during the Third Plan period in introducing a further 19,000 hectares to irrigation and on transferring 22,500 hectares from traditional to modern irrigation methods, may be wasted. The complacency over the use of deep aquifers is not always justified, particularly since artesian wells below 1,000 metres tend to run into saline water. Even in those areas where shallow groundwater resources have been developed — such as the dams in the uplands or in the alluvial areas such as al-Dawasir, al-Sudair, or Aflaj — the essentially private nature of planned water use gives cause for alarm (Moss Helms, 1981, p. 43; Moliver & Abbondante, 1980, p. 58).

The dangers inherent in ineffective water management practices is already well known from the experiences of Qatif oasis and the vast al-Hassa oasis complex. In Qatif, there is an impervious stratum two metres below the surface and thus it is extremely easy to cause waterlogging and salinity. Since most land there is privately owned, the result has been that productivity has steadily fallen, despite considerable concern. The situation is worsened by the fact that 75 per cent of the holdings are below 2 hectares in size and 75 per cent are owner-occupied, while 30 per cent of the land is given over to dates. Such small-scale proprietors will not and cannot invest sufficiently to resolve the problems of waterlogging and salinity (Bowen-Jones & Dutton, 1983, p. 31).

The al-Hassa situation is similar, where the oasis towns of al-Hofuf and Mubarraz, together with 52 villages, covering a total population of 380,000 depend on some 500 artesian wells and springs. However, up to the mid-1960s, salinity had reduced the land available for agricultural use from 16,000 hectares in 1951 to less than 8,000 hectares in 1963 (Schuster, 1979, p. 99). The problem had been worsened by sand movement which had contributed to the rapid loss of land. The fact that two-thirds of the water was privately owned and that most holdings were less than 2 hectares in size made it imperative for the problem to be tackled through a government agency and, in 1967, the al-Hassa Irrigation and Drainage Authority (HIDA) was created. As a result of a sand stabilization programme and a reform of the water regime, together with the reconstruction of distribution and drainage systems, it has proved possible to move towards the goal of expanding the cultivated area in the oasis towards 20,000 hectares (Vidal, 1980, pp. 91–97). The success of HIDA in this project by 1972 meant that it also took over similar projects in Qatif and Aflaj and the old *bedu* resettlement project at al-Haradh, now run on capital intensive lines, has been taken over by a new state-organized company since 1981, the National Agricultural Development Company (NADEC).

The HIDA and NADEC experiences clearly show that Saudi Arabia has not completely rejected the possibility of state intervention, for NADEC, for example, also manages 60,000 hectares in Ha'il, Qasim, and Wadi Dawasir. However, these companies are paralleled by massive private sector companies in Ha'il, Tabouk, and Qassim, which are really responsible for the extraordinary growth in Saudi production, just as companies such as Saadco have developed the dairy sector, often with foreign help. The state seems to be confined increasingly to rescue operations such as that in al-Hassa, and virgin land development in central and western Saudi Arabia (Bowen-Jones & Dutton, 1983, p. 31) with the major emphasis on Jizan, the Tihama, and the Wadi Najran-Wadi Dawasir complex.

THE FUTURE

The Saudi experiment in agricultural development has clearly shown considerable success in terms of production growth — as the recent cereal yields have shown

(Table 11.2 and 11.3). However, it seems that these successes have been bought at the cost of abandoning a large part of the rural agricultural labour force. In addition, the gains have only been achieved at immense cost, particularly in terms of subsidies — an approach only open to an oil-rich economy, where political objectives of self-sufficiency may well outweigh economic considerations. Furthermore, these gains have only been achieved after great changes in policy in the 1970s, whereby Saudi Arabia abandoned evolutionary improvement of the private sector for an advanced high technology enclave solution which will continue to depend on foreign management and inputs (McLachlan, 1984, pp. 126–127). The result must be that the agricultural sector will become simply an enclave activity from which the traditional agricultural labour force will be increasingly excluded through its inability to invest and its lack of appropriate skills. The massive subsidies will, therefore, increasingly pass into foreign hands as they alone offer the skills necessary to maintain the sector and the dream of self-sufficiency will merely become another trap for the loss of reserves. It is difficult to see how, if oil revenues fall, the Saudi authorities will be able to sustain the vast financial inputs into a sector which, although it employs 25 per cent of the indigenous labour force, provides only 1.1 per cent to the country's GDP. In short, the experiment, bold though it is, ultimately offers no sure escape from dependency on food imports from the outside world.

Yet, despite these wider issues, it seems that the very vehicle that the Saudi government has chosen to proclaim its impending success may suffer from fundamental flaws. The major irrigation systems used have proved extremely difficult to maintain and operate, although they are relatively cheap to obtain. Central pivot systems are notorious for high evaporation losses and the danger of salinity build-up. There are fears that the aquifers may eily be over-used and, in any case, crop yields are still low — often only reaching 1 tonne per hectare, rather than the 3 tonnes per hectare that represents a break-even point, and certainly nowhere near the anticipated levels of 9 tonnes per hectare. Elsewhere, too, localized falls in water table have already begun to point out the dangers of overuse of the critical factor in Saudi agriculture — water. There, perhaps, lies the greatest danger, that the fragile agricultural environment in Saudi Arabian may be irreparably damaged by over-rapid and over-extensive exploitation, of which the current enthusiasm over wheat production is the first sign.

REFERENCES

Barker, P. (1982) *Saudi Arabia: the Development Dilemma*, Special Report No. 116, EIU, London.
Bowen-Jones, H. (1980) 'Agriculture and the use of water resources in the Eastern Province of Saudi Arabia', in Ziwar-Daftari, M. (Ed.) *Issues in Development: the Arab Gulf States*, MD Research & Services Ltd, London.

Bowen-Jones, H., & Dutton, R. (1983) Special Report No. 145, *Agriculture in the Arabian Peninsula*, EIU, London.

Bowen-Jones, H., & Dutton, R. (1984) 'Agricultural production and/or rural development?', *Arab Gulf Journal*, **4**, 51–64.

COMET (1983) *Regional Development in Saudi Arabia*, Special Report, London.

Cottrell, A. (Ed.) (1980) *The Persian Gulf States—a General Survey*, Johns Hopkins University Press, Princeton, USA.

El Mallakh, R. (1982) *Saudi Arabia: Rush to Development*, Croom Helm, London.

Fabietti, U. (1982) 'Sedentarisation as a means of detribalisation: some policies of the Saudi Arabian Government towards the nomads', in Niblock, T. (Ed.) *State, Society and Economy in Saudi Arabia*, Croom Helm, London.

Financial Times (various dates) London.

Fisher, W. B. (1978) *The Middle East*, Methuen, London.

Gischler, C. (1979) *Water Resources in the Arab Middle East & North Africa*, Menas, Wisbech.

Hajrah, H. H. (1982) *Public Land Distribution in Saudi Arabia*, Longmans, London.

Lackner, H. (1978) *A House Built on Sand—a Political Economy of Saudi Arabia*, Ithaca Press, London.

McLachlan, K. S. (1984) 'The agricultural potential of the Arab Gulf states', in El Azhary, M. S. (Ed.), *The Impact of Oil Revenues on Arab Gulf Development*, Croom Helm, London.

MEED *Middle East Economic Digest*.

Moliver, D. M. & Abbondante, P. J. (1980) (various dates) *The Economy of Saudi Arabia*, Praeger, New York.

Moss Helms, C. (1981) *The Cohesion of Saudi Arabia*, Croom Helm, London.

Prestley, J. R. (1984) *Saudi Arabia: Achievements and Prospects*, COMET, London.

Schuster, W. (1979) *Wirtschaftsgeographie Saudi Arabiens mit besonderer Berücksichtigung der staatlichen Wirtschaftslenkung*, Dissertationen der Wirtschaftsuniversität Wien 27, VWGO (Vienna).

Vidal, F. S. (1980) 'Development of the Eastern Province: a case study of al-Hasa oasis', in Beling, W. A. (Ed.), *King Faisal and the Modernisation of Saudi Arabia*, Croom Helm, London.

World Bank (1983) *World Development Report*, Oxford University Press, Oxford.

Agricultural Development in the Middle East
Edited by P. Beaumont and K. McLachlan
© 1985 John Wiley & Sons Ltd

Chapter 12

Agricultural Policy and Development: Oman, Bahrain, Qatar, and the United Arab Emirates

RODERIC DUTTON

INTRODUCTION

An examination is made here of the way in which policy in the four countries has and will (or could) affect agricultural development. Only enough physical and general economic background is given to illustrate the main theme; for more detailed statistical appraisals of agriculture see the references.

OMAN

Development policy in Oman only began to be formulated from 1970 when the accession of the modernizing Sultan Qaboos, replacing his very conservative father, and mounting oil wealth made rapid change and development in Oman both politically and financially feasible. From the outset it was clear that the guiding philosophy would be one of encouragement to private enterprise but a clearer formulation of policy had to await the guidelines composed for the preparation of the First Five-Year Plan (1976–80) and the Second Plan (1981–85) (Oman, 1975, 1980a).

However, the ideal of free enterprise has not easily accorded with the fact of oil wealth passing primarily under direct government control, nor with the fact of the great need for the government to develop, from scratch, the physical and social infrastructure required to supply the growing expectations of the population. In this work the new ministries had the very laudable but perhaps over-optimistic and over-ambitious desire to supply as many services and goods to the people as possible.

Thus up to 1977 10 million Omani riyals (OR) were invested in agriculture (excluding water resources). Of this at least half went to large projects which, however, had very little direct impact on the farmers, and therefore on either food production or rural wealth (Dutton, 1980). Indeed, about OR 1 million were spent on state-owned production farms which were not very productive or were totally uneconomic enterprises and therefore operated in wholly unfair competition with the private sector which government policy was pledged to support. In its drive to do everything and to be seen to be helping both food producer and food consumer the government had, to an important extent, muddled its policies.

Furthermore, the value of some of the other government investments, aimed at supporting the private agricultural sector, may also be questioned not only in terms of effectiveness but also in terms of conflict of policy. The highly subsidized tractor service has suffered from the use of unsuitable equipment and a low effective work-rate whilst killing off not only the traditional bull-ploughing service (and all its associated skills) but also any realistic possibility of private tractor services being started. And in Oman, with oil wealth penetrating into the rural areas by remittance money, the private purchase of agricultural machinery was possible in the later 1970s, indeed quite costly diesel irrigation pumps had been privately purchased even before the 1960s.

The crop-spraying service, virtually free of charge but operated by untrained men, encouraged those farmers who used it to blanket spray everything, including beneficial insects. The veterinary service has been poorly managed and has tended to mask poor livestock husbandry with expensive drugs.

Worse, the agricultural extension service which grew rapidly in size during the 1970s has not been, fundamentally, an extension service. It has been a supplier of subsidized services such as those noted above. It has had neither the management support, nor the quality of trained staff, nor therefore the confidence of the farming community, to allow it to play a serious extension role. In this it has not been effectively supported by the research arm of the Ministry of Agriculture and Fisheries. The links between the two have, typically, been tenuous and the quality of research and its proven applicability to any given region of the country doubtful.

The 1970s may, however, be regarded as a learning period, not only for the Ministry of Agriculture and Fisheries but for the government as a whole. The results are reflected in the Second Five-Year Plan (1981–85) which provided the occasion for a re-thinking of policies, and of strategies for enacting them. This process was aided in the Ministry of Agriculture by the happy coincidence of a new Minister determined to make his mark but not tied to previous policies, of a strengthened Planning Unit, and of an experienced consultancy company, Arthur D. Little, employed to formulate the Plan.

The 1981–85 Plan (Oman, 1980b) makes clear that the objectives and targets of the 1976–80 Agricultural Plan 'were not met or even significantly approached'. In the new Plan, for all the productive industries including agriculture the Development Council emphasized the policy of giving 'a strong and stimulating push to the private sector'. This policy was underlined, as far as agriculture was concerned, by OR 100 million allocated to the Ministry of Agriculture (of which OR 25 million was spent on fisheries), OR 15 million on grants to small enterprises, OR 19 million on a new agriculture and fisheries bank, OR 10 million for the Development Bank of Oman, and OR 18.5 million to provide agricultural inputs for farmers.

Consistent with the private sector emphasis the plan also stressed a move away from government investment in projects of a commercial nature which, in agriculture, included the production farms, the date-processing factories, and the cattle farms in Sohar and Salalah.

The plan also recognized that heavily subsidized agricultural inputs were of doubtful benefit and were perhaps tending to inhibit private initiative. One small but very significant development as a result of this policy switch has been the careful formulation and implementation of a scheme to enable farmers to purchase approved, appropriate small tractors (with the emphasis on two-wheel pedestrian-operated machines), financed partly by subsidy and partly by loan, for which agents guarantee spares, maintenance, and training, thereby moving away from the ineffective previous tractor service described above.

The problems of research have been recognized—its continuity, quality, relevance, and dissemination of results—and an outside consultancy appointed to help overcome these problems, for both livestock and crops. The Ministry is also trying to integrate health care and husbandry in its approach to both livestock and crop problems. A greater awareness of the integrated nature of rural development questions is also shown by a Ministry-sponsored feasibility study of livestock-related 'cottage' industries—including skins, milk, poultry, bees, weaving, and red meat.

The wish gradually to move away from input subsidies, as noted above, is paralleled by moves towards greater intervention in the marketing process. An experimental import tariff has helped local producers of bananas, as has guaranteed purchase of bananas grown in Salalah and their shipment to the larger market in northern Oman. The Public Authority for Marketing Agricultural Produce, now coming into being, will provide a large number of collection points and organize sales for a wide range of vegetables and fruits, including fresh limes. There is a danger that this may inhibit local initiative and be cumbersome in practice, with loss of produce. But in principle it provides the Authority with the opportunity to put pressure on producers, through selective purchasing, in order to improve quality and to encourage off-season production.

Throughout the 1981–85 Plan period the Ministry of Agriculture is also continuing to improve its inventory of land and water resources; improving its ability therefore to make long-term decisions about their development. This is the kind of background work which has no short-term benefit for farmers or production but is essential for the future.

However, the Ministry may perhaps be criticized on four counts. Firstly, a high proportion of its field staff have neither the experience, the training nor the ability effectively to help farmers in what should be a period of very rapid change, while oil wealth lasts; there has been a greater emphasis on physical facilities than on quality of staff to man them.

Equally important, the Ministry needs to establish 'action units' on particular crops, stock, or other multi-faceted problems—units that cut across the traditional boundaries between research, services, extension, subsidies, trials, and marketing, and look at the needs of, for example, a given crop, including the economics of its production in the small-farmer context, from start to finish. By concentrating attention on a small number of 'action units' with good field staff, results of relevance to current needs might more readily be attained. As a model, the Ministry might take its own honeybee department. Although this comes under 'research' it is, in practice, research, field trials, demonstration, extension, training, policy-making, and marketing all rolled into one, in the hands of a very capable man based in the centre of Oman's principal bee-keeping region. Similar units (the Ministry could only cope with a very small number at any one time) could look into individual fruits and vegetables, or livestock, or else examine the complexities of a particular traditional production system such as the *falaj* system in the interior of Oman. Under discussion at the present time is a comprehensive approach to livestock requirements, but if a livestock authority is established one person should have special responsibility for each aspect, comparable with the present honeybee specialist.

Thirdly, in Oman agricultural development, as with other aspects of rural and regional development, has been influenced by different regional political considerations. For example, the southern Region (Dhofar) has had a disproportionately high development budget in consideration of its recent history and its location bordering socialist Yemen PDR. Such intensive support is not always successful, as a recently held conference to discuss Dhofar's problems revealed[1].

Finally, in its policies the government has not fully recognized that large (in total) sums of remittance money are available for private investment in agriculture. If government policies can create attractive production systems (possibly using 'action units') and an attractive economic environment during the coming decade, then thousands of small farms throughout the country stand to benefit by self-help, with an equivalent benefit to the farmer and his community, and to the nation as a whole.

BAHRAIN

Bahrain, a minute nation 662 square kilometres in area has the greatest population density of any country in this study and, at about 5 per cent, the highest proportion of its land surface under crop. Modern governmental action has influenced Bahrain's agriculture, both directly and indirectly, for a longer period than in the other states and so a longer historical perspective is here accorded to it.

The earliest settlers in Bahrain were presumably attracted by its location; a small island for which the sea provides natural defences and paths of access to all other Gulf States—the eastern seaboard of Saudi Arabia being only 25 kilometres distant. However, a settled life was made possible by irrigated agriculture using water from numerous springs in the north of the island flowing from aquifers whose origins lie in Saudi Arabia. But Bahrain's location equally encouraged the development of those skills associated with *entrepôt* trading (as well as fishing and pearling), and this, in recent times, has acted in conflict with the interests of the farming community. The ruling family, the al-Khalifa's, until fairly recently the major land owners, were under no pressure to manage their land to ensure high productivity. There was a regime of absentee landlords who exercised oppressive authority over short-term tenants, the latter having little or no incentive to innovate or to make long-term investments in their land.

In this century, Western, notably British, influence has been strong in Bahrain, attracted by the island's strategic location. One important consequence was that formal secondary schooling was started, providing for boys and girls in the mid-1920s, long before it began in neighbouring states and a full 50 years ahead of Oman. The strength of Western contacts led to the successful exploration for oil, and to oil exports from as early as 1932—a first for the Arab Gulf (excluding Iraq) and the Arabian Peninsula.

Oil wealth had complex effects on agriculture, mostly indirect. The urban centres grew and much agricultural labour moved to the oil industry or to urban-based industrial or service sector jobs. Tenants of farmland found it even more difficult to pay their rents as rural labour became more scarce and costly, imports grew, and the relative demand for and value of dates, the principal crop, declined. Land was abandoned, or swallowed up by the burgeoning urban centres of Manama, Muharraq, and Jid Hafs—all situated within the northern agricultural zone (Bowen-Jones & Dutton, 1983).

Education expanded rapidly on the strength of oil wealth, giving rural children aspirations to move into urban jobs, particularly as agriculture was already in decline.

Oil production also peaked much earlier in Bahrain than elsewhere, forcing diversification into the economy. Bahrain has thus a big dry-dock facility, an aluminium smelter, large oil and gas refineries (even refining imported Saudi crude oil), telecommunications enterprises, and offshore banking, as well as a sophisticated marketing system heavily associated with import/export trading of foodstuffs. Thus job opportunities outside agriculture were being created, and farm work became even less attractive to the Bahrainis as imported labour, mostly Omani, prepared to work for low wages, came into the country in the post-war period. Between 1959 and 1965 the migrant population working in agriculture and fisheries doubled to about 1,100 while the numbers of Bahrainis fell from 3,900 to under 3,600. By 1971 the agricultural labour force had fallen to 4,000 in total, and to under 3,000 by 1979/80. After Oman started oil exports in 1967, and Qaboos invited all exiled Omanis back home in 1970, cheap local Arab labour was much harder to obtain in Bahrain.

Meanwhile, Bahrain was suffering increasingly from water problems. Many of the free-flowing springs were running dry, a result of increased water usage in Bahrain and Saudi Arabia, and the pumped water was of worsening quality. Today a high proportion of water available for agriculture is very saline, 3,000–4,000 ppm, though with good management still capable of supporting the more salt-tolerant crops.

But agriculture, in addition to problems of urban-industrial development, shortage of labour, and poor quality of water, has always been threatened by the island's commercial skills and experience. In 1957 the Department of Agriculture noted that local contractors who supplied fruit and vegetables to large customers used the excuse of a local increase in amoebic dysentery to import from Lebanon. These imports undercut local prices and therefore inhibited local production. They helped put an end to a growth from 150 to 576, between 1952 and 1959, of the number of gardens growing vegetables.

Bahrain's international marketing facilities continue to improve, and are being given another boost with the construction of the causeway to Saudi Arabia. Nevertheless, in spite of continuing decline in some sectors of agriculture during the 1970s (as revealed by the agricultural censuses of 1973/74 and 1979/80), notably in date palms, alfalfa, fruit trees, and numbers of cattle, some expansion was recorded.

Poultry is a case of particular interest. Starting as early as the 1950s Bahrain was the first country to start acquiring expertise in intensive poultry production under the difficult local extremes of heat and humidity. The Department of Agriculture was the principal innovator and the government has successfully stimulated private involvement by providing capital for joint enterprises. Production of eggs and poultry has risen rapidly since 1977, with plans for continuing expansion towards self-sufficiency.

Lessons from the success with poultry (now generalized throughout the Arabian peninsula) can be reapplied to other sectors. To be successful, producers need to maximize production per unit labour, per unit land, and per unit water in order to combat problems of costly labour, small land holdings (mostly under 5 hectares), and water shortage, and in order to compete with cheap imports.

Trials by the Department of Agriculture, using simple plastic 'greenhouses', basic drip irrigation and evaporative cooling have produced high yields of various vegetables. The plan is to transfer this relatively inexpensive and unsophisticated technology to a gradually expanded network of local growers. The same approach will have to be applied to sheep/goat production (and associated forage) and to fruit trees. In the former, hormonal control to promote frequent breeding (allied with competent management) is essential to begin to compete with imports from Australasia.

As a result of a greater belief in the potential of Bahraini agriculture, and of the oil price increase of 1979, the government planned a significant increase in investment. If this withstands the more recent drop in oil prices, and if a monitoring eye is kept on the balance of benefits to farmers and food importers, then Bahrain could achieve a greater measure of food self-sufficiency than at present.

QATAR

In contrast with the other three states in this study, in which oil sustains agriculture, oil in Qatar may be said to have created agriculture. Before oil exports began in 1948, Qatar had a pearling and trading population, centred on Doha, and small isolated fishing communities, principally in the northern half of the peninsula, such as Al Khor and Al Ru'ays. The villagers also maintained a small subsistence cultivation of a few date palms and vegetables, but only as a complement to their principal maritime interests.

Pastoralists, of whom there were never a very large number, crossed the neck of the peninsula into and out of Saudi Arabia and utilized the many *roda* for grazing. The *roda*, a feature of peculiar importance to Qatar—though found elsewhere in the peninsula they are not as significant in the overall vegetation picture—are depressions caused by sub-surface limestone solution followed by surface dropping. Gravity then ensures that the low precipitation of about 80 mm per year in the north flows into the depressions where the water is trapped for plant usage by sandy colluvial deposits. The *roda*, of which there are some 850, of diameters varying from tens of metres to several kilometres, were under a loose tribal ownership. Ownership in effect devolved onto Qatar's shaykhly families, a process aided by the tribesmen being pulled into the more remunerative employment opportunities of the urban centres as oil wealth grew. The very greenness of the *roda*, and the fact that water obviously existed in them to support the natural vegetation attracted the *shaykhs* to invest some of this new oil money into installing pumps and fencing land and creating their own green 'oases'—the word oasis here being deliberately used, with its romantic connotations, to indicate the powerful non-commercial motive of much of this development.

By the latter part of the 1960s, there existed the following situation: a new class of land owners, quite inexperienced in soil/water management, was 'developing' a rapidly growing number of land holdings; management and labour were almost entirely expatriate, usually with neither the sophisticated skills required to manage scarce and fragile resources nor the decision-making freedom to do so; the economics of the enterprises were not reckoned, even in terms of output market value against basic input costs, and certainly not in terms of the true cost of the water (surely at least as valuable as oil to Qatar's long-term future?) and the land.

In 1958 there were 40 farmholdings utilizing only 380 hectares of land, but by 1970 there were 365 holdings utilizing 3,400 hectares. By then, agriculture and tree-planting was consuming 30 million cubic metres of water per year, often in a most profligate manner — good land was even going out of production as a result of waterlogging and salination (Bowen-Jones & Dutton, 1983).

By 1970 there was a sufficient realization of the potential dangers for the government to initiate, in conjunction with FAO, a comprehensive water resource, agriculture and land use survey which extended over a decade from 1971 to 1981 (Qatar, 1981), and continues within the Department of Agricultural and Water Research. However, while this survey was producing a more complete picture of local resources and of the urgent need to control their usage than for any other country in the peninsula, a further rapid expansion of land-water usage continued during the 1970s along the same lines as before. By 1980 a maximum irrigated area of 3,300 hectares was absorbing over 57 million cubic metres of water, almost double the safe aquifer yield. Salination and waterlogging had by now put 800 hectares out of production.

Food was, of course, being produced. And it was helping to feed an urban population growing rapidly in numbers (by natural increase and migration) and wealth and therefore demanding more of a wider range of better quality foodstuffs. In particular, up to 70 per cent (over 20,000 tons) of Qatar's vegetable consumption (including watermelons) was being provided locally, with some 4,000 tons exported. In addition, 20 per cent of fruit was produced locally, though some 60 per cent of local production was dates. The date palm was still the dominant fruit crop, occupying half the fruit area of 1,500 hectares though economically it has a very doubtful future. Of the other foodstuffs, including meat, milk, other fruits, and cereals, Qatar produced a small or negligible proportion of local requirements.

Qatar's difficulty, therefore, is that, in spite of agriculture already using double the safe water extraction rate, it is still heavily dependent on food imports, and the food demand is rising and likely to double by the year 2000. But at least the authorities now know, more clearly than in any neighbouring state, the likely physical consequences of allowing the present situation to persist. Qatar is also free from the need to concern itself with an agriculture-dependent indigenous rural population — none exists. It is therefore almost a realistic option greatly to curtail agricultural activity; not many Qatari producers would be adversely affected, water would certainly be conserved, and the security of long-term food imports might be considered safe enough in the light of a 30- to 50-year oil reserve and extremely large deposits of offshore natural gas. The government also demonstrated that it has the political will for this sort of action when in 1980 it prohibited additional exploitation of groundwater reserves for farming, and impounded all drilling rigs.

But perhaps a more positive approach, which has the merit of enhancing local skills for the longer-term future, will be to use the accumulated experience of the research centres and of the better-managed holdings to disseminate knowledge of the techniques required radically to increase yields per unit of water. This movement can be encouraged by financial and legal measures as appropriate. Any combination of capital, technology, imported management, and gas energy can be made to play a part in minimizing water wastage. And undoubtedly, in the not-too-distant future, food production policies in Qatar will be affected by GCC initiatives and by the rapidly expanding agricultural industry of Saudi Arabia, the oases of whose Eastern Province are readily accessible to Qatar.

UNITED ARAB EMIRATES

The United Arab Emirates stand between Oman and the other two states in this brief study in a number of ways of significance for agriculture. The eastern and northern parts of the UAE, from Al Ayn to the border with Omani Musandam, lie astride the Hajjar mountains and are therefore geomorphologically a northern extension of Oman. The mountains are of importance as the origin of the alluvial material that has formed the plains east and west of the mountains, and as the rainwater catchment zone which makes this small eastern area of the UAE relatively well-provided with sub-surface water for irrigation. In the UAE the Batina plain is much narrower than in Oman, and essentially reduced to a number of embayments in which date gardens irrigated from hand-dug wells are the traditional form of cultivation, notably around Fujairah, Khor Fakkan, and Diba. Small mountain settlements use *ghayl falajes*, and in the interior plains larger settlements, such as Falaj al Mu'alla, have substantial *falajes* such as those that water Nizwa or Ibri in Oman. But the mountains here are lower than in Oman, and the rain catchment correspondingly smaller. Other major agricultural settlements in this region include Al Ayn, Mileiha, Hamraniyah, and Digdaga, from south to north, and the coastal settlement of Ras al Khaimah where the mountains meet the Gulf.

In all, these are but slim resources to meet the food requirements of a burgeoning growth of population, centred in Abu Dhabi, Al Ayn and Dubai. To them may be added the Liwa oasis, the only area of agricultural significance in the long western-extending elbow of the UAE whose sands make up 80-90 per cent of the total land surface of the Emirates.

The quarter century from 1955 to 1980 was one of ever-increasing government encouragement to the private sector in farming and of direct action by individual Emirates and by the Ministry of Agriculture and Fisheries in Dubai and the Department of Agriculture in Al Ayn, in both research and production farms. The motivations and drive, with significant British encouragement in the early years, are complex: part desire to stimulate a local industry with use of local skills and the production of food for a growing population; part largesse by individual rulers, as in Ras al Khaimah, handing out parcels of land; part a desire to keep the more rural elements of the population out of the towns; part emulation of the regional drive towards 'modernization' led by Saudi Arabia; and part by Shaykh Zayid's very personal desire to develop Al Ayn, where he was *wali* for twenty years until 1966, and by his enthusiasm for forestry.

Encouragement has taken many forms. There has been a growing input into improving the quality and dissemination of basic data. The first experimental station opened at Digdaga in 1955 and had expanded from 2 to 12 hectares by 1961. Three sub-stations were established in Al Ayn, Ajman, and Umm al Qaiwain, and in May 1969 a new experimental station opened in Al Ayn. There are now three experimental stations on the *UAE*: Batina, one at Al Dhaid, and one at Hamraniyah (replacing Digdaga which closed in 1980), in addition to Al Ayn (Unwin, 1981).

To diffuse the new knowledge there are extension centres at Digdaga, Fujairah, and Al Dhaid (with six sub-offices); and there are eighteen extension centres in the southern region, responsible to Al Ayn, and also government centres in the Liwa oasis.

This large research/extension infrastructure was planned, and largely in operation, by 1978 when the whole vegetable, field crop, and non-date fruit crop area of the Emirates totalled only about 5,400 hectares.

Farmers were also helped by the supply of seeds and plants from a number of government nurseries. These and other services, including wells, pumps and irrigation systems, tractor services, fertilizers, and a crop-spraying service are provided at half price or, in the region dependent on Al Ayn, free, except for half-price fertilizer.

Land, for those without it, has normally been fairly easy to obtain often simply by application to the local ruler. Credit and loan schemes were available to help develop the land from as early as 1957.

All the farmer has to do is grow the crops, and this is increasingly left in the hands (management and labour) of migrants who work for much lower wage rates than would a UAE citizen. It is revealing that about half the owners of agricultural holdings consider themselves as non-farmers and indeed many of their holdings are regarded as private retreats from the main towns—areas of pleasant greenness, using scarce irrigation water without any clear economic aim or irrigation constraint.

Post-harvest, the farmers in Al Ayn have for several years had the support of a subsidized marketing operation and in 1980 it was decided to create an Agricultural Marketing Board for the whole of the UAE.

Most of the above services are for the small farmers but a combination of private, government, and foreign capital has established some large enterprises with production and/or research aims. The former include large-to-very-large poultry units and two major dairy farms. But the hydroponics station on Sadiyat island and the Mazyad station at Al Ayn have a larger research element. The 400 hectares of wheat of the Al Oha project (also in Al Ayn) are also part motivated by 'food security' logic. And the huge foresty project, notably along the Al Ayn/Abu Dhabi highway, has a complex of local justifications which may be summed up as one man's desire to make the desert green.

With the degree of farmer support outlined above, and the contribution of the large enterprises, it is not surprising that the level of agricultural activity rose very rapidly during the 1970s. As additional stimulus, there was a rapidly growing market—the population of 180,000 in 1970 had risen to over one million by 1980. From 1973 to 1979 the land in agricultural holdings almost doubled, from 12,600 to 23,400 hectares; from 1978 to 1979 alone there was a jump of 38 per cent in vegetable area (to 4,300 ha) and of 130 per cent in field crop area (to 2,200 ha) (Unwin, 1983). Farm output value stood at about 446 million dirhams in 1978 and had risen to 773 million by 1981 (MEED, 1983). This compares with a food import bill which, rising rapidly, had attained a value of 3,500 million dirhams by 1980/81.

However, the resource cost (quite apart from the financial outlay) has been high. Land, notably in Ras al Khaimah, has become more saline—with some totally lost to production—irrigation water has also become more saline and groundwater levels in some regions are falling by 2 metres or more per year. A ministry statement in 1981 put the annual water requirement for the country at 565 million m^3 with only 210 million m^3 available; a shortfall of 355 million m^3 (Bowen-Jones and Dutton, 1983).

Local awareness of this problem has been growing for some years—exemplified by the change from furrow to trickle irrigation for the Al Oha wheat project, by the expansion of the cool house vegetable project at Al Ayn and by continued hydroponic work at Sadiyat island. However, national-scale action, agreed by all seven Emirates, has been politically much more difficult to attain. But the recently formed Water Resources General Authority will, it is hoped, keep a much tighter rein on water usage.

The question is really one of political will. Administratively, because the farming system is so tied to government subsidies, it would be simple to cut the rate of support or to make it dependent upon greatly increased efficiency of water usage. Indeed, with only 9,000 holdings and such a large extension network it would be feasible to meter water usage, and even cost it individually. Incentives for types of production which maximized value per unit water could be instigated. The long-term position of the extensive date gardens will have to be reconsidered, and also the forestry programme.

If, in the short-term, such measures reduced output rather than increased water usage efficiency it would not matter. The UAE is already 80 per cent dependent on food imports and an increase will not, therefore, significantly increase the 'security' risk. Nor would there be an undesirable collapse of the farm sector. It is said that many enterprises would flourish without subsidy—if so, why not create the economic conditions whereby they can show the others how it is done?

CONCLUSIONS

On the one hand we are dealing with a very small problem: the total, cultivated area in the four states is less than 70,000 hectares or a mere 1 per cent or less of the cultivated areas of Iraq and Sudan, respectively. Agriculture in these four states will never therefore excite much international interest, except what can be paid for by oil money; but the questions it raises are of real concern locally and the results of development during the next decade or two will be of great importance to the long-term vitality of these countries and the lessons learned will be of great international interest to those concerned with development in hot, arid lands.

In Qatar and the UAE the main problem is one of known water shortage versus a seemingly insatiable appetite by the farming industry for more. The very large oil wealth of these two states (and equally high per capita incomes) has caused the problem; agriculture has expanded very rapidly on the back of oil money. Administratively speaking it would be possible severely to restrict agriculture and rural consumption; very few people consider themselves to be primarily dependent on farming for their livelihood, the countries are heavily dependent on food imports in any case, and large afforestation schemes could be curtailed overnight. However, the political impetus of the 'food security' lobby is strong, and in any case oil wealth can be utilized, in combination with legislative action, to seek a more positive road forward. If water usage can be greatly reduced by the adaption and adoption of appropriate (sophisticated) technology then not only can food continue to be produced and the countries have a far more secure hydrological future, but the rural populations will also develop skills which will have a long-term relevance, and Qatar and the UAE will be very interesting internationally as examples of developing countries which have revolutionized their farming practices, examples which may be of more interest to other developing countries than the arid area practices of a very developed country like the United States. As far as research is concerned, many steps have been taken, the work of the research laboratories of the Ministry of Agriculture at Al Ayn being but a recent example (MEED, 1983); but it is the translation of this knowledge into everyday farming practice which is the real problem.

Bahrain differs from Qatar and the UAE in that its agriculture has been in many respects stagnant or in decline in recent decades. Nevertheless, agriculture accounts for well over half the island's total water consumption (Bowen-Jones, 1980a) but this water is now of such poor quality that, combined with other factors, it is forcing traditional agriculture into a steep decline, for example the great reduction in the date palm area.

Oman is the 'newest' of these countries, in the sense of exposure to oil-induced change, and it is also the best-endowed with land and with both renewable and fossil water. It also has a 'real' rural population, though one steeped in traditional agricultural practices, which are not relevant to today's circumstances, and using greatly increased volumes of irrigation water because of generalized installation of diesel pumpsets. Perhaps learning from its neighbour's experience, Oman already has a Public Authority for Water Resources. But there is an expectation, both privately and governmentally, that new land and water can continue to be brought into production and this is perhaps diverting attention from what must be the main long-term task of maximizing productivity of the few larger units and the thousands of smaller units already in existence by thinking through new and applicable crop and livestock systems.

But if water is the critical physical constraint on agriculture in these countries, the political vision and administrative determination to ensure that farmers use water wisely in intensive production systems will be the determining factors of the agricultural revolution required to bring this wise usage about.

NOTES

1. Oman/IUCN Workshop held in Salalah, September 1983, to discuss conservation and development of the Dhofar mountains.

REFERENCES

Abdul Aziz, M. (1980) *Issues of Development and Agricultural Planning in UAE*, Ministry of Agriculture and Fisheries, UAE.
Bahrain (1974) *Results of the 1974 Agricultural Census*, Ministry of Municipalities and Agriculture, Agriculture Directorate, Bahrain.
Bahrain (1980) *Results of the 1980 Agricultural Census*, Ministry of Commerce and Agriculture, Agriculture Directorate, Bahrain.
Bowen-Jones, H. (1980a) 'Agriculture in Bahrain, Kuwait, Qatar and UAE' in Ziwar-Daftari, M. (ed.), *Issues in Development: The Arab Gulf States*, MD Services, London, pp. 46–64.
Bowen-Jones, H. (1980b) 'Agriculture and the use of water resources in the Eastern Province of Saudi Arabia', in Ziwar-Daftari, M. (Ed.), *Issues in development: The Arab Gulf States*, MD Services, London, pp. 118–137.
Bowen-Jones, H., & Dutton, R. W. (1983) *Agriculture in the Arabian Peninsula*, Economist Intelligence Unit, London.
Cordes, R., & Scholz, F. (1980) *Bedouins, Wealth and Change. A Study of Rural Development in the United Arab Emirates and the Sultanate of Oman*, UNU, Tokyo.
Durham (1982) *Final reports — Research and Development Surveys in Northern Oman*, Vols. 2 & 4, University of Durham.
Dutton, R. W. (1980) 'The agricultural potential of Oman', in Ziwar-Daftari, M. (Ed.), *Issues in Development: The Arab Gulf States*, MD Services, London, pp. 170–184.
Dutton, R. W. (1984) 'Interdependence, independence and rural development in Oman: the experience of the Khabura development project', in *J. Oman Studies*, **6**.
Jaidah, A. M. (1980) 'Prospects for gas prices and the development of the natural gas industry in Qatar', in Ziwar-Daftari, M. (Ed.), *Issues in Development: The Arab Gulf States*, MD Services, London, pp. 154–159.
MEED (1983) *U.A.E.*, MEED Special Report, Middle East Economic Digest, London.
Oman (1975) *The Five-Year Development Plan, 1976–80*, Development Council, Oman.
Oman (1980a) *The Second Five-Year Development Plan, 1981–85*, Development Council, Oman.
Oman (1980b) *Second Five-Year Agricultural Development Plan (1981–85)*, Ministry of Agriculture and Fisheries, Oman.

Qatar (1981) *Water Resources and Agricultural Development Project Reports, 1972–1981*, Ministry of Industry and Agriculture & FAO/UNDP, Doha.
UAE (various dates). *Annual Statistical Abstracts*, Ministries of Planning, Agriculture and Fisheries, UAE.
Unwin, T. (1981) *Agriculture in the United Arab Emirates: A Preliminary Report*, University of Durham (unpublished).
Unwin, T. (1983) 'Agriculture and water resources in the United Arab Emirates', in *Arab Gulf Journal*, **3** (1), 75–85.

Agricultural Development in the Middle East
Edited by P. Beaumont and K. McLachlan
© 1985 John Wiley & Sons Ltd

Chapter 13

Yemeni Agriculture in Transition

SHEILA CARAPICO

Throughout history Yemen has been the bread-basket of the Arabian Peninsula. Drenched twice annually by Indian Ocean monsoons, the southwest corner is ecologically distinct from the arid interior of the peninsula. The interior Yemeni highlands, with elevations of up to three thousand metres, enjoy a relatively moist, temperate climate; the Red Sea coastal plain, the Tihama, is hot and dry but watered by runoff from mountain rains into a dozen major wadis. To the north and east, cultivation recedes into the Arabian sands. Traditional farm techniques were finely tuned to micro-environmental variations in terrain and water availability, but exhibited certain common features. Arable land — some 15 per cent of the total land surface — was overwhelmingly devoted to drought-resistant grains. The agricultural labour force was mostly organized at the house-hold level. Most families kept some livestock and subsisted on a diet of cereals and dairy products. This primary agricultural regime was supplemented by garden-scale production of vegetable, fruit, spice, and non-food crops on some of the better-watered parcels. Until the mid-twentieth century, nine out of ten Yemeni families supported themselves by farming.

In the past two decades, rapid changes in the political economy of the Yemen Arab Republic have begun to transform the system of agricultural production. The recent trends and the prospects they portend for the future of Yemeni agriculture are the subject of this paper. In particular, two salient factors are examined in some detail. The first is the effect of heavy male out-migration from the rural areas, especially labour migration to the oil-rich, labour-short nations of the Arab Gulf. After 1973, migrant remittances were a major element in GNP and contributed cash income to a majority of farm families. The second factor is investment in agricultural innovations and technology. Modern methods have only been incorporated into farm practice on a limited basis and have had a correspondingly modest impact on overall productivity. The Yemen AR Government, foreign donors, and even agricultural cooperatives have had only minimal success in encouraging the investment of remitted income into the farm sector. Rather, the basic outlines of the traditional system remain intact with the exception that the country has lost its relative self-sufficiency in food.

The first section below examines the pre-capitalist mode of production to highlight both environmental considerations and social relations of agricultural production. The second part of the paper discusses economic changes which have occurred since the republican revolution of 1962 as they pertain to agricultural development. The focus is on the impact of labour scarcity and the economic open door on the traditional use-oriented agricultural regime, and incentives or disincentives to commercialization of farm production. Finally, the conclusion briefly reviews real and projected performance of the agricultural sector during three national plan periods: an experimental Three-Year Plan (1974–76), and the recent and current Five-Year Plans (1976–81 and 1981–86).

THE TRADITIONAL SYSTEM OF CONSUMER PRODUCTION

Traditional agriculture was an environmentally sensitive system of production oriented towards consumption. An ecological balance of farmers with their land, crops, and livestock was approximated by minimizing waste and maximizing utilization of all by-products of the system. Hardy grains — varieties of sorghum, millet, maize, barley, and wheat — were favoured because cereals can be stored for year-round consumption and because of the value of their grasses as fodder and tinder. Domestic livestock — principally cattle, goats, and sheep, but also chickens, donkeys, and camels — were fed from the fields and reciprocated by providing traction and fertilizer for crop production. Where wood was scarce, dung mixed with straw provided fuel. Bread, porridge, sour milk, clarified butter, and meat or eggs represented the staples in the family diet. The system of grain and animal production was economical in utilization of scarce resources and well-suited to management by an extended family household. Tools were simple hand-held or animal-driven implements. Though never completely self-sufficient, families producing within this system provided most of their own regular consumption requirements (Carapico and Tutwiler, 1981).

Within the grain-farming family, men specialized in maintenance and cultivation of the land. Highland Yemeni agriculture is mainly terrace cultivation. Generally, terraces are stone-walled; in the wetter southern uplands, grass sod is used instead. Narrow mountainside terraces require annual or semi-annual repair. In the lower elevations where wadi flows were channelled to cultivated fields via a spate irrigation system, restoration of irrigation ditches was a major chore before and after each flood. In addition to maintaining the physical structures, men did most of the heavy field work associated with ploughing, planting, and some of the harvest tasks. Although they were very busy in the spring and autumn, in the traditional farm system men were prone to chronic seasonal unemployment.

Farm women, by contrast, specialized in animal-husbandry and procurement activities which must be done daily. Women fed, milked, and otherwise tended the livestock, going frequently to the fields to cut fresh fodder. They collected fuel and water. At harvest they cut, sorted, and stored the grains and stalks for later use. The temptation to classify these tasks as 'housework' should be resisted; in the traditional farm system they were essential productive activities. Whereas the system could tolerate the absence of one or more men from the household for part of the year, women tended the fields and the animals on a day-to-day basis all the year round.

In addition to cereal and livestock production, a range of indigenous crops was produced in limited quantities for local or foreign exchange. Potatoes, onions, garlic, beans, horseradishes, grapes, almonds, apples, apricots, and peaches lent occasional variety to highland diets. Alfalfa was grown as a fodder crop. The two most important highland non-grain crops were coffee and qat (*Catha edulis*). Yemeni 'Mocha' coffee was famous in the eighteenth century but was later eclipsed on the international market by colonial plantation production in East Africa and elsewhere. Qat is a tree or shrub peculiar to the high mountains flanking the lower Red Sea. Chewing of fresh qat leaves produces a pleasant sensation of physical well-being and mental alertness. Coffee beans were once Yemen's main export. Within the country there was also a market for the husks, which were brewed with spices for a special hot drink. Qat dries and loses its narcotic effect after twenty-four hours; hence its market was limited to the radius of a day's donkey trip from where it was grown.

Wadi lands in the Tihama yielded a wide range of tropical fruits and vegetables — papaya, bananas, okra, sesame, watermelon, and especially dates — but the two most important non-grain crops were tobacco and cotton. Yemeni tobacco was marketed mainly within the country. Like qat, it was something of a luxury available to common people only on special occasions but enjoyed by the well-to-do. Along with coffee and animal hides, cotton was a major commodity in the small export sector of the traditional economy. Cotton and straw weaving were cottage industries utilizing farm products.

With the exception of qat, non-grain crops generally require some irrigation to supplement rainfall and were thus limited to a tiny fraction of the total land area.

Landholding and tenure relations varied from one region to the next. At one point in the distant past, perhaps, every family owned and operated its own property; but the trend during the past few centuries has been towards consolidation of prime lands and fragmentation of smallholdings. Before 1962 there were two basic forms of land tenure relations, a smallholder form and a landlord form. Family-owned and -operated smallholdings predominated in regions of low or intermediate water supply and productivity. These included the semi-arid north-central plateau, the steep terraced slopes of the western mountains, and dry stretches between the Tihama wadis. Most rain-fed smallholdings in the highlands produced summer sorghum and sometimes a winter crop of barley or wheat. Some highland smallholdings also included a tiny irrigated plot of alfalfa or a few fruit, coffee, or qat trees. In the Tihama, the average 'small' holding was larger, but without access to spate irrigation would still produce only a single modest harvest of sorghum or millet. Smallholdings were typically not contiguous farm properties but scattered fields within a region; fragmentation of holdings among numerous parcels was the result of centuries of dividing lands among heirs.

Consolidation of holdings was greatest where the highest surpluses were realized from the land (al-Mujahid, 1978). True largeholdings of many thousand hectares were accumulated mainly in the two regions of highest productivity, the southern uplands and the Tihama. In the southern mountains heavy monsoons have helped create wide, fertile, temperate valleys where multiple crops of sorghum, maize, legumes, potatoes, qat, and even coffee can rely on rainfall alone. In these rich valleys the majority of farm families share-cropped from a landlord, although some tenants also owned small parcels and a few smallholders survived. Ownership patterns were comparable adjacent to the major Tihama wadis, where semi-annual floods provided ample spate for two or three grain crops plus tropical fruits and vegetables. The prime land was nearest to the wadis; most of this was owned by a few landlord families. At the perimeter of the spate system were some irrigated smallholdings. There were some large land owners in the western mountains and the central plateaux as well.

Cash rarely entered the cycle of agricultural production. Large and small holdings alike relied on a labour force organized at the household level. Production techniques were essentially the same; there were no economies of scale in the cultivation of large holdings. Landlord's properties were subdivided and rented out in household-sized parcels. Tenant families worked the land in the same fashion as smallholders, except that tenants surrendered a share of their harvest to the landlord. Share-cropping contracts varied according to a complex of inter-related factors: terrain, water supply, crop suitability, and arrangements for the provision of inputs such as seed and traction. Depending on the specifics, landlords might claim one-third to two-thirds of the harvest, and sometimes additional labour or special products such as honey. Wage labour was virtually unknown in agriculture. Landlords, tenants, or smallholders requiring workers to supplement the household labour force normally bartered or contracted for a share of the crop. Few if any inputs were purchased. For instance, blacksmiths traded implements for grain. Vegetable-growers traded their produce with grain farmers. Agricultural tax, *zakat*, the major source of state revenues, was also most frequently paid in kind, although collectors could demand coin; tax-men reserved a share of the revenue as commission.

This was not a fully closed subsistence system at either the household or the national level, despite the autocratic policies of the ruling Imams. Imports and exports were controlled by the royal family through licensing of monopoly traders. Political and economic isolationism helped protect the existing social order. However, barriers to commercialization of the agrarian system were gradually eroded. A major source of change was the British fuelling and trading colony at the Indian Ocean port of Aden, established in 1839. To protect its commercial interests, the British Empire extended Protectorate administration over southern Yemen, separating it politically from the Mutawakkilite Kingdom in the north. But socio-economic ties endured. The expansion of the Aden port coincided with increasing pressure on tenants and smallholders to sacrifice a larger share of their harvests to landlords, exporters, and royal tax collectors (Messick, 1978). In response to these pressures, thousands of young men, especially from the southern uplands, laid down their farm tools and walked to Aden in search of wage employment or petty trading opportunities. The vibrant port also represented a new and lucrative market for north Yemeni grain and fresh produce. Initially, most of the Yemeni farm products reaching the British colony were surpluses collected as rent and taxes, export of which enriched landlords and the ruling classes without altering the actual relations of production. Large land holders and tax collectors stored a portion of their grain receipts for their own families' consumption and sold the remainder. By the Second World War, however, even some smallholders and share-croppers were selling their crops on the market and beginning to consume cheaper imported substitutes. Aden was the main link to the international market, but the Italians in the Red Sea also purchased Tihama produce.

THE COMMERCIAL TRANSFORMATION

The pace of change in the agricultural sector was accelerated during the 1960s and early 1970s by national and international political events. First, liberal republicans overthrew the Imamate in September 1962. Yemeni royalists, backed by Saudi Arabia, resisted the revolution; Egyptian troops were sent to defend the fledgling republican government. A civil war raged until 1968; in 1970 all parties finally recognized the government of the Yemen Arab Republic. In the meantime, the combined effects of a radical anti-imperialist revolution in South Yemen and the 1967 Arab–Israeli war over the Suez Canal closed the Aden port to international commerce. Lastly, another Arab–Israeli war in 1973 precipitated unprecendented rises in the price of oil and a concomitant expansion in the economies of the Arabian petroleum-exporting nations. All of these events had an impact on Yemeni agricultural production.

The political revolution removed extra-economic constraints on commercialization. The republican government issued its own currency, the Yemeni riyal, to replace the old Austrian Maria Theresa thaler used in exchange under the Imams. A market for wage labour developed; in the early 1970s manual farm labour earned two to five riyals per day. Land also went on the market for the first time—first around the cities, and then in some rural areas.

Some old royalist landlords sold their holdings to the incipient urban bourgeoisie, and the Imam's holdings were taken over by the republican government. Import restrictions were lifted. The war and the presence of foreign troops increased demand for food from abroad. After the civil war the new government turned its attention to the creation of transport, communications, educational, and administrative facilities; international donor agencies pledged assistance in the construction of infrastructure and the reconstruction of the national economy. Agricultural experimentation and extension stations were established on prime state lands and some 'green revolution' technology was imported. Finally, the geometric expansion of Arab Gulf economies created seemingly unlimited opportunities for Yemeni wage labour and petty entrepreneurship. Within five years of the first rise in oil prices, a million and a half Yemeni men, out of a total population between six and seven million, spent some time working in Saudi Arabia and other Gulf states. Remittances peaked in about 1979, when over a billion US dollars entered the Yemen AR economy as unrequited transfers. Imports soared.

The contribution of agriculture to the national economy declined relative to other sectors. Initially this was more a function of rapid expansion in the tertiary sector, construction, transport, trade, finance, and services, than absolute declines in farm output. But there are signs also that the labour shortage, inflation, foreign imports, and competition from the tertiary sector have contributed to the stagnation of agriculture. Farming represented 70–90 per cent of employment, GNP, and state taxes before the revolution; 20 years later it provided about two-thirds of domestic employment, half of GDP, and only a tiny fraction of government revenues and overall capital investment.

The effects of the absence of a third of the male rural workforce on agriculture is difficult to assess. The foregoing discussion of traditional farm conditions suggests that certain levels of migration may simply absorb or counteract perennial male underemployment. The absence of young men is most likely to be felt during harvest and planting periods—when some migrants do return from abroad to work their fields. During the year, women in the household and one or more men remaining at home can maintain productive activities. Hence the direct impact of migration may not be detrimental.

Indirectly, however, migration undermines agriculture by fuelling inflation and rapid expansion in the off-farm sectors. Wage rates have increased astronomically. Agricultural day-labourers who were happy with five riyals per day in 1972/73 would not accept less than fifty (US $11) in 1978/79. Unskilled construction workers demanded twice as much. Relatively high wages in the farm sector discouraged the hiring of day-labour, while even higher earnings off-farm drew more labour from agriculture into the 'modern' sector. Similarly, the rapidly growing commercial and services sectors offered a much higher rate of return on investment than agriculture. Rapid inflation in import commodities and urban real estate encouraged speculation at the expense of productive investments in either industry or agriculture. Entrepreneurs shunned all labour-intensive enterprises.

The lion's share of remitted income financed consumption rather than investment in any sector. Foreign food, yardgoods, gadgets, vehicles, and televisions flooded the market, and most households spent their discretionary income on these commodities. The hypertrophy of a consumerist tertiary sector has indirectly contributed to agricultural stagnation. In many ways, the transformations of the past two decades have reinforced the time-honoured system of use-oriented production. This is contrary to the optimistic predictions of some development experts and strategies that remittances invested in agriculture would stimulate both import substitution and production for the export market. Since the majority of families still farm but also have access to some remitted cash, it was hoped that capital investments could offset the loss of labour. By converting to pump irrigation, utilizing modern technology, and producing high-yield crops, farmers could increase overall output. Rather, both households and the private sector have consistently under-invested in the still-predominant farm sector. To understand this trend, it is necessary to examine recent developments and investment opportunities in agriculture more closely.

Many mechanical and chemical inputs have limited applicability in the Yemeni environment. A significant share of highland cultivation takes place on mountainside terraces which are too narrow, weak, or inaccessible to be ploughed with a tractor. Mechanical pumping of groundwater is similarly limited by geophysical conditions. Two regions have surface or subterranean water resources which may be tapped for irrigation purposes. Pump irrigation is most prevalent in the lowlands, where shallow hand-dug wells have always provided drinking water. For the cost of a small pump and generator, farmers can draw groundwater to the fields. Using modern technology, deep wells are being drilled for irrigation in the central plateau as well, although costs for both drilling and pumping are considerably higher. In sections of these two flat regions it is possible to modernize agriculture by replacing the oxen-and-plough with tractors and by pumping groundwater. However, these innovations are technologically feasible on less than a quarter of the nation's farmland and actually employed on less than 5 per cent. Moreover, other innovations are limited by the same constraints. Fertilizer, for instance, burns the soil unless ample water is applied at the correct times, and is therefore useful mainly on pump-irrigated land. Likewise, most garden vegetables and imported 'miracle' grains require controlled irrigation.

Existing social relations of production may not favour innovations even where they are technologically feasible (Walker et al., 1983). A disincentive to the utilization of wells in some areas is the fragmentation of landholdings. Particularly in the central plateau region, the costs of installing a well and pump are only justified if a household works several contiguous hectares to be served by the irrigation system. Even in the Tihama, where average holdings are larger and less prone to be divided into several scattered parcels, the expense of a motor, pump, and maintenance may not make economic sense for smallholders. Rental and water sharing arrangements have only partially mitigated these constraints.

Nor have landlords taken a strong lead in the capitalization of agriculture. Share-cropper contracts which divide inputs and harvests between owner and producer are closely related to the system of grain and animal production for consumption. A major abrupt shift towards cash cropping implies repudiation of share-cropping contracts in favour of wage labour. Although tenant crop-shares have risen slightly in response to the labour shortage, landlords favour the share system because it maintains an equivalence between labour costs and total output. They fear higher costs for day-labour and also increased labour requirements for high-yield and garden variety crops. Tenants, on the other hand, may fear the loss of security inherent in share-cropping. Some landlords have invested in pump irrigation, and renegotiated tenant farmer contracts to reflect the costs of new inputs. If, for example, the crop was formerly divided equally between landlord and tenant, the renegotiated contract might divide the increased harvest in thirds; one-third to the share-cropping household, one-third to the land owner, and one-third to the pump owner (either the landlord or, less commonly, a third party). The complexities of crop-sharing are too intricate to be examined fully here; the point is that the introduction of a new source of irrigation does not necessarily entail a revolution in the social relations of production.

Even on pump-irrigated land, production remains oriented towards consumption. Grains account for half to two-thirds of hectarage. Alfalfa and potatoes often represent a significant fraction of the remainder. Sorghum, maize, barley, wheat, potatoes, and alfalfa have in common that their value does not depend on exchange; they may either be used or sold. Cereals and potatoes may be stored for family consumption. Grain stalks and fodder crops support livestock, which provide milk or eggs and sometimes meat. Fresh or processed dairy products and young livestock may be consumed or kept by the family, but also command high prices on the market. In short, production of these staples neither demands nor denies access to the market. Rather, they offer maximum flexibility in the decision of when and whether to sell. This may be contrasted with the market situation for most perishable garden vegetables, such as tomatoes, which must be sold immediately after harvest to protect a whole season's investment and which suffer from low prices during the harvest season. On irrigated land most farmers experiment with cash crops on a few fields, but grain and animal production are still primary.

Several factors, then, conspire to preserve the basic outlines of the old system of production for use. Drought-resistant crops are favoured by the majority of farmers still relying on rainfall or rain-determined spate irrigation. Additionally, the high cost and limited supply of agricultural day-labour discourages hiring of workers; instead, rational farm management maximizes utilization of imputed (unpaid family) labour or payment in kind (crop-shares). Labour-intensive crops and procedures are avoided. Kulakization is further inhibited by an uncertain domestic market for cash crops and competition from cheaper imported foodstuffs. Within the country, Iraqi dates, Canadian wheat, Ecuadorian bananas, Norwegian eggs, Chinese honey, Sudanese cotton, and Brazilian coffee are all less expensive than their unprotected Yemeni counterparts. Despite their inflated cost, native products are preferred by Yemeni consumers for special occasions; however, more and more households purchase imported food for daily use. Coffee, cotton, and date farmers are particularly disadvantaged.

The most lucrative cash crop during the past decade has been qat. The crop is attractive to highland farmers for several reasons. First, it benefits from but does not require artificial irrigation; unlike coffee, the qat tree goes dormant when deprived of water, but will revive if irrigated. Secondly, compared with other cash crops, qat demands far less labour. Finally, the market value for fresh qat has been consistently high, even during late summer when it is most plentiful. The profit margins are higher for qat than for any other product of Yemeni agriculture. The expanded cultivation and chewing of qat is further explained by other socio-economic factors. Before the revolution, qat-chewing was something of a luxury enjoyed primarily by wealthy mountaineers; only in qat-producing areas did ordinary families have an opportunity to chew regularly. Since the civil war, the national market has been integrated via state highways and rural tracks. Labourers, farmers, and shopkeepers can afford to spend cash on qat, which is now marketed daily throughout the country. Qat-chewing is an immensely popular form of relaxation, too widespread to be limited politically. Despite its popularity with both producers and consumers (and the fact that it is the only luxury consumable originating within the domestic economy), planners feel that it contributes little or nothing to national and agricultural development.

Apart from qat cultivation, relatively few farmers converted from production of subsistence crops to production expressly for the market during the 1970s. Indeed, modest increases in production of some fruits and vegetables may have been offset by declines in coffee, dates, and cotton. The strategy favoured by most farm families has been to supplement grain cultivation and animal husbandry with cash from off-farm activities. Women remain occupied full-time in agriculture while one or more men migrate abroad or find employment in the modern sector. The consumption pattern which results from this mixed strategy is optimal in the eyes of most rural people. Grain is stored for family consumption, a supply of fresh milk is assured, and cash is reserved for luxuries such as qat, televisions, and home improvements.

Yet the imperatives of farming have softened. The nation as a whole was less reliant on domestic agriculture during the oil boom than ever before. Now that the market supplies white wheat, vegetable ghee, and tinned milk, and given high cash earnings off-farm, households are no longer obliged to grow food in order to eat. Reduced reliance on the land for subsistence has been reflected in modifications of the agricultural regime (Swanson & Herbert, 1981). Some marginal lands, especially terraces on high steep slopes, where the ratio of maintenance effort to crop production is very low, have gone out of production. Some of the painstaking labour-intensive steps in sorghum cultivation have been eliminated. Since the market offers alternative sources of income and food, farmers are less inclined to seek the last increment of marginal utility. As a consequence, there has been some erosion in overall productivity.

NATIONAL STRATEGIES FOR AGRICULTURAL DEVELOPMENT

The agricultural policies of the Mutawakkilite Imams were a conservative strategy designed mainly to preserve the social relations of production and surplus appropriation by the ruling classes. No effort was ever made to increase overall or specific commodity production, although the state relied heavily on agricultural *zakat*. In the first ten years of the republic the government likewise pursued a non-strategy with respect to agricultural development. Recent administrations have been concerned with productivity in the farm sector and its contribution to the national product, but have only limited policy instruments at their disposal to influence agricultural investments. Analysis of the national development plans shows a philosophical commitment to private sector agricultural innovation but lagging progress in the farm sector.

The first National Development Plan, covering the three years 1974–76, legitimately and necessarily emphasized investments in basic infrastructure, communications, health, education, and utilities, but also gave an important role to agriculture, which was scheduled to receive 15 per cent of total development spending over the three years. Only transport and communications and education were to claim higher shares of investment than agriculture. Investment records reviewed at the end of the plan period, however, show that agriculture took a back seat not only to communications and education but also to the industry and power and the social services sectors. Agriculture attracted about 8 per cent of actual investment, mostly from the private sector. The rate of growth in the agricultural sector averaged 5 per cent per annum, considerably lower than projected. But it was noted that growth in this sector 'is in fact clearly related to the rainfall seasons, thus the rate of growth in the first year declined by − 10.8 per cent from the previous year, and in the second year it compensated its previous decline by having a growth rate of 28.4 per cent on the first year; while in the third year it declined from the second year by about 7.4 per cent' (sic: CPO, undated). Growth and decline indicated not investment decisions but the effects of fluctuations in rainfall.

Architects of the Five-Year Plan inaugurated in 1976 viewed these developments with some alarm. All analyses and speeches underlined the critical need for real increases in farm productivity. The Second National Plan did not revise prescriptions for agricultural development but admonished public agencies to redouble efforts to see that targets were achieved. The plan targeted a real growth rate of almost 5 per cent for the agricultural sector which was scheduled for 14 per cent of capital expenditure. Planners estimated that this could be achieved through private investment chiefly in irrigation technology to reduce reliance on highly variable annual rainfall. Vegetables, potatoes, wheat, cotton, fruit, and poultry production were targeted for high rates of expansion, with modest increases for all other crops (except qat). To encourage capital formation in the agricultural sector, the previously inoperative Agricultural Credit Bank would offer low-interest credits to individual farmers and agricultural cooperatives. The Tihama Development Authority, the Southern Uplands Rural Development Project, and other experimentation and extension programmes designed by multilateral technical assistance would show the way to modern farming.

Again the analysis at the end of the plan period was sobering, as the previous discussion has indicated. Engorged by remittances during the five-year period, the private sector surpassed all targets in the tertiary sector; but private investors continued to neglect the farm sector. Agriculture claimed 8 per cent of total investment between 1976 and 1981, much of this, clearly, either foreign project development of state lands or private investments in qat production. Agriculture, which accounted for over three-quarters of GDP when estimates were first made towards the end of the civil war, now provided about 45 per cent. Imports of food and live animals doubled between 1977/78 and 1979/80, and the total value of imports was thousands of times greater than exports.

A number of reasons for the failure of farm practice to conform to planned cash cropping objectives have already been identified. Cash from abroad enabled families to continue subsistence production while consuming imports: to have their cake and eat it too. Under the peculiar conditions which obtained in the mid-to-late 1970s, there were a number of disincentives to private investment in agriculture. In addition, national strategies for agricultural modernization failed to address the real conditions of most farmers. Experimental stations employed a system of heavy water, mechanical, chemical, and labour inputs on prime land, all practices principally suited to irrigated plantations. Even Western agricultural experts admitted they had little to teach Yemeni farmers on rain-fed land. Farmer interest in agricultural cooperatives for purchasing and marketing was limited to a few areas where politicians seized the idea. Some credits were extended to farmers for pumps or other innovations, but few illiterate farmers understood or trusted bank loans.

While imports continued to climb, remittances had levelled off. Reversal of the trend in world oil prices contracted job markets in the Gulf petro-economies.

Some migrants returned to Yemen and fewer young men went abroad. In 1981 the widening gap between production and consumption contributed to a balance of payments deficit estimated at $350 million. The IMF dismally predicted continuation of these trends (Harvey, 1982). In unveiling the 1981–86 Plan, the government called for greater self-sufficiency in food, less consumerism, and more productive capital formation.

The new plan stresses that agriculture remains the primary sector in the economy, but that its development is a private-sector responsibility. Farm productivity is again optimistically targeted for a growth rate of nearly 5 per cent or 16 per cent of fixed capital formation. Despite the lobbying efforts of some coffee and date producers, no protective tariffs which might offend international benefactors have been seriously contemplated. The public sector will continue to concentrate on research and extension. An enlarged role is indicated for farmer's cooperative societies in addressing the needs of producers within particular micro-environments, and in fostering capital formation for productive investments. The Agricultural Credit Bank has been merged with the Cooperative Bank to rationalize credit policies while encouraging farm modernization.

There are some prospects for a revitalization of agriculture in the 1980s. Some of the economic constraints on productive investment have been partially alleviated. The declining frequency of out-migration and the Yemen AR's admission of several hundred thousand labour immigrants and political refugees (from East Africa, South Yemen, and Vietnam) have increased the ranks of the domestic workforce. At the same time, the slowdown in inflation throughout the peninsula has contributed to stabilization of Yemeni wage rates and commodity and real estate prices. These trends suggest a levelling off speculative investment in the tertiary sector. Furthermore, households deprived of their external source of cash may be more motivated to maximize farm production, and more judicious in their consumption. These new market factors could favour a greater commitment of labour and financial resources to agriculture and agro-industry.

The prognosis for the future of Yemeni agriculture is mixed. Tendencies already apparent within the system could suggest decline, maintenance, or growth in the farm sector. In conclusion, therefore, three different hypothetical scenarios are offered. The first is agricultural involution through steady deterioration of the existing productive system. If the reduction in the agricultural labour force and diminishing returns on coffee, dates, and cotton continue to exceed rates of new investment in farm production, then stagnation will occur. Traditional production will continue, but in a diminished, distorted, and less efficient form. The agricultural sector will contribute less and less to national product and domestic consumption. The outcome of this underdevelopment in agriculture would be deepening dependency on imports, and, ultimately, international indebtedness.

A less dismal prognosis is based on the rejuvenation of agriculture through appropriate intensifications of existing cultivation, storage, and distribution techniques. Farm production in this instance is oriented towards consumption at the household level and import-substitution at the national level: in short, food self-sufficency. This implies that the farm sector not only feeds itself but produces a surplus for urban consumption. Households produce both for storage and for the domestic market; grain and livestock production remain primary, but there is some expansion of cash cropping on irrigated land. This could happen if the gap between earnings off-farm and agricultural incomes narrows in response to changing labour and commodity market conditions. If private and cooperative investments in agriculture are higher than in the previous two plan periods but still fail to meet planned targets, then the recent declines in the contribution of agriculture to GDP may be reversed. Net growth, however, would require high levels of investment.

The most optimistic scenario envisages capitalist production for the international market. Production for export requires well-considered initial capital outlays in technology to increase production. Remittances still represent an important, though widely dispersed, source of capital. The problem remains to mobilize this capital for appropriate agricultural investment. Changes in administration of credit facilities and agricultural cooperatives, and increased farmer appreciation for these institutions, could raise investment to levels indicated in the plan. The very considerable recent improvements in rural roads and market access offer an additional incentive to cash cropping. Finally, there exists a very proximate and potentially lucrative market for Yemeni grain, livestock, and/or fresh produce in the Arab Gulf cities. The potential for the emergence of agrarian capitalism is thus perhaps more auspicious than in many peripheral economies.

REFERENCES

Carapico, S., & Tutwiler, R. (1981) *Yemeni Agriculture and Economic Change*, Yemen Development Series No. 1, American Institute for Yemeni Studies, Sana'a.

Central Bank of Yemen (1980) *Financial Statistical Bulletin*, July–September.

CPO (Central Planning Organization) (undated) *The Yemen Arab Republic Five-Year Development Programme 1976–81*, Sana'a.

Harvey, N. (1982) 'Development demands more investment and less consumerism', *Middle East Economic Digest*, 8.1.82, London.

Messick, B. M. (1978) *Transactions in Ibb: Economy and Society in a Yemeni Highland Town*, unpublished Ph.D. dissertation, Princeton University.

Al-Mujahid, A. (1978) *Al-Ta'awun Al-Zara'i Madkhal Lil-Tanmiyya fi Al-Yaman*, Kitab al-Ghad, 5, Sana'a.

Swanson, J., & Herbert M. H. (1981) *Rural Society and Participatory Development: Case Studies of Two Villages in the Yemen Arab Republic*, Cornell University Rural Development Committee, Yemen Research Program, Ithaca, N.Y.
Walker, C. T., Carapico, S., & Cohen, J. M. (1983) *Emerging Rural Patterns in the Yemen Arab Republic: Results of a 21-Community Cross-Sectional Study*, Cornell University Rural Development Committee, Yemen Research Program, Ithaca, N.Y.

Agricultural Development in the Middle East
Edited by P. Beaumont and K. McLachlan
© 1985 John Wiley & Sons Ltd

Chapter 14

Agricultural Development in Syria

IAN R. MANNERS
and
TAGI SAGAFI-NEJAD

INTRODUCTION

One of the characteristic features of Middle Eastern society has been the persistence of agricultural patterns and practices. Firmly rooted in the physical environment, these are the outcome of centuries of ecological experimentation and adaption and represent the best efforts of Middle Eastern farmers to manage the resources of a semi-arid environment (Manners, 1980). In Syria, as elsewhere in the Fertile Crescent, the predominant means of livelihood outside urban centres has been the cultivation of cereals wherever winter rainfall and terrain provided the opportunity for arable farming (Weulersse, 1946). Yields in these dry-farmed lands vary markedly from year to year and there is always a risk of crop failure as a result of drought. Irrigated production of fruit and vegetables based on the withdrawal of groundwater or the diversion of perennial streams has provided a more stable source of agricultural production. Yet even irrigated farming may be subject to the vagaries of the rainfall regime, since most traditional diversion systems, lacking storage facilities, are subject to marked fluctuations in supply. Moreover, the seasonal distribution of streamflow, concentrated as it is into the winter and spring months of least irrigation need, is ill-suited to the needs of the farmer. In this difficult environment, complex social and economic networks have evolved, linking the farmer with the nomadic pastoralist and the urban merchant in an urban trilogy that has only recently begun to disintegrate under the pressures of nationalism, population growth, and technological change (English, 1967, 1973).

For all its durability, Middle Eastern agriculture has been affected by extremely low levels of productivity, whether measured in terms of actual yields or per capita output. This is intricately related to physical constraints, notably the variability of winter rainfall, to cropping practices that emphasize labour inputs over energy inputs, and to tenurial systems that discourage innovation. Overcoming these constraints on agricultural performance represents an unprecedented challenge. In this context the efforts of the Syrian government parallel those of other Middle Eastern countries seeking to modernize the agricultural sector and achieve a rapid increase in agricultural output. These efforts are occurring against a background of technological improvements that offer the opportunity for more intensive use of land and water resources, of population growth and rising per capita incomes that mean greatly increased food requirements, and access to regional and global markets that provide expanded opportunities for commercial farming. While the obstacles to agricultural improvements are deeply rooted in the physical environment and in the social and political fabric of the state, the conventional wisdom has been that these obstacles can be overcome only through direct government involvement in agricultural planning, water development, technology transfer, and agrarian reform.

How successful have government efforts been in addressing and resolving those underlying problems that are usually identified as limiting factors on Syria's agricultural development? Although information on Syrian agriculture is frequently sketchy, this chapter will attempt to answer this question by (1) identifying major trends in land and water use over the past quarter century, and (2) evaluating those policies and programmes designed to remove or moderate physical and social constraints on agricultural development. In the process, it will be suggested that the Syrian experience raises a broader set of issues that are relevant to agricultural development throughout the region. Such issues include the appropriate level of investment in water development, the relative benefits to be derived from vertical as opposed to horizontal expansion of agriculture, whether emphasis should be placed on diverse, small-scale improvements to existing production systems as opposed to single, large projects, and the priority that should be given to achieving food self-sufficiency. In general, there has been a tendency to think that a few dramatic changes in technology—a large dam or a new grain variety—will radically transform agricultural performance. Yet experience suggests the need for careful appraisal of alternative investment opportunities and for greater attention to the social and ecological impacts of development decisions. Nowhere is such an appraisal more important than in the arid lands of the Middle East for, as numerous writers have emphasized, these are lands of risk and uncertainty where ecological processes are easily disrupted and where engineered solutions may consequently prove counter-productive (Dregne, 1970; Hills, 1966; Walton, 1969; White, 1966). An underlying concern must therefore be whether recent gains in agricultural productivity can be sustained over the longer term without degradation of the resource base.

THE AGRICULTURAL ECONOMY OF SYRIA

A useful point of departure in any assessment of the performance of Syrian agriculture over the past quarter century is the World Bank mission to Syria in the early 1950s (IBRD, 1955). The mission's task was to prepare a development plan for the government's consideration and its final report provides some insight into the structure of the Syrian economy at a time when official statistics are unreliable. The mission's best judgement was that between 65 and 75 per cent of the population derived their living from agriculture and that agriculture and animal husbandry together accounted for 45 to 50 per cent of national income. Nearly 90 per cent of the cultivated area was dry-farmed which, as the mission noted, meant sharp fluctuations in cereal output and farm income. Nevertheless, the two decades preceding the mission's visit had been characterized by rapid economic growth based primarily on the performance of the agricultural sector. 'The availability of idle land (in the Jezira) combined with food shortages attributable to the war provided a tremendous incentive to an expansion of agricultural output' (IBRD, 1955, p.18). In sum, the agricultural sector was satisfying a high proportion of the country's food requirements and contributing to foreign exchange earnings as a result of the export of cotton and cereals. Indeed, despite the growing importance of irrigated cotton, cereals still accounted for around 20 per cent of the total value of Syria's exports.

The period since the World Bank mission has seen structural changes in the Syrian economy similar to those experienced in other developing countries. In particular, the relative contribution of agriculture to the GDP had fallen to around 20 per cent by the late 1970s, although the total value added by the agricultural sector (measured in constant prices) grew at an annual average rate of 3.5 per cent between 1963 and 1978 (USAID, 1980). Similar changes have occurred in the structure of the labour force, recent estimates by the World Bank suggesting that only just over one-third of the labour force is now employed in agricultural production (World Bank, 1982). Arguably, agriculture's primary role is to contribute to the food requirements of the population. In Syria, despite a continuing emphasis on cereal cultivation and increases in total food production, per capita production has lagged in the face of rapid population growth. As a result the value of imported foodstuffs rose at an annual average rate of around 20 per cent during the 1970s in order to meet the country's expanded requirements (USAID, 1980).

Throughout this period the Syrian government has pursued several major goals for the agricultural sector. These are framed within the context of such broader national development goals as sustained economic growth, increased national self-sufficiency, full employment, and greater social equity and economic well-being. Specific agricultural goals relevant to this chapter include (1) sustained and rapid increases in agricultural production (defined, for example, in the Fourth Development Plan for 1976–1980 as a real annual average growth rate of 8–10 per cent) and (2) achieving greater self-sufficiency in major agricultural commodities. Other goals, rephrased and redefined in successive development plans, include provision of raw materials for industrial processing, raising the nutritional standards of the population, and increasing agricultural exports to reduce trade balance deficits (USAID, 1980). Against this background we now turn to the specific policies and programmes pursued by the government in order to achieve its development goals.

EXPANSION OF IRRIGATION

Expansion of the irrigated area to reduce dependence upon rain-fed cultivation has been the stated goal of successive Syrian governments. As the World Bank report emphasized, 'irrigation is the principal means for expanding the cultivated area, increasing and stabilizing yields and diversifying agricultural production by developing summer crop cultivation' (IBRD, 1955, p. 41). In these circumstances development of the waters of the Euphrates basin came to be perceived by many as the panacea for Syria's agricultural problems.

The immediate post-war period had seen a considerable increase in irrigated farming (Figure 1). Most of this expansion resulted from improvements in existing diversion systems through lining of distribution canals to reduce seepage and from the introduction of advanced pumping technologies (Hudson, 1968).

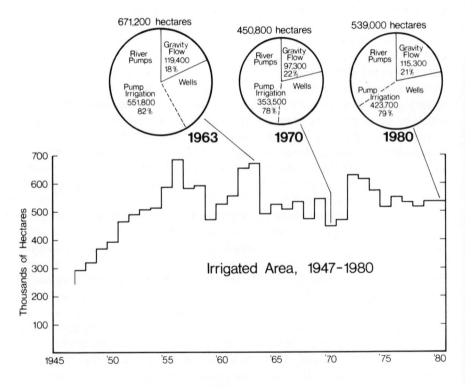

Figure 14.1   Irrigated area, 1947–1980

Table 14.1   Syria: irrigated area by source of water, 1955–80 (hectares)

1955[a]		1963[b]	1980[b]
	Pump irrigation		
	River pumps	393,700	186,000
	Groundwater wells	158,100	237,100
250,000		551,800	423,700
	Gravity flow		
	Water wheels	8,100	1,200
	Diversion from		
	rivers, springs	111,300	114,100
150,000		119,400	115,300
	Total		
400,000[c]		671,200	539,000

[a]Estimated by the International Bank for Reconstruction and Development.
[b]Syrian Arab Republic, 1982.
[c]This figure reflected the World Bank opinion that the government's estimate of 375,000 hectares was too conservative. Both estimates differ from that of the Food and Agriculture Organization which reported that the irrigated acreage in the early 1950s exceeded 500,000 hectares.

Perhaps the greatest change during this period occurred along the Euphrates where the use of river pumps enabled large areas of land to be brought under irrigation for the first time. At the time of the World Bank mission, the irrigated area in Syria probably amounted to around 400,000 hectares. Of this area approximately 250,000 hectares were believed to be irrigated by pumping and the balance by gravity flow from rivers, springs, and underground *qanats* (Table 14.1). Only 48,000 hectares were covered by government irrigation schemes and in many instances these projects were receiving an inadequate supply of water.

The post-war expansion of irrigation was characterized by small-scale, privately financed development. However, the World Bank concluded that there were only limited opportunities for further expansion of irrigation in the private sector. Installation of new pumps seemed to be declining and this was attributed to the higher lifts involved in bringing additional land under irrigation in the Euphrates basin and to the poor quality of groundwater resources. In these circumstances there was some consensus that future agricultural expansion was dependent upon more efficient use of surface water resources. The capital investment requirements of such large-scale storage and distribution networks clearly implied much greater public sector involvement in irrigation development.

In its final report the World Bank recommended a number of projects for detailed study. Together with projects already under construction, the mission felt that an additional 90,000 hectares could be added to the irrigated area of Syria during the following decade. In making its recommendations, however, the mission emphasized the need for careful planning to determine soil quality

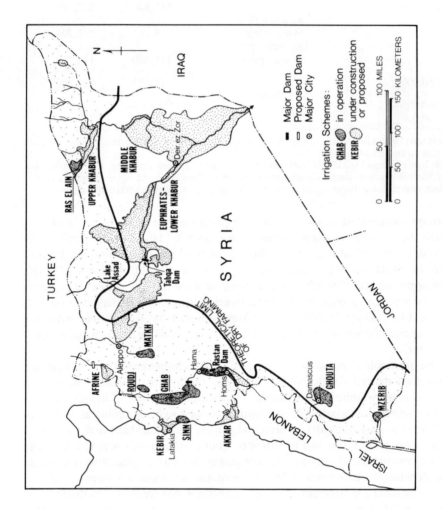

Figure 14.2   Syria: major irrigation projects

and water duty requirements within the area to be irrigated. The lack of such preparatory work led the mission to recommend deferring construction of the proposed Youssef Pasha dam on the Euphrates (IBRD, 1955, p. 201).

Several of the projects recommended in the World Bank report were completed within the next 15 years and by 1970 around 140,000 hectares were reportedly irrigated by government schemes (Figure 14.2). Of particular importance was the Ghab reclamation project which transformed an area of marsh and swampland in the Orontes valley into some of the most productive farmland in Syria. With the completion of the Ghab project in 1968, attention turned once again to making fuller use of the Euphrates and its tributary the Khabur. Although the Euphrates accounts for around 85 per cent of the nation's surface water resources, in the 1960s probably less than one-fifth of the country's irrigated area was located within its basin. The subject of repeated engineering studies, financial and technical assistance was finally secured from the USSR for a large-scale, multi-purpose project involving hydro-electric power generation and irrigation of an area officially estimated at 640,000 hectares.

The central element in the Euphrates project is the Tabqa Dam, a large earth-fill structure, sixty metres high and four-and-a-half kilometres long. Begun in 1968, the dam was finally completed in 1973 at which time Lake Assad began to fill, in the process inundating some 25,000 hectares of irrigated cotton land (Pitcher, 1974). Ultimate storage capacity of Lake Assad is 11.6 billion cubic metres, of which 7.4 billion will be live storage. With completion of the dam and power portions of the project much greater emphasis was placed on the associated irrigation and drainage programmes. How successful these programs have been remains unclear. There are, however, growing indications that both the lead time required for land development and the technical problems of irrigation agriculture in the Euphrates basin have been seriously underestimated by the government. The initial goal was to complete the development of all irrigable land within a 20-year period. To this end the Fourth Five Year Plan for the period 1976–80 targeted 240,000 hectares of land to be prepared for irrigation. This goal proved unattainable. By 1978 only 7,400 hectares of land had actually been prepared, and the 1980 target for land preparation had been revised downwards to 43,200 hectares (USAID, 1980, Vol. 1, p. I–31). Despite this experience, the Fifth Five Year Plan (covering the period 1981–85) has as one of its goals the development of 188,000 hectares of irrigated land in the Euphrates basin, a goal that seems unrealistic in view of progress to date.

In large part the delay in implementing proposed irrigation projects appears to be due to inadequate planning, leading to unexpected problems with gypsum subsoils in the project area. Particularly disturbing is the evidence that the area of land suitable for irrigated crop production may be substantially less than originally anticipated. Land classification data are presently available for only about 400,000 hectares, but examination of these data 'suggests that less than half of the 640,000 hectares is reasonably good land for irrigation purposes'

Table 14.2   Syria: proposed irrigation projects, 1980

Basin/project	Storage capacity (million m³)	Irrigated area (hectares)	Description
Orontes			
Kremish Dam	275	25,000	Reservoir upstream from Hama to provide water supply for remaining irrigable land in the Ghab.
Afrin River Dam	230	20,000	Dam on a northern tributary of Orontes to irrigate areas outside Ghab.
Ghab redevelopment	—	15,000	Reclamation of land currently flooded or poorly drained.
Khābūr			
Upper Khābūr	600	35,000	Storage reservoirs on the
Middle Khābūr	320	19,000	Upper Khābūr; initial feasibility studies revealed some gypsum problems in area to be served by lower dam.
Kabir			
Shamali Dam	214	14,000	Dam on Al Kabir Al Shamali to irrigate land south of Lattakia.
Janubi Dam	150	10,000	Proposed.
Akkar Plain			
Arouse River Dam	77–89	23,000	Integrated development of surface and groundwater resources.

Source: USAID, 1980.

(USAID, 1980, Vol. 2, p. II–4). Clearly any major revision of the original estimate of irrigable land could have a significant impact on the economic viability of the entire project.

The experience with the Euphrates programme poses something of a dilemma for the government with respect to future water development policies. At the present time detailed feasibility and design studies have been completed or are under way for an additional series of storage reservoirs that would bring an estimated 161,000 hectares of land under irrigation (Table 14.2). Yet the high level of public investment in large-scale water development projects in recent years has so far produced rather meagre results. Indeed, while estimates of the irrigated area are unreliable in view of inaccuracies and inconsistencies in reporting procedures, the government's own figures suggest that there has actually been a net loss of irrigated land since the early 1960s (Figure 14.1). This decline is variously attributed to inundation, waterlogging, and salinization, and to the impact of a land reform programme that set a ceiling on the ownership of irrigated land. Regardless of the particular explanation, it is something of an indictment of water development policies that in the quarter century since the World Bank mission, the total irrigated area of Syria has been expanded by no more than 150,000 hectares (Table 14.1).

In these circumstances it is perhaps timely to question whether the government's preoccupation with large-scale water development schemes, particularly the Euphrates project, has not led to a neglect of other needs and opportunities that might offer a better return on investment. Little attention has been given, for example, to alternative approaches towards increasing the effective water supply, notably through improvements in the efficiency of water use. Specific measures to achieve such an improvement would include the renovation and lining of older canals, the reorganization of distribution networks, and, perhaps most important of all, the upgrading of field application techniques. That such measures are urgently needed is evident from studies that show that 54 per cent of the water diverted into the Homs-Hama canal is lost before it reaches the farmer (USAID, 1980, Vol. 2, p. II–24). Perhaps the most serious problems involve groundwater development. Thus, contrary to the World Bank conclusion that private pump irrigation had reached the limit of its expansion, the number of pumps used for agricultural purposes has continued to increase, rising from 20,990 in 1963 to 47,206 in 1980 (Table 14.9). As a result the area irrigated from groundwater wells has increased quite markedly and in 1980 accounted for 44 per cent of the total irrigated area (Table 14.1). Yet information on actual rates of groundwater withdrawal and recharge is sketchy and incomplete. In theory all pump operators are required to obtain a licence, but monitoring and enforcement measures to ensure compliance with the terms of the licence are perfunctory. A serious overdraft situation now exists in several parts of the country, notably in the vicinity of Damascus and Aleppo and in the Orontes basin, yet pumping licences continue to be issued for these areas in large numbers. As a USAID study concluded, '. . . information is badly needed as a basis for groundwater resource planning . . . with improved information, a much needed integration of surface and groundwater supplies and uses could be at least partially achieved' (USAID, 1980, Vol. 2, p. II–22).

In sum, performance in the development of the country's water resources has not matched expectations. A careful re-evaluation of government priorities with respect to the future direction of water resource development appears essential. In any such re-evaluation, consideration should be given to assigning higher priority to strategies that emphasize better water management rather than large-scale development.

EXPANSION OF CULTIVATION

Perhaps the most striking development in Syrian agriculture over the past quarter century has been the net expansion of the cultivated area. By the end of the 1970s the Ministry of Agriculture and Agrarian Reform (MAAR) was reporting a cultivated area of around 5.7 million hectares, a considerable increase over the 3.5 million hectares reported by the World Bank. Such an expansion was certainly not anticipated by the mission, which 'gained the impression . . . that the amount of land still available for cultivation without irrigation is extremely limited' (IBRD, 1955, p. 36).

From a historical perspective this expansion of the cultivated area represents the culmination of a process of agricultural settlement and colonization begun during the nineteenth century (Lewis, 1955). By the 1940s the agricultural frontier had reached the Jezira, an 'island' of slightly higher elevation in northeastern Syria where winter rainfall is sufficient for dry farming. The strong regional demand for cereals during and immediately after the Second World War provided a stimulus for large-scale development, with private investment capital playing a key role. Urban merchants and entrepreneurs were quick to lease uncultivated land from tribal shaikhs and to introduce mechanized methods of cultivation that by bringing extensive areas of rain-fed land into production promised a rapid return on investment. By the 1960s, the Jezira had become the country's granary with 1.4 million hectares under cultivation (Beaumont et al., 1976, p. 354). Yet the speculative nature of agricultural operations in the Jezira was already a matter of concern at the time of the World Bank mission, which commented on the apparent failure of many urban entrepreneurs to adequately consider the long-term productivity of the land then being brought into production. In the mission's judgement, much of the land that was being ploughed under in the Jezira was too marginal for rain-fed cultivation (IBRD, 1955, p. 297). Despite these reservations, development in the Jezira expanded the cultivated area in Syria to a peak of around 6.5 million hectares in the early 1960s, since when there appears to have been a fairly steady decline (Table 14.3).

The 'loss' of nearly a million hectares of cultivated land since the 1960s is open to a variety of interpretations. It is possible that this represents a regional pull-back from more marginal lands because of declining yields, a trend that became more pronounced during the drought years of the late 1950s (Beaumont et al., 1976). An alternative explanation may lie in the difficulty of defining cultivated land in an area of limited and unreliable rainfall. In this regard, it is not unusual to see land officially classified as uncultivated brought into production in years of exceptionally good rainfall. Moreover, in 1974 the Supreme Agricultural Council of Syria adopted a revised set of land use definitions that restricted the term fallow to lands already prepared for cropping or lands in rotation and uncultivated for a period not longer than two years. Under this revised definition, land left uncultivated for more than two

Table 14.3 Syria: land use ('000 ha)

	1963–66	1974	1975ᵃ	1976	1977	1978	1979	1980
Forests	465	446	445	457	452	455	459	466
Steppe and pasture	5,803	6,393	8,631	8,459	8,535	8,413	8,274	8,378
Uncultivable	3,470	3,627	3,487	3,630	3,671	3,702	3,727	3,520
Cultivable	8,780	8,052	5,955	5,882	5,863	5,948	6,058	6,154
Uncultivated	2,261	2,025	479	338	355	360	372	470
Cultivated	6,519	6,027	5,476	5,544	5,509	5,588	5,686	5,684
Fallow	3,143	2,493	1,776	1,295	1,642	1,597	1,847	1,791
Rain-fed	2,826	2,956	3,184	3,702	3,336	3,472	3,300	3,354
Irrigated	550	578	516	547	531	519	539	539
Total	18,518	18,518	18,518	18,518	18,518	18,518	18,518	18,518

Source: Compiled from Syrian Arab Republic, *Statistical Abstracts*.
ᵃAdoption of new land use classifications.

agricultural seasons, a common conservation practice in the drier interior of Syria, is no longer classified as fallow. Thus the loss of cultivated land since the mid-1960s may be partly a statistical aberration, earlier estimates having included, as fallow, land left uncultivated for more than two years.

As the preceding discussion illustrates, a persistent difficulty in assessing Syria's agricultural potential has been uncertainty over the area of land actually suitable for cultivation. At the present time, there appears to be a consensus that the cultivated area of around 5.5 million hectares represents the maximum that can be brought into sustained production in the absence of further irrigation development. Yet until quite recently much higher estimates of the arable area were widely circulated and officially endorsed (Table 14.3). Throughout the 1960s, the Central Bureau of Statistics' *Statistical Abstract* reported a cultivable area in excess of 8.7 million hectares, of which as much as 2.9 million hectares was described as uncultivated. Yet this optimistic view received little support in other independent studies. Indeed, the World Bank mission registered its dismay over the widespread belief, reinforced by official estimates, 'that agricultural land resources were virtually limitless' (IBRD, 1955, p. 36).

Uncertainty over exactly how much of Syria's total land area of 18 million hectares is suitable for cultivation is the result of (i) inadequate data on land capability; (ii) inconsistencies in criteria used to define arable land; and (iii) variations in rainfall. Early estimates of the cultivable area included some land with apparent irrigation potential in areas of meagre rainfall without any consideration as to whether or not an irrigation supply could actually be made available. More critically perhaps, the extreme variability of the winter rainfall makes it virtually impossible to define with any degree of precision that area wherein dry farming can be practised on a regular basis.

Of particular interest in this regard are the findings of a recent USAID study undertaken to assist the Syrian government in defining agricultural priorities and goals for the 1981–85 Five Year Development Plan. Using Landsat imagery supplemented by detailed field work, the study provides the most comprehensive estimate to date of land use potential in Syria. By comparing existing land use with estimated capabilities, the study team was able to identify (1) land which can safely be retained in or brought into cultivation, (2) cultivated land which should be shifted to less intensive use, such as range, and (3) land with irrigated crop production potential (USAID, 1980, Vol. 1, p. II–4). The purpose of this comparison was to identify adjustments in land use that would enhance long-term agricultural productivity.

The results of the USAID assessment, summarized in Table 14.4, confirm that a critical feature of contemporary Syrian agriculture is the amount of marginal land that has been brought under the plough. Of the 5.2 million hectares currently devoted to rain-fed crops, fallow land, and unirrigated orchards (defined as extensive agriculture), about *1.5 million hectares are unsuited to their present use*. Most of the land unsuitable for extensive

Table 14.4   Syria: land use suitability (ha)

Land use category	Current land use	Land use suitability	
1. *Extensive* (includes all rain-fed cultivation, fallow land,	5,211,000	High Medium Low	253,000 1,948,000 1,853,000
and unirrigated orchards)			4,054,000[a]
2. *Intensive* (includes all land		High Medium	2,919,000 3,293,000
receiving full or supplementary irrigation	683,000		6,212,000
3. *Other* (includes land under range, pasture, forest)	12,825,000		8,453,000
	18,719,000		18,719,000

Source: USAID, 1980, Vol. 1, Table II.1.
[a] This net figure is arrived at as follows: currently used for extensive agriculture (5.2 million hectares) less land considered unsuitable for this use (1.5 million hectares), plus presently uncultivated land suitable for this use (0.3 million hectares).

agricultural use is found in the interior steppe where rainfall is limited and variable, but the report also identified erosion-prone land with slopes in excess of 15 per cent under cultivation in the more mountainous regions of western Syria. The study recommends converting this land 'as rapidly as possible' to less extensive, more appropriate uses — range or pasture in the interior and reafforestation on the better-watered mountain slopes. An additional 1.8 million hectares were considered to have a low potential for rain-fed cultivation where 'over time some producers may find the low and irregular yields and income to be derived from these areas to be unacceptable' (USAID, 1980, Vol. 1, p. II–53). While the study recommends retaining this land in its present use, it felt little could be done to increase or stabilize yields. 'With the greater than average increase in demand for livestock products, the prices of these products may be expected to increase relative to prices for crops now produced on this marginal land. At that time government programmes to assist in converting such lands to range uses would be desirable' (USAID, 1980, Vol. 1 p. II–53). In effect, of the 5.2 million hectares currently dry-farmed, only just over 2.0 million hectares were judged to have reasonably good potential for rain-fed crop production. Moreover, in inventorying uncultivated lands the study was only able to identify an additional 296,000 hectares scattered throughout Syria that could be converted to rain-fed crop production.

The implication of the USAID analysis are striking. If the study's recommendations with regard to land now extensively cultivated were to be implemented — in effect restricting dry-farming to those lands considered suitable for such activity — there would be a net reduction of 1.2 million hectares in the area dry-farmed. From a historical perspective, the overall assessment of the USAID team, that around 4 million hectares is suitable for extensive cultivation, tends to confirm the more intuitive observations of the World Bank mission that the viable limits of dry-farming had already been reached a quarter of a century ago.

LAND USE

The experience of the past quarter century suggests that the limits of viable rain-fed cultivation have been exceeded and that the expansion of Syria's irrigated area is likely to be slower and more costly than originally envisaged. In these circumstances, increases in agricultural output will be dependent upon raising productivity through improved cropping practices.

One of the notable trends in Syrian agriculture over the past decade has been an increase in the area actually cropped each year (Figure 14.3). In the late 1970s the cropped area was just under 4 million hectares, exceeding the cropped area reported during preceding decades by nearly 1 million hectares. More intensive use of the land is being achieved through an increase in multiple cropping on irrigated land and a reduction of the fallow period on rain-fed land. Cultivation of an irrigated winter wheat crop, usually high-yielding Mexican varieties, has

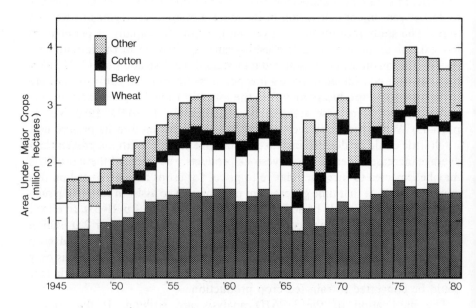

Figure 14.3   Syria: cropped area 1945–1980

increased to the point where one quarter of the country's wheat crop is produced under irrigation. However, by far the greatest change has occurred on unirrigated land where land that traditionally remained fallow for at least one year has been brought into production. According to the USAID analysis of land use trends in Syria, nearly two-thirds of the increase in the cropped area between 1968 and 1978 can be attributed to expansion of the area under non-irrigated winter crops. As a result, the rain-fed crop acreage actually rose from 45 per cent to 62 per cent of the total cropped acreage (USAID, 1980, Vol. 1, p. III–1).

Clearly, changes in fallowing practices may be highly desirable. In other arid zones, however, the trend has been towards replacing cereals with a form of mixed farming that includes a leguminous crop in the rotation cycle to provide forage for livestock and to maintain soil fertility. In Syria, however, the cereal monoculture on rain-fed land has been intensified with the land diverted from fallow being cultivated with cereals, particularly barley.

Table 14.5 Syria: cropped area, 1940–80 ('000 ha)

Period/ average	Wheat	Barley	Cotton	Fruit	Total	Cereals as a % of total
1946–1950	884	368	33	184	1,821	76%
1951–1960	1,392	602	220	217	2,784	75%
1961–1970	1,249	707	275	259	2,862	70%
1971–1980	1,490	919	194	380	3,515	70%

Despite more intensive use of the land, there has been little change in the underlying structure of Syrian agriculture. Cereals continue to occupy by far the largest acreage, a steady 70 per cent of the cropped area throughout the 1960s and 1970s (Table 14.5). Dry-farming remains the dominant mode of production with all that this implies in terms of uncertainty and instability. Despite the trend towards cultivation of a winter wheat crop on irrigated land, 88 per cent of the wheat acreage and 99 per cent of the barley acreage is still entirely dependent upon rainfall (USAID, 1980, Vol. 2, p. II–60). As a result cereal production continues to fluctuate quite markedly, although such fluctuations are now occurring around a norm somewhat higher than in the 1950s and 1960s owing to the increased contribution from irrigated land (Table 14.6).

The importance of rainfall to cereal production is well illustrated by comparing the 1969/70 and 1971/72 crop years. Although approximately 1.35 million hectares of wheat were harvested in both years, the 1970 crop was one-third that of 1972 (625,000 tons as compared with 1.8 million tons). As a result, Syria in 1972 was able to export around 200,000 tons of wheat after having been a

Table 14.6   Syria: cereal production[a], 1943–80

	High	Low	Average	Standard deviation	Coefficient of variation
Wheat					
1943–1950	909	404	595	± 179	30%
1951–1960	1,354	438	784	± 277	35%
1961–1970	1,374	559	930	± 264	28%
1971–1980	2,226	593	1,445	± 485	34%
Barley					
1943–1950	357	169	281	± 57	20%
1951–1960	721	137	365	± 203	56%
1961–1970	798	203	541	± 204	38%
1971–1980	1,587	102	629	± 423	67%

[a] Quantities in thousand metric tons.

net importer during the 1960s (USDA, 1975). The 1972 bumper harvest was not surpassed until 1980 (2.2 million tons) when winter rainfall was again abundant and well distributed throughout the growing season. Even record harvests can cause problems, however, and a serious shortage of labour and mechanical harvesters in 1980 delayed harvesting and contributed to substantial crop losses (USDA, 1981).

The expansion of the area planted with grains during the 1970s has occurred against a background of increased government emphasis on national self-sufficiency in agricultural commodities. Among the goals of the Fourth Five Year Plan (1976–1980) was full self-sufficiency in major food commodities as well as increased levels of self-sufficiency in other agricultural commodities. The goal of increased domestic agricultural production in order to reduce dependency on imported products is based, at least in part, on the desire to secure the country's supply of food and other essential commodities against unforeseen emergency situations. It also reflects the poor harvests of the 1960s and the implications of increased food imports for Syria's balance of payments. As a result, the government has sought to giver higher priority to grain cultivation by encouraging use of high-yield varieties and a shift to wheat cultivation on irrigated land formerly planted with cotton. The steady improvement in the food production index during the 1970s suggests that farmers have responded to government policy, but this improvement has occurred during a period of relatively good rainfall. The USAID analysis notes that while there was a significant upward trend in the area planted with rain-fed winter crops between 1968 and 1978, this was *not* matched by significant upward trend in yields on rain-fed land (USAID, 1980, Vol. 1, p. III–4). In these circumstances it seems questionable whether the record harvests of recent years can be sustained when the amount and timing of winter rainfall is less favourable.

LAND REFORM

As suggested earlier in this chapter, obstacles to the improvement of Syrian agriculture have been encountered in the social fabric of the country as well as in the physical environment. In particular ownership of the critical factors of production had become concentrated in the hands of a comparatively small number of families.

> In areas where rainfall is inadequate the pump owner, who is frequently not the landowner, can exact 45% to 60% of the crop simply for the water he supplies. Where labor is plentiful and rainfall adequate, the landowner can get up to half the crop simply by allowing others to cultivate his land. On the other hand, in the Jezireh land is plentiful by comparison with labor and rents for only a 10–15% share of the crop. Here the owners of tractors and combines . . . can command the lion's share of the output. (IBRD, 1955, p. 37)

The wide disparities in agricultural income, the abuses of share-cropping, and the inefficiencies of fragmentation combined to produce growing pressure for land reform.

In general the government's reforms have sought to (1) redistribute ownership of the means of production, (2) resolve uncertainties over title and tenure, and (3) promote improved methods of production through a system of agricultural cooperatives. While it is necessary to give some brief account of the progress of land reform, evaluating its success is problematic in view of the limited information released by the government. Moreover, any such evaluation must take into account the political goals of land reform. Thus land reform in Syria provided the government with a vehicle for breaking the social and political power of the land owning–merchant elite while mobilizing a peasant base of support.

The Progress of Land Reform

The first serious attempt to implement a reform programme followed the union of Egypt and Syria in 1958 (Warriner, 1962). However, the reforms, which were modelled after those introduced in Egypt, provoked widespread opposition. As noted by Issawi (1963, p. 61):

> The application of most of these laws to Syria, where the landlord class was still powerful and the native bourgeoisie traditionally much more vigorous; where state intervention was a much more recent and restricted phenomenon; and where a remarkable rate of growth between 1945 and 1957 had been succeeded by a depression—partly

owing to poor rainfall—following the union with Egypt in 1958 had dramatic consequences. Discontent, which had been smoldering for some time, flared into the September revolt which dissolved the union.

With the rise to power of the Ba'ath Party, land reform again emerged as a major government priority. A government decree issued in 1963 reinstated the original reform law but incorporated a number of important modifications intended to adapt the law to the particular circumstances of Syrian agriculture. Of particular significance was the greater flexibility allowed in the permitted ceiling on land ownership. Thus the new ceiling on ownership of irrigated land varied from 15 hectares in the Ghouta to 55 hectares in those areas irrigated from groundwater wells in northeastern Syria. Similar flexibility was allowed with respect to ownership of rain-fed land with ceilings fixed according to region and cropping pattern. As in the 1958 law, expropriated land was to be redistributed to small farmers, tenants, share-croppers, and agricultural labourers in holdings not to exceed 8 hectares of irrigated land or 30 to 45 hectares of unirrigated land per family. Beneficiaries of the redistribution programme were required to form state-supervised cooperatives but the 1963 law granted operatives the right to own land in excess of the members' individual ceilings if that land could be used productively.

According to the government, 1.4 million hectares of land have been expropriated under the reform programme, equivalent to around one-quarter of the cultivated area. However, the disposition of this land remains unclear. The 1982 *Statistical Bulletin*, for example, only includes data for the period up to 1975. At that time an area of 466,100 hectares had been distributed to individuals while a further 254,000 hectares had been allocated to cooperatives and other government agencies. Of the remaining sequestered land, 329,800 hectares is reported as 'excluded and sold', leaving a balance of 351,400 hectares still to be distributed.

The Impact of Land Reform

Although documentation is difficult, the reform programme does appear to have been successful in breaking up the very large estates. According to the 1970 *Agricultural Census* only 18 per cent of the surveyed area was in holdings larger than 100 hectares (Table 14.7). This compares with a figure of 49 per cent reported by the World Bank. The same data suggest an increase in the number of smaller holdings with around 70 per cent of the area surveyed in 1970 farmed in holdings of less than 50 hectares. However, the limitations of the data must be emphasized; thus there is considerable ambiguity about the area and type of holdings covered in the agricultural census, while the World Bank figures were representative only of the area covered by the cadastral survey up to the

Table 14.7   Syria: land holdings, 1952–70

Size of holding (hectares)	1952[a] Area (% of total)	1970[b] Area (% of total)	('000 hectares)	Land holders ('000)	(% of total)
<2	1%	3%	112	128	32%
2–10	12%	20%	757	170	43%
10–50	28%	47%	1,746	88	22%
50–100	10%	12%	431	68	2%
>100	49%	18%	663	3	1%

Sources: [a] IBRD 1955, p. 354. Unfortunately the World Bank mission was unable to secure data on the number of holdings in the various categories.
[b] Syrian Arab Republic, Ministry of Agriculture and Agrarian Reform, *Agricultural Census, 1970.*

end of 1952. Nevertheless the overall impression is that reform has produced a wider distribution of land ownership.

Critics of the Syrian land reform programme have suggested that the break-up of large estates was responsible for the reduced use of farm machinery and fertilizers during the 1960s, thereby adversely affecting agricultural output. Such a conclusion would not be out of line with the experience of other Middle Eastern countries. Certainly the 1960s did see a decline in irrigated and cropped acreage in Syria and a sharp reduction in agricultural contribution to GDP. However, one must be very cautious about attributing the relatively poor performance of the agricultural sector during the 1960s to any single factor. The reforms were undertaken during a period of exceptionally poor rainfall when repeated crop failure undoubtedly discouraged many farmers. Moreover, the expropriation of large, private estates was only one element, albeit a crucial one, of an interlocking set of policies and programmes for the countryside. Other economists attribute the stagnation in agricultural production during the 1960s at least in part to government inexperience in coordinating and implementing rural development programmes. From this perspective efforts to increase farm production were hampered by (1) a lack of trained personnel that slowed down the redistribution of appropriated land, (2) a failure to provide sufficient credit, particularly on the long-term basis necessary to finance capital improvements and purchase farm machinery, and (3) contradictory pricing policies for farm produce (Nyrop, 1978, pp. 120–122). Illustrative of these problems were efforts to encourage cereal production (for example, by shifting from cotton to wheat on irrigated land) while simultaneously keeping grain prices low relative to other crops in order to minimize basic food costs for urban consumers. Nyrop (1978, p. 121) observes that the government's performance improved during the 1970s, noting in particular the increased availability of credit through the Agricultural Co-operative Bank, and more realistic pricing policies that favour grain, soybeans and sugarbeet over cotton. In this respect pricing policies now appear to be consistent with national development goals that emphasize self-sufficiency in basic agricultural commodities.

The promotion of agricultural cooperatives has been an integral part of the government's rural development programme. In effect cooperatives were expected to replace and expand on the assistance and support formerly provided by landlords and managers of large estates. More specifically it was anticipated that cooperatives would play a key role in modernizing Syrian agriculture by (1) introducing improved cropping practices, (2) facilitating mechanization on small farms through cooperative ownership of equipment, and (3) encouraging the consolidation of small holdings into more efficient units of operation (Nyrop, 1978).

Undoubtedly the cooperative movement got off to a very shaky start. Beaumont et al. (1976, p. 147), for example, note early difficulties and attribute these to the failure to adapt the Egyptian model to Syria's very different administrative traditions. However, the number of cooperatives more than doubled during the 1970s and by 1980 approximately 25 per cent of the cultivated area was being farmed cooperatively (Table 14.8). The remarkable growth of the cooperative movement during the 1970s suggests that any initial difficulty was indeed due to organization and staffing problems rather than any deep-rooted aversion to cooperative farming on the part of Syrian farmers.

Table 14.8   Cultivated area by type of farm (ha)

		1975	1980
1.	Private farms	3,041,000	2,866,000
2.	Cooperatives	593,000	1,001,000
3.	State Farms	66,000	26,000
	Total	3,700,000	3,893,000

Source: Syrian Arab Republic, 1982.

Table 14.9   Syria: indicators of technological change in agriculture

	Irrigation		Mechanization		Fertilizer
	No. of sprinkler units	No. of water-raising pumps	No. of combined harvesters	No. of tractors (>50 h.p.)	Total reported application (metric tons)
1963	205	20,990	1,566	2,093	61,900
1970	1,163	29,042	1,328	2,929	111,780
1975	1,325	40,416	1,607	9,030	189,935
1980	1,081	47,206	2,244	21,145	305,365

Source: Syrian Arab Republic, 1982.

How successful cooperatives have been in bringing about the desired transformation of Syrian agriculture is a matter of conjecture. According to one assessment, little has been accomplished in the way of farm consolidation (USAID, 1980, Vol. 5, p. 9). On the other hand, the more intensive use of both irrigated and rain-fed land has already been described. According to Nyrop (1978, p. 122) extension agents and cooperatives were particularly active during the 1970s in persuading farmers to reduce the amount of land left fallow. In addition the rate of technological change in Syrian agriculture has clearly accelerated over the past decade, with significant increases in the use of mechanized equipment and fertilizer (Table 14.9).

CONCLUSION

The vulnerability of Syrian agriculture to the vicissitudes of the rainfall regime has been a recurrent theme in this chapter. Despite the considerable investment in irrigation over the past two decades — 80 per cent of all development funds allocated to the agricultural sector during the Fourth Five Year Plan (1976–80), for example, were for irrigation and land reclamation — the World Bank's characterization of Syrian agriculture as rainfall-dependent is as accurate today as it was a quarter of a century ago. Indeed the proportion of the cultivated area that is irrigated has actually declined since the World Bank mission. As a result of the predominance of rain-fed cultivation, agricultural production (which in view of the emphasis on cereals in the cropping pattern essentially means food production) continues to fluctuate markedly from year to year (Figure 14.4). The experience of the past two decades suggests that this basic pattern will not be easily changed and that extension of irrigation, particularly in the Euphrates Basin, will be costly and time-consuming.

This instability should not disguise the better performance of the agricultural sector during the 1970s relative to the preceding decade. In contrast to the 1960s when both irrigated and cropped acreage declined, agricultural and food production indices showed significant gains during the 1970s, culminating in the record harvest of 1980. Even during the 1970s, however, production increases barely kept pace with population growth, which averaged in excess of 3 per cent per year. The nadir was reached in 1973 when food production on a per capita basis was half that achieved in the early 1960s. As a result Syria was a net importer of cereals during much of the 1960s and early 1970s. The improved harvests of the late 1970s merely meant that at the end of the decade output per capita had just about returned to the level achieved 20 years earlier.

This situation underscores serious long-term difficulties in developing Syria's land and water resources. Historically, agricultural practices in Syria and throughout the Middle East have been geared to survival, albeit at a subsistence level. Despite low levels of productivity they represented a sensible, low-risk way of coping with environmental uncertainty. Such a strategy is no longer

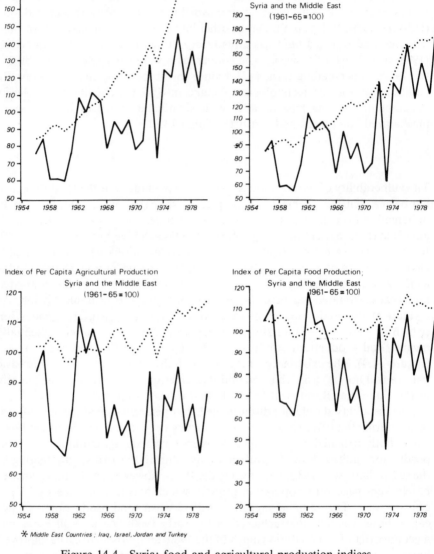

✳ Middle East Countries ; Iraq, Israel, Jordan and Turkey

Figure 14.4   Syria: food and agricultural production indices

tenable, given the heightened expectations and expanded needs of a growing population. The only alternative to sustained increases in agricultural output is higher expenditures on food imports. But whether the upward trend in total and per capita production can be maintained over the long term remains to be seen. The improvements recorded during the 1970s were primarily achieved through an expansion of dryland crop production and occurred during a period when rainfall was reasonably abundant and well distributed. In this regard the production gains of recent years may be attributable to meteorological good fortune as much as to improved planning for the agricultural sector. Moreover, intensification of production on rain-fed land has meant the conversion of fallow to cereals rather than the introduction of a leguminous crop that would help maintain soil fertility and integrate crop and livestock husbandry.

In light of the experience of the past two decades, achieving real annual growth rates in agricultural GDP of 8–10 per cent, the goal of the Fourth Five Year Plan, seems unrealistic. Few countries have been successful in sustaining an annual growth rate of 5 per cent for any length of time, let alone accomplish this in an environment characterized by uncertainty. The experience of the 1960s and 1970s should have produced a more realistic assessment of the opportunities for and constraints on agricultural development in Syria. The parameters to any development programme remain essentially the same as those outlined in the World Bank report. These include an intensification of production on those lands best suited to dry-farming and a reduction in the exploitative cropping of marginal lands that are better suited to range and pasture. The establishment of the International Center for Agricultural Research in Dry Areas (ICARDA) near Aleppo is an encouraging indication that improvement of dry-farming production techniques is now receiving much higher priority in the government's development plans. Irrigation will continue to play a key role, although the construction of new storage facilities and diversion canals will not by itself suffice. Much more can be accomplished through improved management of land and water resources within existing projects. In this sense there is no panacea for Syrian agriculture, only careful planning that is sensitive to human needs and ecological realities.

REFERENCES

Beaumont, P., Blake, G. H., & Wagstaff, J. M. (1976) *The Middle East: A Geographical Study*, John Wiley, London.
Clawson, M., Landsberg, H. H., & Alexander L. T. (1971) *The Agricultural Potential of the Middle East*, American Elsevier Publishing Company, New York.
Dregne, H. E. (Ed.) (1970) *Arid Lands in Transition*, American Association for the Advancement of Science, Washington DC.
English, P. W. (1967) 'Urbanites, peasants, and nomads: the Middle Eastern ecological trilogy', *Journal of Geography*, **66**, 54–59.
English, P. W. (1973) 'Geographical perspectives on the Middle East: the passing of the ecological trilogy', in Mikesell, M. W. (Ed.) *Geographers Abroad: Essays on the Problems and Prospects of Research in Foreign Areas*, Department of Geography Research Paper No. 152, University of Chicago, pp. 134–164.
Hills, E. S. (Ed.) (1966) *Arid Lands: A Geographical Appraisal*, Methuen, London.
Hudson, J. (1968) 'The role of irrigation in Syria', *Focus*, **18**, 10–12.
IBRD (International Bank for Reconstruction and Development) (1955) *The Economic Development of Syria*, The Johns Hopkins Press, Baltimore.

Issawi, C. (1963) *Egypt in Revolution — An Economic Analysis*, Oxford University Press, London.
Lewis, N. E. (1955) 'The frontier of settlement in Syria: 1800–1950', *International Affairs*, **31**, 48–60.
Manners, I. R. (1980) 'The Middle East', in Klee, G. (Ed.), *World Systems of Traditional Resource Management*, John Wiley, New York, pp. 39–66.
Nyrop, R. F. (Ed.) (1978) *Syria: A Country Study*, The American University, Washington DC.
Pitcher, S. (1974) 'Syria's Euphrates Dam promises rapid agricultural development', *Foreign Agriculture*, **13**, 14–15.
Syrian Arab Republic, Ministry of Agriculture and Agrarian Reform (no date) *Agricultural Census, 1970*, Damascus.
Syrian Arab Republic, Ministry of Agriculture and Agrarian Reform (1981) *The Annual Agricultural Statistical Abstract*, Damascus.
Syrian Arab Republic, Office of the Prime Minister (1976) *Fourth Five Year Economic and Social Development Plan of the Syrian Arab Republic, 1976–1980*, Damascus.
Syrian Arab Republic, Office of the Prime Minister (1982) *Statistical Abstract, 1982*, Damascus, Cultural Bureau of Statistics.
USAID (United States Agency for International Development) (1980) *Syria: Agricultural Sector Assessment* (5 volumes), Washington DC.
USDA (United States Department of Agriculture) (1975) *The Agricultural Situation in Africa and West Asia* (Foreign Agricultural Economic Report No. 108), Washington DC.
USDA (United States Department of Agriculture) (1981) *Agricultural Situation: Africa and the Middle East* (Economic Research Service Supplement No. 7), Washington DC.
Walton, K. (1969) *The Arid Zones*, Hutchinson, London.
Warriner, D. (1962) *Land Reform and Development in the Middle East: A Study of Egypt, Syria and Iraq*, Oxford University Press, London.
Weulersse, J. (1946) *Paysans de Syrie et du Proche-Orient*, Gallimard, Paris.
White, C. F. (1966) 'The world's arid areas', in Hills, E. S. (Ed.), *Arid Lands: A Geographical Appraisal*, Methuen, London, pp. 15–30.
World Bank (1982) *Annual World Development Report*, Washington DC.

Agricultural Development in the Middle East
Edited by P. Beaumont and K. McLachlan
© 1985 John Wiley & Sons Ltd

Chapter 15

Kuwait: Small-scale Agriculture in an Oil Economy

MUHAMMAD RASHID AL-FEEL

INTRODUCTION

Kuwait has the distinction of being among the nations least able to feed themselves. In the modern period it has been entirely dependent on external supplies of cereals and is heavily reliant on the outside world for every other agricultural commodity. The reasons for Kuwait's position lie in an amalgam of outcomes of rapid growth generated by oil revenue and the poor natural endowment of the state in land and water.

Kuwait, in the northeastern quadrant of the Arabian peninsula, is bounded to the north and north-west by the Republic of Iraq, to the east by the Gulf and to the south and south-west by Saudi Arabia. It covers an area of about 18,000 square kilometres and is a generally flat or gently undulating plain, broken by low hills, shallow depressions and escarpments. The land falls gradually to the east in the southern half of the state and to the north-east in the northern sector, falling some 275 metres (900 feet) in about 145 kilometres. The highest point, about 305 metres above sea level, is in the south-west corner of the country.

In a country with a population of 1,355,827 at the 1980 census, agricultural production contributes less than 1 per cent to GDP and the employment opportunities afforded by it constitute approximately 1 per cent of the total gainfully employed adults. To a degree, the official statistics might be misleading. Within Kuwaiti borders there is a constantly fluctuating number of bedouin herders, who are to a large extent ignored in the census since they do not possess Kuwaiti citizenship but are not enumerated as immigrants. In the outer low-income housing and shanty areas, it is a common sight to see sheep, goats, and even camels still in the ownership of current or settled bedouin families. If the bedouin herders, part-time farmers and full-time gardeners are taken into account, agricultural employment in Kuwait takes on a more substantial aspect, though still not one to rival neighbouring states.

The comparatively insignificant role played by agriculture in the Kuwaiti economy is related *inter alia* to a number of environmental constraints, including limited water resources, a harsh climatic regime and poor soils, as well as to difficulties of an inexperienced and poorly trained labour force, low productivity per worker employed and the ease with which Kuwait can pay for food imports.

SOILS OF KUWAIT

The surface of Kuwait is formed of sedimentary rocks ranging from early Miocene to recent times. The soil was formed under arid climatic conditions, is permeable, sandy in texture and very low in organic matter (under 1 per cent) with a cemented calcareous subsoil. It lacks important nitrates and phosphates but contains a high percentage of other salts. Sand covers the eastern half of Kuwait, while sand dunes exist to the north and south of the country. Playas occur and impermeable sandy lime or *gatch* is common. Richer alluvium covers only 7 per cent of the land.

Kernick (1966), Ergun (1968), and other pedologists recognized four soil categories:

(1) Desert soils such as sandy desert soil, gypsiferous desert soil, gravelly saline desert soil. This group constitutes 78 per cent of the total surface area.
(2) Desert regosol soils such as sandy desert soils of the type found at Sulaibiyah and dunes at Umm Al-Negga and in the south of Kuwait.
(3) Lithosols, escarpments such as Jal Al-Zur. These are very shallow soils and cover one per cent of Kuwait.
(4) Alluvial soils at Bubyan Island and along the sea shore are contributed by continuing deposition of the Shatt Al-Arab.

The soil of Kuwait is characterized by an impermeable *gatch* layer, which is a consolidated calcareous sandy matrix of high silica content, slightly gypsiferous and highly saline (Al-Feel, 1981). The disabilities provided by the presence of *gatch* may be partially overcome by using special crop rotations (Ueda & Ueda, 1968), which increase the nitrogen content of the soil, add natural fertilizer, and produce green fodder. Ergun (1968) adds that water of less than 800 parts per million of dissolved solids must be used in order to protect the soil and prevent salinization.

CLIMATE

Climate has been unkind to Kuwait, situated in a zone of extremely high temperatures, where cloud cover is rare and where negligible precipitation leaves the state for eight months of the summer without rainfall. Yearly average sunshine is about 3,200 hours. Temperatures show a considerable diurnal range, despite the proximity of the water of the Gulf. Climatic conditions have imposed a four-season rhythm on agricultural activities. Summers are long and hot in contrast to much milder winters: but spring and autumn seasons are very short. Weather conditions are exacerbated by high relative humidity and frequent dust storms, which together make labour in agriculture difficult and at times intolerable.

The typical hot desert weather gives average summer temperatures in the 35–39°C range, with extremes reaching 50°C. Winters are cooler but zero temperatures have (very rarely) been recorded. Relative humidity is highest in the winter season, averaging about 65 per cent, while average summer humidity is generally less than 30 per cent. The prevailing wind comes from the north and north-west, frequently bringing dust and sand. Annual rainfall is mainly confined to the winter months and amounts to 110 mm, though the yearly totals vary. Potential evaporation varies from 2 mm/day in January to 12 mm/day in July. In conditions such as these, agricultural activity can survive on any scale only with irrigation, in which area Kuwait is severely deficient.

WATER RESOURCES AND REQUIREMENTS

The major sources of water supply in Kuwait are from brackish subterranean aquifers, though in recent times these have been supplemented with other supplies. Water distillation plants produce 106 mn gallons/day, which is used

Table 15.1   Water field developments

Name of field	Existing yield $(10^6 \times IG/D)^a$	Potential yield $(10^6 \times IG/D)$	Salinity (ppm)^b	Life expectancy (in years)
Shagaya project field A.B.C.	32	40	3,000 to 4,000	More than 50
Sulaibiyah	15–20	20	4,500 to 5,500	More than 25
Rawdatain and Umm Al-Aish	1.5	4	Fresh 700 to 1,200	More than 20
Abduliya	Approx. 5	Unknown	Approx. 4,500	Not known
Abdali Umm Nigga farms	20–25	Limited	3,000 to 7,000	More than 15
Wafra farms	20–25	30	4,000 to 6,000	More than 15
Shagaya project field D.E.	—	25	3,000 to 4,500	More than 50

^{a}IG/D = imperial gallons/day.
^{b}ppm = parts per million.

mainly for domestic consumption. Recycled sewage water can make available up to 85 mn gallons/day and re-use of water from the industrial areas gives a further 30 mn gallons/day. A long-planned project to bring water from the Shatt Al-Arab system in Iraq to supply 350 mn gallons/day together with 300 mn gallons/day from the Euphrates River has failed to materialize, since Kuwait and Iraq have been unable to reach a final agreement on the scheme, though this might be an option for the future.

Underground water is present in abundant quantities, but both quantity and quality vary with location. Table 15.1 shows the water fields that have been developed or are under construction. Use of water for agricultural purposes is still at a low level, amounting to 243 mn gallons/day as outlined in Table 15.2. Demand for water for irrigation is growing and should reach some 471 mn gallons/day for all forms of cultivation (Table 15.3). Should the demand estimates prove to be correct, Kuwait will face a shortage of water in this sector by the year 2000 unless a sizeable increase in supplies, such as that proposed in the Shatt Al-Arab project, is realized. Given rates of increase in domestic use of water, the state's planned additions to desalinization capacity will be altogether inadequate to cater for irrigation, which has always had a lower priority than the urban sector.

It is estimated that approximately 20,000 ha of land are cultivable in Kuwait's 18,000 km^2 surface area (Bowen-Jones & Dutton, 1983). In practice, the area under cultivation in any one year is much less than this. Only 36,000 donums were in use for field and orchard crops throughout the 1970s, representing a negligible proportion of total land. The pasture area is significant, taking in 1,340,000 donums (134,000 ha). In all, the most generous assessment puts agricultural land at a mere 7.7 per cent of total surface area and even with

Table 15.2   Water consumption in Kuwait, 1980

Types of use	Location	Sources	Salinity (ppm)	Quantity (10^6 G/D)
Agriculture	Wafra	Underground	5,000–10,000	50.0
	Abdali	Underground	2,500–10,000	30.0
	Sulaibiyah	Sewage	2,000–2,500	6.5
		Underground	4,500–7,000	5.0
	Scattered farms	Underground	3,500–10,000	10.0
Public and private gardens	Kuwait	—		110.0
		Underground	500	18.0
		Distilled	—	92.0
	Al-Jahra	Underground	—	27.0
	Al-Ahmadi	Underground	3,500–4,500	5.0
Total				243.0

Source: Al-Sharhan, 1982, p. 56.

Table 15.3   Projected water demand in Kuwait — year 2000

Types of use	Volume (10^6 G/D)	Type of water
Private gardens	227.0	Potable
Public gardens	70.0	Potable
Afforestation	16.0	Treated water
Cultivation	22.0	Potable
	65.0	Low salinity
	71.0	Treated water
Total	471.0	

Source: Al-Sharhan, 1982.

potentially cultivable land only 14 per cent. Bearing in mind the smallness of Kuwait and the tiny fractions of it that are used for agriculture (Table 15.4), the state is all but bereft of real agriculture. In recent years, the development of industry, works related to the oil field systems, and the spread of housing have all eroded the area that might be considered to be of agricultural value.

Cultivation is undertaken in three basic forms, winter cropping, specialist horticulture in the summer, and the farming of semi-perennial crops. Winter vegetables are the main concern in the agricultural calendar, with 833 ha in 1979/80 season. Principal items were tomatoes (12,275 tons production), radishes (5,595 tons) and onions (2,526 tons) together with other salad crops. Summer crops took in 303 ha in 1979/89, with the cucurbit group — watermelon and melon being particularly important. Output from summer cultivation amounted to 8,109 tons in 1979/80. Semi-permanent crops include clover and leeks, in all covering 2,830 ha.

According to the latest official data, there are 501 agricultural holdings in the state. Of these, it is reported (Ministry of Planning, 1982) that 318 are concerned with vegetable growing, 58 with poultry rearing, and 42 with dairy farming, and 1 is used as a multi-purpose unit. Of the land in farms, approximately 95 per cent is under cultivation for vegetables and more than 60 per cent of farm holdings are engaged in vegetable production.

It is a slight irony that the two main agricultural areas of Kuwait, Al-Abdali and Wafra, are both situated on the geographical fringes of the country, Al-Abdali in the north adjacent to the Iraqi frontier and Wafra in the extreme south. Between them, they account for 822 ha or 94 per cent of total agricultural land. Sulaibiyah in Hawalli Governorate, with 492 ha under cultivation, is the only other appreciable area under crops.

Table 15.4 Land utilization in Kuwait ('000 donums)

Type of Land Use	1971	1972	1973	1974	1975	1976	1977	1978	1979
Vegetables	7	8	8	9	10	11	13	11	13
Fruit & others	23	23	23	23	23	23	23	23	23
Pasture	1,340	1,340	1,340	1,340	1,340	1,340	1,340	1,340	1,340
Fallow/unused	169	168	168	166	166	165	165	161	162
Uncultivable land	16,279	16,279	16,279	16,280	16,279	16,279	16,279	16,283	16,280
Total	17,818	17,818	17,818	17,818	17,818	17,818	17,818	17,818	17,818

Source: Ministry of Planning, 1981.

MANPOWER IN AGRICULTURE

Employment in agriculture in Kuwait, as in many other Middle Eastern countries, fluctuates with the level of seasonal activity. Spring is the time of greatest activity and high summer one of comparative quiet. In the 1979/80 agricultural year, employment in March was 4,280 persons, falling to 4,162 in June and rising again in September to 4,235. There are few self-employed farmers, the bulk being short-term employees in the ratio 1:25. In September 1980 there were only 169 farm proprietors actively working in the sector. Labour supply posed difficulties in so far as Kuwaiti nationals do not normally offer themselves for work in the manual occupations and, other than the bedouin, have little or no interest in cultivation or herding. It is immigrant workers who are therefore employed, often reluctantly and rarely with real commitment. It was found, for example, during field studies in 1983 that several foreign workers on the farms had been brought to Kuwait under the false impression that they would be given urban or managerial work, only to discover that their employers, to whom they were contractually attached, allocated them to farm labouring tasks (Al-Moosa & McLachlan, 1985).

Certainly, there are great problems in managing a workforce of mixed nationalities, some with no understanding of farming, in extremely arid conditions such as those in Kuwait. The total number of persons engaged in agriculture in 1979/80 included only 2–3 per cent of Kuwaitis but approximately 40 per cent Iraqis and 15 per cent Iranians. Low wages and arduous conditions meant that immigrants were far from keen to stay on the land. Once established in Kuwait, foreign workers tended to gravitate towards better-paid and more congenial employment in Kuwait City. In consequence, labour suffers from constant turn-over and keeps only the least skilled and least enterprising of a mixed and generally poor quality workforce.

Plans to enforce contracts for those immigrants in agricultural employment have foundered on difficulties of control in a country where 70 per cent of the workforce is foreign. Plans to introduce severe penalties for employers poaching employees from agriculture have also been considered. Neither measure would solve the basic problem, that agriculture does not pay well enough to compete satisfactorily with other sectors such as construction or services. The most reliable and productive labour will remain Kuwaiti, especially those with traditional interests in livestock-rearing and cultivation in the Al-Jahra Governorate of the north. These small groups, aided by family and foreign labour, are highly efficient but not apparently easily augmented. Not so acceptable to the Kuwaiti authorities are the large numbers of bedouin and others who keep livestock in the main housing areas, though particularly in low-budget and shanty housing areas. Part-time farming of this kind produces remarkably high levels of output despite the disapproval of the regional administrations.

Food self-sufficiency in Kuwait is low. The state produces no cereals, about 35 per cent of its vegetable requirements, almost no fruit, some 20 per cent of egg demand, and, if semi-official estimates are to be believed, 40 per cent of the meat required. Imports of agricultural goods have risen steadily. The value of imports of food and animals rose from $234 mn in 1974 to $681 mn in 1979, with total agricultural imports rising from $276 mn to $798 mn during the same period and continued to rise thereafter. By 1982 imports were valued at $987 mn against re-exports of $119 mn.

There is an element, too, of disguised import-dependence. Kuwaiti livestock-rearing outside the scavenger flocks in the housing areas relies on imported feeds. Feed for the large-scale poultry farms also comes in from abroad, while technologies for long-running experiments in hydroponics are all imported, often with the management to go with them. Virtually all labour in agriculture, other than the special groups already noted, is foreign, so that, in all, even the internal sector is deeply dependent on imports of goods and people.

PLANNING FOR AGRICULTURE

Kuwaiti planners have long faced the problem of matching a felt need to expand domestic production of basic foodstuffs with the limited resources available with the state. In the first development plan of 1967/68 to 1971/72 (Planning Board, 1968), it was noted that constraints of water supply, land availability and provision of trained manpower made all aspirations for agriculture seem unrealistic. At best, it was foreseen that surveys could be put in hand to determine which areas might be developed once water supplies could be assured and that production units in poultry and livestock could be begun using modern equipment and methods.

World Bank estimates of the agricultural potential of Kuwait in the 1960s were also lacking in optimism (IBRD, 1965). Recommendations for agricultural development were limited to some expansion of vegetable cultivation, improvement of dairy farming to enable the greater part of domestic demand to be met and an increase in meat output. At that time, it was considered advisable for Kuwait to concentrate on fattening imported livestock before slaughter. Such recommendations were modest and have in many ways formed the basis for such government policy towards the sector as has existed.

At various periods, planners have been tantalized by the prospects opened up by hydroponics and high technology intensive cultivation as demonstrated at the government experimental farm. This originally 20 ha site showed that relatively high yields could be attained using slightly brackish water but under closely controlled conditions. There has been the temptation to try to increase the number of farm units of this kind as a means of increasing agricultural output rapidly. In reality, conditions at the farm have not been without significant problems of both a technical kind and in terms of labour supply.

Despite these difficulties, the 1982–86 plan for agricultural expansion devised by the Department of Agriculture proposes that the area under glass be increased and that cultivation by hydroponic techniques be exploited to a greater extent. In all, it is estimated that glasshouse cultivation will be extended to 450 ha. Adoption of more intensive vegetable-growing and an expansion of the area under vegetable crops to 3,500 ha by 1986 is hoped to lead to a doubling of vegetable production (Bowen-Jones & Dutton, 1983). Output should reach 98,000 tons/year by 1986.

More grandiose plans for a rapid reclamation of new lands for fodder-growing is expected to permit the construction of new livestock units. This expansion would depend on the availability of irrigation water (Table 15.3), projects for which are under construction but the output from which may not be allocated to agriculture.

However brave the plans formulated, they will ultimately have to take account of the two great realities of the situation in Kuwait. First, the state has a highly developed oil-based economy. In addition to obvious factors such as very high standards of living, with income per head estimated at $20,900, it must be borne in mind that all Kuwaiti citizens are guaranteed employment in white-collar jobs by the government. Other than the bedouin, whose interests are in livestock rather than cultivation, and the small number of cultivators in the Al-Jahra oases, Kuwaitis have no traditional or current interest in agriculture. At best, some Kuwaiti families have developed gardens where some fruit and vegetables are grown, though these are retreats for leisure and amenity purposes rather than agricultural holdings. Gardens represent a form of consumerism little different from the purchase of other luxury commodities. At the same time, one or two Kuwaitis, as much out of eccentricity as anything else, have spent considerable sums of money investing in agricultural enterprises. But these farms, often well run, are subsidized by profitable occupations elsewhere in the economy and are not viable as independent units.

Kuwait, too, has every reason to believe that its position of comparative wealth will remain for many years. Oil reserves in place are estimated to have a life of up to 100 years. There is no pressure in the state to act in the immediate future to protect its food supply against the day when it can no longer afford to import its requirements. The Kuwaiti population is small — slightly over half a million persons. It remains a Kuwaiti strategy to keep the numbers of registered citizens to a minimum and the three-quarters of a million foreigners living in the State are almost entirely denied rights to nationality. Although the indigenous population is growing, its small size means that there is no threat of population numbers outstripping available financial resources and the volume of imports of food to feed it are relatively small. It cannot be expected in such circumstances that Kuwaiti development will be oriented towards self-sufficiency in any economic sector, least of all in agriculture. The planners fight, therefore, against a deep conviction in the Kuwaiti section of the population that costly and urgent measures to develop agriculture are unnecessary and wasteful.

Second, agricultural expansion is inhibited by more than inertia and complacency. Water supplies are poor, except from brackish sources, the utility of which is limited. Expansion of cultivation can be intensified through use of slightly more good quality water; but a rapid and general expansion that would genuinely reduce Kuwait's import dependence would have to come through major land reclamations underpinned with the provision of substantial new sweet water supplies. The only near source of water on this scale would have to be brought in from Iraq by resuscitation of the Shatt Al-Arab project. There are severe political obstacles in the way of achieving this scheme. Despite their complexity, it might be observed that the Kuwaiti authorities are unwilling to acquire water from Iraq if this might put into jeopardy Kuwaiti sovereignty over the northern borderlands and the islands of Bubyan and Warba. While this fear remains, no progress can be expected in bringing to Kuwait abundant volumes of new water and for the foreseeable future it must be assumed that this situation will persist, with all its negative implications for the outlook for agricultural expansion in Kuwait.

Agriculture in Kuwait appears to be condemned to be a tiny sector of the total economy, whether measured by its contribution to national wealth or its function as an employer of labour. Minor improvements are feasible, despite losses of land to urban growth, but in general Kuwait seems destined to be a food-importing country and one heavily dependent on the outside world for other provisions for farming activities.

REFERENCES

Al-Feel, M. R. (1981) *Water and Food Security in Kuwait*, Kuwait University, Kuwait.

Al-Moosa, A. A., & McLachlan, K. S. (1985) *Immigrant Labour in Kuwait*, Croom Helm, London.

Al-Sharhan, A. (1982) *Water and Agriculture in Kuwait*, Government of Kuwait.

Bowen-Jones, H., & Dutton, R. (1983) *Agriculture in the Arabian Peninsula*, Economist Intelligence Unit, London, p. 127.

Ergun, H. N. (1968) *Reconnaissance Soil Survey: Report to the Government of Kuwait*, Government of Kuwait.

IBRD (1965) *The Economic Development of Kuwait*, Johns Hopkins University Press, Baltimore, pp.–130–131.

Ministry of Planning (1981) *Annual Statistical Abstract 1981*, Central Statistical Office, Kuwait.

Ministry of Planning (1982) *Annual Statistical Abstract 1982*, Central Statistical Office, Kuwait, p. 132.

Planning Board (1968) *The First Five-Year Development Plan 1967/68–1971/72*, Kuwait, pp. 74–75.

Ueda, H., & Ueda, T. (1968) *Preliminary Study on Agriculture in Kuwait*, Kuwait. Mimeograph.

Agricultural Development in the Middle East
Edited by P. Beaumont and K. McLachlan
© 1985 John Wiley & Sons Ltd

Chapter 16

Progress in Turkish Agriculture

BRIAN W. BEELEY

Rapid changes are evident among Turkey's farmers and in the way in which they operate their farms. Key features of this change are migration and mechanization, but there are also major new developments in other aspects of agricultural production and marketing. A national planning organization is now in its third decade of coordinating the increased involvement of state agencies in agricultural and rural change. Some of this initiative has been directed towards the modernization of farming practice, from mechanization and the provision of fertilizers and new seed types to irrigation projects and a variety of infrastructural projects—including roads. At the same time, Turkish agriculture has moved still further from locally orientated, largely self-sufficient production towards integration into a nationwide commercial system.

Most important of all have been the changes in Turkish rural people themselves. Many have left their villages as migrants for Turkey's burgeoning cities or have gone to work in eastern Europe or Arab countries. Non-farming activities are increasingly in evidence in villages and remittances sent home by migrants' finance investment in the village—or in leaving it. Yet agriculture remains the largest employment sector with 55 per cent of the labour force (including livestock-rearing, forestry, and fishing) in 1980 compared with nearly 75 per cent in 1960. One in four of the 10.5 million people in Turkish agriculture are listed as employed on their 'own account' (93 per cent of these are men). Unpaid family workers account for all but a few of the remainder in the official figures (and over 70 per cent of them are female). Though the contrasts in these figures are clear, they understate the 'paid employee' category since payment is made in kind and they fail to reflect the growing numbers of farmers and farm-workers who are now only working 'part-time' in agriculture because they are developing other skills in the village—or spend most of their time away from home as migrant workers.

The very highest levels of socio-economic development are associated with Turkey's four major urban areas—Istanbul, Ankara, Izmir and Adana. These cities are now linked to each other and to middle-rank and smaller towns across the country by a comprehensive road network: improvements in the rail system and in inter-city air services have further contributed to the infrastructure of communications. Thousands of villages are now linked to nearby market towns by dependable feeder roads, frequently with truck, bus, or shared-taxi services. Some of these local roads have cut journeys which might once have taken hours to little more than minutes. Turks now seem to be perpetually on the move through a country where distinctively separate regions were once connected only by lonely and often difficult roads.

Indeed, the movement of people has, arguably, become the principal factor affecting Turkish agriculture and the rural people upon whom it depends. Migration, rural–urban and international, has quantitative and qualitative implications. As to the former, it is clear that more than one-tenth of Turkey's districts or sub-provinces (*ilçe*) were losing rural people during the 1960s. By 1980 30 per cent of the districts in the country were reporting a net loss of rural people—although birth rates remained high. Most of these districts are located in the upland fringes of the Anatolian plateau but some are coastal and some in Turkey's European corner in Thrace. During the 1970s a number of provinces began to show net losses in urban as well as rural population. Between 1975 and 1980 there was a drop in the total population (allowing for reproduction as well as migration) in five of Turkey's 67 provinces. Seventeen other provinces meanwhile gained population by migration. All but two of these are in the western half of the country and four are provinces of the very largest Turkish cities where urban growth is at its most dramatic. In the capital, Ankara, for example, most of the inhabitants were born outside the province—and most are rural and small-town Turks living in squatter (*gecekondu*) housing.

Few immigrants from farming backgrounds find full employment easily in the formal economy of the cities: many remain underemployed at best. Sometimes groups move together into a city—perhaps from a simple village. Links with the home village or small town persist: money may be sent home and visits made. Such links may be critical to the maintenance of relatives in a rural area, especially where local resources cannot measure up to aspirations. These links and the attitudes developing along with them contribute to the qualitative change which is taking place in rural as well as urban Turkey. The farmer increasingly feels that he is part of a national, urban–rural society rather than of the narrower world of the much more limited horizons which confined his parents. To the personal links of ties with relatives who have left farming—or to the subsidies which are sent back—are added the physical links which are now possible with the improvements in travel and communications from the centre to the rural hinterland through the media, agricultural extension, and other branches of state service.

As a result of the development of a variety of urban–rural links, more—perhaps most—of Turkey's 36,000 villages are now 'market-seeking' in that they show signs of losing their strictly village status (Kolars, 1967). They are becoming less different from urban areas as the expansion of production of surpluses for sale beyond the village leads to higher income levels. Meanwhile, village labour forces develop a variety of non-farm skills and the range of local amenities grows. Except where resource potentials are minimal, the number of 'market-ignoring' subsistent villages—long characteristic of rural Turkey and much of the developing world—dwindles fast.

One of the most visible changes in Turkish agriculture in the 1970s and 1980s has been the appearance of tractors and other modern farm machines in place of the traditional energies of animals and people and old-style implements and techniques. In 1948 there were fewer than 2,000 tractors in the whole of Turkey: now there are more than half a million. By 1960 the number had reached 42,000 and there was a growing range of other machinery such as sprayers (nearly 3,000), pumps, ploughs, seed cleaners, and many more. Greater use of tractors accounted for much of the additional area—much of it on sloping ground and pasture—brought under cultivation during the 1950s. Since 1960, however, tractors and other farm machines have tended to replace oxen and horses rather than to contribute to area expansion. Figures for selected farm implements (and draft animals) since the beginning of the period of national planning in 1963 are found in Table 16.1.

Between 1968 and 1981 there was a five-fold increase in the number of tractors: in the Black Sea coastal zone it was more than ten-fold. By the mid-1970s, two-thirds of all cropland was being cultivated by tractor: back in 1950 the percentage had been less than ten. In 1983 mechanization had reached the point where the Turkish Industrial Development Bank was publishing studies of the prospects for the export of Turkish-made agricultural machinery to Algeria, Egypt, Iran, Iraq, and elsewhere in the Middle East. Production of machines within the country was increasingly substituting for imported items which had soaked up large amounts of Turkey's limited foreign currency reserves as well as imposing immense burdens of import-duty payment—and consequent indebtedness—on farmers.

Perhaps machines have been more than farm implements where they have brought higher status as well as providing a focus for investment for rural Turks. During the first rush to mechanize, the import of farm machines—especially

Table 16.1   Numbers of farm implements and draft animals
in Turkey (in thousands)

	1963	1974	1982
Tractors	51	200	491
Sprayers	65	234	416
Farm pumps (centrifugal and motor)	52	161	295
Wooden ploughs	1,964	1,541	824
Pairs of draft animals	2,652	1,953	1,692

Source: State Institute of Statistics.

tractors—was often ad hoc. Different makes of machines would arrive in a village, with incompatible needs for spare parts and specialist servicing and repair. Rural roads and tracks were frequently quite unsuitable for large tractors which, moreover, were capable of demolishing field and terrace walls with consequent damage and soil loss. But despite difficulties of cost and maintenance, the tractor was seen not only as a means of working more land for one's own account but also as a basis for extra earnings from hiring to others.

Farm mechanization at the scale experienced in Turkey seems ill-suited to a largely labour-surplus enterprise, but it continues. One can debate the extent to which machines have contributed to the 'push' of rural people out of the villages: certainly machines have displaced the labour of women and children more directly than that of men. Equally, the nature of agricultural implements is such that they work more efficiently on larger and less fragmented holdings than most of those characteristic of traditional agricultural practice in Turkey. Except in the south-east, medium-sized farms predominate, though these may tend to diminish in size where fragmentation (brought about by inheritance) is not balanced by consolidation or where cultivation is interrupted by disputes over title as more and more agricultural land is surveyed and registered.

Between 1960 and 1980 the area under the plough (whether animal- or tractor-drawn) in Turkey increased by 7 per cent but production expanded by much more—even doubling in some cases. Prior to 1960, increase in land cultivated had contributed more to greater output than had improved yields. In 1950 less than 19 per cent of Turkey's land area was classified as farm land: by 1960 nearly 30 per cent was so described (compared with 31.4 per cent today). Within the arable total, fallow accounted for 32 per cent in 1950 and 34 in 1960 and again in 1980. Forests accounted for 13 per cent of Turkey in 1950. According to official figures this had doubled by 1980 but the increase is more the result of the re-classification of land (much of it not rich in trees) than of afforestation (see Table 16.2).

Wheat, Turkey's principal agricultural product, has shown a nearly five-fold increase in output since the years following 1945. Other crops have increased

Table 16.2   Land use in Turkey, 1950–80 ('000 hectares)

		1950	1960	1970	1980
Total arable land		14,540	23,264	24,296	25,560
Sown		9,870	15,305	15,591	16,372
of which:	vineyards	561	782	845	820
	olive groves	297	548	731	813
	orchards	608	730	1,019	1,492
Fallow		4,670	7,959	8,705	8,188
Forest		10,418	10,584	18,273	20,199

Source: State Institute of Statistics.

even more: cotton, potatoes, and sunflowers are examples, as are sugarbeet and
oranges with twelve- and nineteen-fold increases respectively. Figures for selected
crops are shown in Table 16.3.

Whereas increases during recent years in the output of field crops have been
primarily due to improved yields (with increases of as much as 60 and 70 per
cent registered, respectively, by barley and wheat since the early 1960s), tree
cultivation occupies much more space than before, with 39 per cent more
capacity for olives and no less than 71 per cent more non-citrus fruit in 1981
than in 1962—though some of this apparent expansion was the result of re-
classification.

Livestock and animal products accounted, in 1981, for nearly 33 per cent
(by value) of Turkey's total agricultural sector output. Some indication of
trends in the overall numbers of farm animals in the country is shown in
Table 16.4.

Of the 204,000 tons of meat production reported for 1980 from official
slaughterhouses, 53 per cent was accounted for by beef and veal and nearly
33 per cent by mutton and lamb. However, 'unofficial' slaughtering accounts
for most of the meat consumed across rural Turkey. Certainly cattle and sheep
account for the growth in total animal numbers in Turkey in recent decades,
a trend which seems likely to continue as it is supported by a policy of progressive
conversion to pure-bred—in place of domestic—breeds along with official

Table 16.3   Output of selected crops, 1946–80 ('000 tons)

		1946–50[a]	1956–60[a]	1966–70[a]	1976–80[a]
All cereals		7,068	14,140	16,410	24,546
of which:	wheat	3,629	7,510	9,924	16,770
	barley	1,725	3,430	3,630	4,988
	rye	404	682	800	744
	maize	634	901	1,000	1,293
Pulses		328	555	588	775
Cotton lint		78	170	402	500
Cotton seed		139	301	642	800
Tobacco		95	125	160	263
Potatoes		431	1,334	1,833	2,854
Sugarbeet		723	2,937	4,454	8,593
Sunflower		61	108	269	566
Grapes		1,361	2,741	3,562	3,371
Hazelnuts		64	103	165	279
Oranges		33	162	405	642
Olives		233	375	629	875
Figs		102	142	218	191

Sources: 1976–80 data from State Institute of Statistics; others from:
Dewdney, 1981, p. 218.

[a] Five-year mean.

Table 16.4   Numbers of selected livestock in Turkey, 1955–80 (in thousands)

	Total	Cattle	Sheep	Goats	Angora goats	Horses	Camels
1955	62,706	11,059	26,444	16,217	4,816	1,219	72
1960	76,120	12,435	34,463	18,637	5,996	1,312	65
1965	72,065	13,203	33,382	15,305	5,500	1,199	46
1970	73,031	12,756	36,471	15,040	4,443	1,049	31
1975	77,612	13,751	41,366	15,216	3,547	871	18
1980	87,067	15,894	48,630	15,385	3,658	794	12

Source: State Institute of Statistics.

Table 16.5   Regional shares in total annual production by value

	1965 (%)	1981 (%)
Coastal regions		
Mediterranean	13.1	13.5
Aegean	17.1	16.8
Marmara	9.2	9.8
Black Sea	10.2	10.7
Inland regions		
Central–north	14.2	14.3
Central–south	14.8	12.1
Central–east	8.0	6.9
North–east	5.4	5.9
South–east	8.1	10.0

Source: Ziraat Bankası, *Annual Statistics*.

emphasis on the need to expand the production of feed crops and on the control of animal diseases.

The increases in area, yields, and output add up to a major expansion in Turkish agriculture since 1945. The total increase appears to have been remarkably evenly spread across the country — especially in more recent years. Regional figures for 1965 and 1981 show similar patterns of distribution of total agricultural output. For both years, moreover, the coastal and inland regions each account for about one-half of the totals shown in Table 16.5.

Regional contrasts have persisted in Turkey since the establishment of the national commitment to planning in 1961, but there have been important changes. The three zones on Turkey's Mediterranean, Aegean, and Black Sea coasts have generally retained their lead over the Anatolian interior and the eastern uplands in levels of development in general and of average income in particular. But certain advantaged areas stand out: the grain-growing plains of inland Konya and those parts of the centre and east which have been able to benefit from irrigation schemes in the Euphrates valley are examples. Certainly the breakdown of Turkey into supposedly richer western and poorer eastern

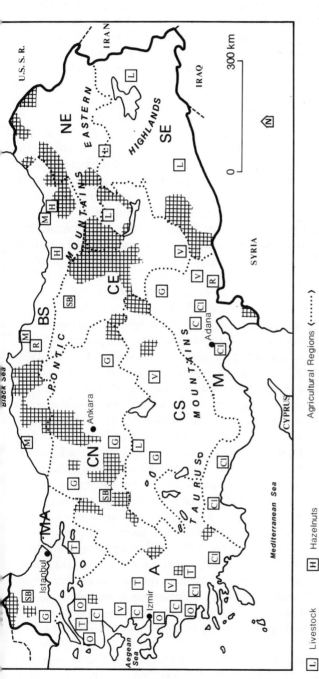

Figure 16.1   Turkey: agriculture and agricultural regions

halves which has figured in much thinking about the development of the country is as misleading now as before. The real pattern is more complex, with patches of backwardness in the west and relatively advanced areas in the east.

Environmentally, the patterns of agriculture in Turkey reflect the country's varied landscape and climate. The Mediterranean and Aegean coastal areas have hot, dry summers and mild, wet winters. The Black Sea zone is less mild but precipitation is greater throughout the year. The interior plateau and eastern uplands are cut off from much of the moderating maritime influence by mountain ranges — the Taurus in the south and the Pontic in the north. This mountain rim and the eastern highlands are better watered because of their height but are cold in winter — especially in the east. Traditionally, the interior of the Anatolian peninsula has depended on seasonally rain-fed farming, while the southern and western coasts have been able to develop irrigation in response to summer drought. A major thrust of Turkish agricultural policy is to extend the irrigable area. This, together with other developments in the structure and support of Turkish agriculture should reduce its vulnerability to the elements.

Despite state initiatives, individual enterprise and the evident expansion in areas, numbers and productivity, the *relative* position of Turkey's agriculture has slipped. Farm products now account for only some 37 per cent of exports compared with 60 per cent for industrial goods (though one-sixth of these are processed agricultural products) and 3 per cent from mining. Agriculture has fallen into third place behind both industry and services in its contribution to national income, having accounted for a 66 per cent share back in the 1920s and for 42 per cent in 1961. Currently a review of national policy regarding the role of the agricultural sector is under way, with implications for its place in both international trade and within the country. The background to this review includes both the recovery of Turkey from the worst extremes of its economic fortunes in the late 1970s and the very rapid development of Turkish trade in the Middle East. Meanwhile, uncertainties about future links with the European Economic Community have a special relevance to Turkey's agriculture.

The winter of 1979–80 marked a low point for the Turkish economy with foreign debt and exchange crises, escalating rates of inflation, the search for external support from the International Monetary Fund and elsewhere and then — in September 1980 — a military take-over in Ankara. Now the country's current account deficit is falling markedly and, partly in response to a sharply devalued Turkish lira, exports are rising impressively — though the traditional pattern of food and food products, plus some minerals such as chrome, going to European and other Western industrialized countries is changing. In 1977 the EEC countries accounted for 50 per cent of Turkey's total exports, by 1980 their share had dropped to 43 per cent and, by 1982, it was below 33 per cent (Table 16.6).

The new export opportunities for Turkey in the Middle East have helped the country's recovery from its recent economic problems — other than those

Table 16.6   Turkish exports (by destination) (per cent)

	1980	1982
European Economic Community	43	30
of which, West Germany	21	12
Middle East and North Africa	22	47
of which, Iraq	1	11
Iran	3	14
Other countries	35	23

Source: *The Economist*, April 16, 1983.

surrounding the future of its links with the EEC. Trade with both Iran and Iraq has grown most rapidly, even with these two countries at war with each other, and Saudi Arabia is now Turkey's fourth largest export market *after* these and West Germany. Many new Turkish exports to Middle Eastern neighbours are manufactured goods, while the agricultural element consists of non-traditional export products such as fresh meat, live animals, fresh fruit (notably citrus and vegetables), and cereals. Meanwhile the EEC, which now takes less than one-third of all Turkish exports, continues to buy 'traditional' luxury foodstuffs (figs, raisins, and hazelnuts) and industrial crops such as cotton and oil seeds and fodder but its demand for ordinary foodstuffs has been slow to rise.

Along with the quite dramatic expansion of Turkish trade to Middle Eastern countries goes another innovation in the form of contract working by Turkish firms and migrant workers in several countries. By 1981, some 38 Turkish companies were operating in Libya, with another 15 in Saudi Arabia, eight in Iraq and three in Kuwait. Little more than a year later the total of Turkish firms in Arab countries had exceeded 150 and there were some 150,000 Turkish migrants on work contracts (66,000 of them in Libya and 60,000 in Saudi Arabia). Further possibilities for Turkish enterprise in the Middle East range from Algeria and Egypt to war-ravaged Lebanon.

A dilemma for Turkish agriculture is that markets are as hard to penetrate in the Middle East as they are in Europe, if not necessarily for the same reasons. In terms of closer association with the EEC, Turkey presents a competitive threat in food exports to the Community's existing Mediterranean members. Apart from limited quantities of traditional sales such as nuts, figs, and tobacco, Turkey has been able significantly to expand exports to Europe only of cereals and cereal preparations and of meat; but even these show sharp fluctuations from year to year. In both European and Middle Eastern markets, Turkey's prospects for the export of grains, fruit, vegetables, and livestock products will depend in part on improved yields and quality control.

Remittances to Turkey from workers are having an unforeseen and quite direct effect on Turkish agriculture. This is truer of the new wave of migrant workers going to construction and other sites in the Middle East than those who went to Western Europe in earlier years — for two reasons. First, those migrants who went to West Germany, the Netherlands, Sweden, and elsewhere in the 1960s and 1970s helped to establish *gastarbeiter* communities which looked increasingly permanent as time passed and the proportion of dependent relatives grew. By contrast, the overwhelming majority of Turkish workers in Saudi Arabia, Libya, and other Arab countries have limited-term contracts and, with very few exceptions, are unaccompanied males. Secondly, the typical migrant to West Germany was from an urban Turkish background or, at least, was a rural Turk who went to Europe *via* the city. Thousands of the migrants working in Arab countries, by contrast, have gone there directly from their home villages in Anatolia where their families remain. In these circumstances remittances may be wads of banknotes sent or brought home, rather than the 'officially' transferred funds which figure in published statistics. In 1982 such statistics showed one billion dollars from workers in the Middle East — nearly as much as was sent back by Turks working in Europe. Moreover, most Turks in Western Europe were sending only about one-quarter of their savings, while those in Arab lands were remitting nearly half.

The impact of migrant work on farming in Turkey is evident in three ways: capital, skills, and more general changes in attitudes and aspirations. Money brought back to rural Turkey has paid for consumer goods such as cars and appliances but has also provided investment funding in farming itself (machinery, land, livestock, etc.) and in non-farming enterprises in the village. Skills developed on migrant labour contracts (plastering, carpentry, and metalwork are examples), together with the experience of non-family-orientated paid employment, are resulting in the appearance in many of Turkey's villages of a range of activities where previously the immediate needs of family-based farming contributed to a less varied economic scene. Along with the growing range of skills in rural areas go changes in the division of labour. Non-wage field work by women and children declines rapidly where machines take over. Older villagers may lack the literacy and technical skills demanded by appliances of many kinds. Even the younger male, literate and mechanically minded, may make the further transition to part-time farming where he has a second, specialist skill as a craftsman or a driver — or where he spends most of his time as a migrant worker in some distant factory or construction site.

Turkish agriculture is changing in response to national as well as individual initiatives. Imported money is increasingly important at all levels. But both activity and commitment remain vulnerable to changes in the availability of finance. Western Europe is no longer seen as the cornucopia which it was before economic depression and rising unemployment restricted the opportunities for migrants. Equally, Turks have no guarantee that prospects in the Middle East will continue to expand as rapidly as they have in the first years of the 1980s. Exports of agricultural commodities, whether to Europe or to the Middle East, remain vulnerable to competition and to protectionism. Internally, Turkish agriculture continues to be subject to environmental factors, to changes in development priorities in Ankara, to repercussions of trends at the international level, and to new abilities and aspirations among rural Turks. More pessimistic forecasters see approaching limits to agricultural expansion in Turkey, both in the total area farmed and in the extent to which Ankara can influence the progress of that sector.

The military government established in Turkey after a coup in 1960 reintroduced planning as a guiding principle in the economy and the new State Planning Organization's First Five-Year Development Plan for 1963–67 indicated official priorities. It stressed the improvement of the technological basis of agriculture. More than 45 per cent of the state's investment was to be for the irrigation projects, 15 per cent for tractors and other farm equipment, 6 per cent for land improvement, nearly 9 per cent for forestry, 4.2 per cent for animal-husbandry and 2.6 per cent for fisheries. The remaining 18 per cent was to go to other agricultural initiatives and to the support of the provincial administration and extension services of the Ministries of Agriculture and Forestry. Further small sums were earmarked for agricultural training purposes. Such investment outlays represented 12.8 per cent of total annual investment in 1963 rising to over 20 per cent by the end of the period of the First Plan in 1967.

During the Second Plan period (1968–72), the emphasis on the main investment priorities was strengthened, with 54 and 21 per cent of total state investment going, respectively, to water and soil resource developments and to tractors and equipment. The Plans in the late 1970s and the 1980s have maintained the stress on improvements in irrigation, some of this in the context of a score of major dams and water tunnel projects with multiple regional development aims. The Fourth Plan (1979–83) has also emphasized a range of further yield improvements from new seed strains to measures to limit the extent of land allowed to lie fallow. These are consistent with calls for higher growth rates for the output of both crops and livestock than those achieved previously, though the overriding commitment to economic development based primarily on industrialization persists. The greatest optimism for export expansion centres on wheat. Somewhat lower growth levels are hoped for in cotton, tobacco, nuts, and, especially, fruit and vegetables where the scope for improvement is considerable. The Plan recognizes that agricultural exports face not only problems of market penetration but also the persistent Turkish shortcomings in matters of standardization, quality control, and the coordination of marketing initiatives.

Overall, the National Plan envisages a continuing change in the production mix away from traditional patterns dominated by subsistent, market-ignoring village farming towards coordinated, market-seeking, commercial agriculture providing a basis for agro-industries and expansion. Such a national plan formula in fact recognizes what is in any case happening in Turkish agriculture and stresses the need for integrated development with consistent and linked changes in sectors and in demographic and socio-economic trends. The principal problem facing the decision-makers and planners in Ankara is that movements of people within the country and changes in their expectations threaten to overwhelm the efforts of the state. Along with these problems is the reality of an agricultural system in which most production has been in the hands of hundreds of thousands of small-scale producers, while marketing, financial support, and extension services have been developed by a broad structure of governmental agencies and ministries.

Since the 1950s, Turkish farmers have become accustomed to subsidies and price support, to low-interest credit, and to help from the extension services. Taxation has hit them far less harshly than other sectors—notably people in state service employment. This state support role in agriculture is now questioned and re-assessed by policy-makers in Ankara while farmers themselves are developing new demands and expectations. In short, there is a changing state–individual 'mix' in Turkish agriculture with initiative and response on both sides.

A major arm of state involvement in agricultural development, low-interest credit, has been distributed primarily through small cooperatives which, it is hoped, will help to maximize the effectiveness of the loans and, at the same time, get farmers used to formal cooperation itself. The durability of such cooperatives varies greatly, as does the readiness of Turkish farmers to invest in those areas preferred by the Agricultural Bank. Inevitably, cheap credit on such a scale becomes politically sensitive as entrepreneurs and agricultural policy makers press their different priorities. Similarly, support prices for agricultural products have become sensitive elements in the relationship between producer and state—which now increasingly recognizes a need for them to be set more strictly in regard to domestic and world conditions.

The prominence of tractors and other farm machinery in recent agricultural development in Turkey has itself been subject to government initiative through import duty and quota fixing. Similarly, the state can manipulate the supply of increasingly popular chemical fertilizers through tax, price support, and credit arrangements. In 1980, six million tons of chemical fertilizers were applied to nearly 12 million hectares (more than 72 per cent of the area sown). The expectation was that more than three times as much chemical fertilizer would be used in 1983 as in 1973.

State involvement in irrigation expansion in Turkey combines both credit funding to farmers for small-scale activities and major water development projects. Plan targets called for 11.7 per cent of the total arable area to be irrigable by 1983, compared with 7.2 per cent a decade earlier. Substantial areas of potentially irrigable land, however, await the completion of irrigation works. One problem here, as elsewhere in Turkish agriculture, is the lack of efficient coordination between state agencies.

The state agricultural extension service has been one important catalyst for change during the past two decades. Administratively, it has had an identity separate from the many other government ministries and agencies responsible for work with farmers in agricultural or related infrastructural developments. But shortcomings of coordination have been less serious than shortages of the qualified manpower, transport, and equipment necessary for work with farmers. At the beginning of the planned period in the early 1960s, there were fewer than 500 extension agents for the whole country and most of these were under-qualified. By the mid-1970s there had been more than a ten-fold increase, though the problems of under-qualification remained along with personal reluctance to indulge in field work, continuing vehicle and other shortages, and the use of extension personnel for data collection and other routine tasks. Notwithstanding these limitations, the Fourth Plan (1979–83) aimed at a further doubling of extension staff. Probably the *most* important single achievement of the rural extension service, as the government's training arm in agriculture, has been to make Turkey's farmers aware of the ways in which the state can help them, from the provision of credit or seed for new varieties to pest control measures or advice on cultivation and marketing.

In a brief review of the main features of progress in Turkish agriculture it is easy to list persistent operational constraints and resource shortages. One can point also to the relative decline of the sector within the national economy. Yet there are spectacular changes to report. Some sub-sectors have expanded rapidly: certain export developments in recent years have been a rapid as they have been innovative. Planned expansion rates for field crops, animal-husbandry, forestry products, and water schemes have not been achieved in most cases, but the reality has nevertheless been impressive, with annual growth rates at around 4 per cent in most cases (greater than increases in the demand for food) promising to hold or even increase. Ultimately, the most important changes in agriculture in contemporary Turkey involve the farmers and stock-breeders themselves.

REFERENCES

Agricultural Bank (Republic of Turkey) (various dates) *Agricultural Production Value of Turkey*, annual volumes, Ankara.

Aresvik, O. (1975) *The Agricultural Development of Turkey*, Praeger, New York.

Dewdney, J. C. (1981) 'Agricultural development in Turkey', in Clarke, J. I. & Bowen-Jones, H., (Eds), *Change and Development in the Middle East*, Methuen, London, pp. 213–223.

Hale, W. (1981) 'Agriculture, fisheries and forestry', in *The Political and Economic Development of Modern Turkey*, Croom Helm, London, pp. 74–189.

Kolars, J. F. (1967) 'Types of rural development', in Shorter, F. C. *et al.* (Eds), *Four Studies on the Economic Development of Turkey*, Frank Cass, London, pp. 63–87.

State Planning Organization, *First Five-Year Development Plan 1963–67* (and subsequent Five-Year Plans, Annual Plans and associated documents), Ankara.

Part III
Conclusions

Agricultural Development in the Middle East
Edited by P. Beaumont and K. McLachlan
© 1985 John Wiley & Sons Ltd

Chapter 17

Trends in Middle Eastern Agriculture

PETER BEAUMONT

INTRODUCTION

Any work on agriculture has to recognize the profound social changes which have occurred in all the countries of the region over the last fifty years. As populations have grown and the impact of Western technology has increased, rapid urbanization has occurred. Many people have migrated from rural to urban areas, but natural rates of increase in urban areas have also been high. As a result rural populations now make up a much smaller proportion of the total than was the case at the time of the Second World War, though in terms of absolute numbers rural areas have registered increases as well. Throughout the region underemployment and unemployment are rife, in both urban and rural settings, with the exception of those countries still experiencing the impact of an oil boom.

Massive differences in standards of living exist within countries as well as between them, though data are hard to come by, certainly at the regional level. Two useful indicators of how nations are faring are GNP per capita and energy consumption per capita (Table 17.1). These permit the countries to be put into various groupings. At the top of the league are the small, prolific oil-producers such as Qatar, Kuwait, and the United Arab Emirates. These are followed by the other main oil-producers including Bahrain, Libya, Oman, and Saudi Arabia, together with Israel. Then there are a large group of countries — Algeria, Jordan, Syria, Tunisia, Turkey, and probably Lebanon, Iran, and Iraq. These last two countries are currently showing signs of upward mobility. Finally, the poorest countries include Egypt, Morocco, the Sudan, Yemen AR, and Yemen PDR. It is interesting to note that the countries with the greatest agricultural production and potential fall in the middle range in terms of GNP and energy production per capita. This would suggest that they may well posess enough funds for a certain amount of agricultural investment, but obviously will not be able to compete with those states with massive oil revenues and small populations.

In the period since the Second World War the agricultural sectors of all Middle Eastern countries have experienced great changes. Almost all of them have undergone major land reform programmes which have often changed the balance of power within the country, though very rarely have all the reform objectives

Table 17.1 National indicators of change

	GNP per capita, 1981[a] ($)	Energy consumption per capita, 1979[b] (kg of coal equivalent)
Algeria	2,129	645
Bahrain	7,490	—
Egypt	654	539
Iran	—	1,141
Iraq	—	664
Israel	5,450	3,513
Jordan	1,623	522
Kuwait	25,850	6,159
Lebanon	—	1,028
Libya	8,560	2,254
Morocco	869	302
Oman	5,924	—
Qatar	27,790	—
Saudi Arabia	12,720	1,984
Sudan	380	133
Syria	1,569	952
Tunisia	1,417	590
Turkey	1,511	771
UAE	25,560	4,451
Yemen AR	459	58
Yemen PDR	512	509

[a] World Population Reference Bureau, 1983.
[b] World Bank, 1981.

been achieved. Certainly many more people own land than was previously the case, but often these new owners do not possess sufficient land or funds to sustain their families at a reasonable standard of living. The net effect of land reform has been a disruption of the social life of rural areas. Although this disruption has often been beneficial it has inevitably meant that the long-standing pattern of agriculture has been disturbed, with consequent losses of production.

With the advent of development planning, which began in certain countries as early as the 1950s, the industrial sector received large injections of capital, as governments attempted to Westernize their economies. In contrast, the agricultural sector, although not devoid of funds, has undoubtedly suffered in relative terms. Partly this has been because agriculture has been thought of as a traditional sector of the economy, where change would not come easily. Unfortunately this has often led to the superimposition of large, high technology agricultural projects, with little consideration of the effects which these would have on the local economy. Until recently few attempts have been made to improve agriculture at the 'grass roots' level. By the late 1970s, as population numbers have grown rapidly and food imports mushroomed, many governments have come to realize that agriculture plays a vital role in their national economies. Because of this, recent development plans have placed greater emphasis on strengthening the agricultural sector and in particular on increasing food production. Whether these attempts will produce the desired effect is still unknown, but attention is at last being focused on the structural weaknesses which have limited agricultural expansion.

There can be little doubt that physical factors, and in particular water shortage, exercise the greatest constraint on agricultural production in the Middle East. Insufficient precipitation means that large areas of the region are incapable of supporting crop growth without irrigation. Within the areas where water is available or can be supplied, topographic features, such as excess slope, further limit the cultivated area. In the highlands lack of soils or in the lowlands salination of soils also prevent cultivation. Low temperatures may limit the length of the growing season on the high plateaux, while other crops have a specific need for high temperatures over long periods. Late frosts in spring can cause considerable damage to blossom on fruit trees and so reduce yields substantially.

In the Middle East most countries have a long history of arable cultivation based on both dry-farming and irrigation. The only exceptions to this are Libya, Saudi Arabia, and the small nations around the Gulf where aridity and the lack of easily accessible water supplies have meant that, until recently, agricultural production has been on a relatively small scale. With the larger countries there are major differences in agricultural practices. For example, in Turkey most arable land is cultivated using dry-farming techniques, whilst in Egypt crops can only be grown with irrigation. Elsewhere, a mix of dry-farming and irrigated agriculture is found.

THE CAPITAL INVESTMENT SOLUTION

In recent years what has tended to happen in the richer oil-producing states is that large sums of capital have been invested to reduce the impact of the naturally occurring physical environmental constraints. Large dam schemes have been built and deep well fields sunk to provide water for crop growth, while in areas with soil salinity limitations complex drainage schemes have been installed. The benefits derived from such schemes, although often considerable, are made to seem even more impressive by the fact that the cost of the capital investment is often written off as a charge, or else costed with a low interest rate. As a result the net benefits will appear to outweigh the costs by a large margin, although with a more rigorous accounting system the opposite may in fact be the case.

In other cases money has been used not for capital investment programmes, but rather to provide subsidies for particular crop or animal programmes. The objective is the same, namely, to help reduce the difficulties of agricultural production under harsh environments. The subsidies can take many forms but the net effect is that the total cost of producing a particular crop or animal product (including subsidy) is often far greater than the cost of the same product on the open world market. It should be admitted, though, that for some commodities world prices are artificially low as a result of internal subsidies paid directly to farmers in other countries. Many countries in the Middle East have taken the view that certain aspects of agricultural production are so important that they warrant a large and continuing investment of funds to provide a measure of self-sufficiency with regard to certain key products.

In Saudi Arabia during the 1970s the tremendous growth rate of the economy brought an increasing reliance on imported food to supply essential needs. Then, with the war between Iraq and Iran, Saudi Arabia became concerned about her dependence on foreign countries for strategic imports. As a result, in its Third Five-Year Plan, 1980–85, Saudi Arabia chose to move towards self-sufficiency in food supply as one of its major priorities, despite the high costs involved. The price of meeting this objective is enormous, with a budgeted figure in excess of 20 billion dollars for agriculture and water resource development. On the other hand, the results have been impressive and by 1984 the country was already self-sufficient in eggs and dairy produce. The support for agricultural development is through the Saudi Arabian Agricultural Bank, which provides interest-free loans for stock, seed, and machinery purchase, as well as land reclamation work. In 1981/82 the Bank lent more than 830 million dollars, of which 25 per cent went to the poultry industry, 14 per cent to water well drilling, and 14 per cent for the provision of agricultural machinery. Besides these loans, subsidies of various kinds are available for obtaining land, seed, stock, chemicals, and machinery, with some of these subsidies amounting to 100 per cent of costs.

The case of wheat production in Saudi Arabia also provides an interesting example. Over the last few years Saudi Arabia has experienced a massive increase in wheat production from about 150,000 tonnes in 1979 to 700,000 tonnes in 1983. One of the main reasons for this growth is that the government-owned grain organization is paying farmers a price of approximately 1,000 dollars a tonne for their wheat, which is about five times the current world price O'Sullivan, 1984). It is not surprising, therefore, that farmers are moving into wheat production when the financial benefits can be so considerable. The financial drain on the Saudi Government is enormous. The net cost of the subsidy is currently of the order of 500 million dollars per annum and if wheat production continues to rise could well reach the 1 billion dollars level in a few years. While it is true that Saudi Arabia might well become self-sufficient with regard to wheat production, it seems that this might be at the expense of other crops.

The UAE also provides an excellent example of the kind of explosive growth in agricultural activity which has occurred in small, oil-rich nations around the Gulf. With the rapid rise in oil revenues from 1971 onwards, coupled with a marked population increase, the UAE used capital to overcome the natural constraints imposed by an arid environment. As surface water resources were almost non-existent, the UAE was forced to rely on its groundwater resources. Many deep wells were sunk and extensive irrigation systems built. The results, in terms of crop production, have been impressive, though one must not forget that agricultural output before the 1970s was very low. In the decade up to 1982 vegetable production had increased by more than 60 per cent per year and the cultivated area has increased by five times. The current five-year plan is hoping for an annual increase in agricultural production of around 10 per cent.

Whether these growth rates can be maintained still remains to be seen. By far the most important limiting factor is water. During the agricultural boom of the 1970s wells were sunk indiscriminately with little or no thought about the safe yield of the aquifer. To begin with, the farmers used traditional flood irrigation methods which consumed large volumes of water and it was not until the late 1970s that more efficient systems such as sprinkler irrigation were introduced. During most of the 1970s the attitude of the farmers was that the water resources were unlimited. Reality was different and by the late 1970s the water table was falling markedly, in some places by over a metre a year, leading to sea water intrusion and the production of more saline groundwater.

By 1983 cultivated land in the UAE was about 23,500 ha. It has been estimated that probably 75 per cent of all the nation's water use is for agricultural production. Actual volumes of water used are unknown but thought to be somewhere between 235 and 850 million m^3/yr. Groundwater on the other hand is believed to be being recharged at a rate of about 120 million m^3/yr. There can be no doubt, therefore, that current water use by agriculture is at least double the safe yield of the aquifer and may be as high as seven times that value. Although water storage volumes in the aquifer are not known with any precision, it would seem reasonable to suppose that the resources will be seriously depleted in a number of years unless something is done to limit water use. In recognition of the seriousness of the water problem a number of dams have been commissioned over the last few years to capture runoff along the wadis draining into the Gulf, which will then be diverted for agricultural purposes.

The two examples of Saudi Arabia and the UAE show that very impressive bursts of agricultural development can be achieved if huge capital sums derived from oil revenues are expended on water resources infrastructure provision. However, it does seem that this type of development will have a limited life expectancy for, in the case of both Saudi Arabia and the UAE, the water resources which are being used are from groundwater reserves which are not being replenished at anything like the rate at which they are being exploited. In effect these waters are being mined and, as is the case with all mineral extraction, there will come a time when the reserves are exhausted or become too expensive to mine. What seems likely to happen in these two countries is that cropped areas will expand to a maximum value and then contract as water resources become more difficult to obtain. The intriguing and as yet unanswerable question is: when is the period of maximum production going to occur?

POST-SECOND WORLD WAR TRENDS IN CROP PRODUCTION

Having looked at two special cases it is important to study the region as a whole and to discover the changes which have occurred in agricultural production in the post-Second World War period.

As cereals form the staple diet of all people in the Middle East it is instructive to examine these crops in detail. In the late 1960s and early 1950s total cereal production averaged around 25 million tonnes (Table 17.2). At this time the

Table 17.2   Cereal production in the Middle East

	Cereal production, 1948/49– 1952/53[a] (1,000 metric tons)	Cereal production, 1982[b] (1,000 metric tons)
Algeria	1,961	1,935
Bahrain	N.D.	N.D.
Egypt	4,103	7,768
Iran	3,087	9,189
Iraq	1,412	1,797
Israel	75	138
Jordan	194	47
Kuwait	N.D.	N.D.
Lebanon	101	31
Libya	(79)	235
Morocco	2,729	4,154
Oman	N.D.	3
Qatar	N.D.	N.D.
Saudi Arabia	N.D.	538
Sudan	811	2,540
Syria	1,197	2,276
Tunisia	689	1,331
Turkey	8,800	26,387
UAE	N.D.	N.D.
Yemen AR	N.D.	763
Yemen PDR	N.D.	94
Total	25,238	59,226

[a] FAO *Production Yearbook*, 1965.
[b] FAO *Production Yearbook*, 1982.

Table 17.3   Cultivated area of major cereal crops in the Middle East ('000 ha)

	Wheat 1948/49–1952/3 [a]	Barley 1948/49–1952/3 [a]	Maize 1948/49–1952/3 [a]	Rice (paddy) 1948/49–1952/3 [a]	Wheat 1982 [b]	Barley 1982 [b]	Maize 1982 [b]	Rice (paddy) 1982 [b]
Algeria	1,597	1,166	7	—	2,000	850	1	—
Bahrain	—	—	—	—	—	—	—	—
Egypt	605	64	660	256	577	45	816	411
Iran	2,080	934	6	220	6,000	1,400	45	330
Iraq	936	—	20	17	1,200	760	35	80
Israel	34	52	8	—	78	16	3	—
Jordan	182	62	—	—	100	50	—	—
Kuwait	—	—	—	—	—	—	—	—
Lebanon	70	20	7	—	18	5	—	—
Libya	24	204	1	—	303	280	1	—
Morocco	1,287	2,013	518	2	1,699	2,046	392	5
Oman	—	—	—	—	—	—	—	—
Qatar	—	—	—	—	—	—	—	—
Saudi Arabia	18	15	15	2	200	7	3	—
Sudan	13	1	13	—	253	—	67	4
Syria	994	369	24	6	1,300	1,350	22	—
Tunisia	917	589	—	—	800	425	—	—
Turkey	4,770	1,972	599	31	9,250	2,950	590	75
UAE	—	—	—	—	—	—	—	—
Yemen AR	—	—	—	—	66	56	38	—
Yemen PDR	—	—	—	—	10	2	10	—
Total	13,527	7,461	1,878	534	23,854	10,242	2,023	905

[a] FAO Production Yearbook, 1965.
[b] FAO Production Yearbook, 1982.

Table 17.4 Yields (kg/ha) of major cereal crops in the Middle East

	Wheat 1948/49–1952/53[a]	Barley 1948/49–1952/53[a]	Maize 1948/49–1952/53[a]	Rice (paddy) 1948/49–1952/53[a]	Wheat 1982[b]	Barley 1982[b]	Maize 1982[b]	Rice (paddy) 1982[b]
Algeria	620	690	910	—	600	765	1,100	2,857
Bahrain	—	—	—	—	—	—	—	—
Egypt	1,840	1,920	2,090	3,790	3,496	2,681	3,321	5,571
Iran	900	1,010	1,030	1,930	1,083	857	1,190	4,242
Iraq	480	770	680	1,170	750	724	2,571	3,135
Israel	690	860	970	—	1,282	813	5,000	—
Jordan	700	840	—	—	200	400	5,000	—
Kuwait	—	—	—	—	—	3,077	6,667	—
Lebanon	730	1,230	1,890	3,860	1,278	1,200	696	—
Libya	290	310	820	—	528	253	950	—
Morocco	610	740	580	4,000	1,074	929	804	2,200
Oman	—	—	—	—	3,778	—	—	—
Qatar	—	—	—	—	2,182	2,250	—	—
Saudi Arabia	890	890	1,380	1,050	2,000	1,714	1,600	10,000
Sudan	1,180	740	830	—	593	—	781	1,899
Syria	770	870	1,290	3,580	1,188	489	2,273	5,000
Tunisia	490	370	—	—	1,250	706	—	—
Turkey	1,000	1,150	1,240	3,510	1,908	2,034	2,373	4,667
UAE	—	—	—	—	2,500	20,000	28,000	—
Yemen AR	—	—	—	—	1,021	946	1,558	—
Yemen PDR	—	—	—	—	1,500	1,533	1,500	—

[a] *FAO Production Yearbook*, 1965.
[a] *FAO Production Yearbook*, 1982.

individual countries of the region were largely self-sufficient in cereals except during periods of drought. By 1982 production had more than doubled to a figure of almost 60 million tonnes, but at the same time dependence on food imports had sharply increased. How then has this production increase been achieved? The two possibilities are increasing the cultivated area or increasing yields (or a combination of the two).

Tables 17.3 and 17.4, reporting data for four of the main cereal crops, show that in fact both possibilities have tended to occur. Looking at the same countries for which data are available in the two years, it can be seen that wheat shows by far the most spectacular increase in absolute area, rising 76 per cent over the 30-year period (Table 17.3). Barley and maize also record increases in cropped areas of 25 per cent and 5 per cent, respectively. Rice cultivation reveals a 69 per cent area increase, which is impressive in relative terms, but in 1982 it still only covers about 4 per cent out of the area given over to wheat. A comparison of yields is more difficult to interpret as a result of the climatic variations which can occur from year to year. However, there is no doubt that significant increases in yields have been registered with all crops (Table 17.4).

When dealing with agricultural production, one of the most important questions is how much food is being produced for each inhabitant within a particular nation. In the Middle East it is well known that population has increased rapidly in the post-war period and so it is instructive to discover the extent to which indigenous agricultural production has been able to supply local needs.

In 1950, when most countries were still largely self-sufficient in basic foodstuffs, data show that cereal production varied from a minimum of 60 kg/capita in the case of Israel to a maximum of 420 kg/capita for Turkey (Table 17.5). Using the same approach for the 1980 data set, a much wider variation is seen from 2 kg/capita in Oman and the United Arab Emirates to 539/kg capita for Turkey. This difference is largely to be attributed to the fact that information is available for many of the smaller countries in 1980 which was not the case in 1950. Comparing the data it can be seen that certain countries, for example Turkey and Syria, record very high values in both years, as well as a significant absolute increase over the period concerned. The Sudan reveals a major increase as well, but from a much lower base figure. At the other end of the scale a number of countries, including Algeria, Iraq, Jordan, Lebanon, and Morocco show marked declines in per capita cereal production values.

To meet these growing production deficits many countries have had to resort to large-scale import of cereals since the early 1970s. By 1980 these cereal imports came to 27 million tonnes, or about 45 per cent of indigenous production in the region in the same year (Table 17.6). It will be noted that the imports in 1980 approximated the region's total production of cereals in 1950 (Table 17.6).

Analysing the cereal imports and production in per capita terms permits the build-up of a picture of total cereal consumption. In terms of imports, Israel

Table 17.5 Cereal production in kilograms/capita in the
Middle East

	1950	1980
Algeria	224	128
Bahrain	—	—
Egypt	201	195
Iran	190	213
Iraq	268	170
Israel	60	78
Jordan	153	58
Kuwait	—	—
Lebanon	70	17
Libya	77	72
Morocco	305	220
Oman	—	2
Qatar	—	4
Saudi Arabia	—	30
Sudan	94	160
Syria	351	433
Tunisia	194	188
Turkey	420	539
UAE	—	2
Yemen AR	—	155
Yemen PDR	—	63
Total	241	245

Source: Figures on national population and cereal production obtained
from various issues of FAO *Production Yearbook*.

reveals the largest figures, followed by the oil-rich states of Kuwait, Qatar, Saudi
Arabia, Libya and Bahrain (Table 17.6). With many of these states, large imports
are to be expected, given the very arid nature of many of their environments.
What perhaps is somewhat surprising is that many states which possess
considerable lands capable of producing cereals still have to import cereals in
excess of 100 kg/person. These include Algeria, Egypt, Iraq, Syria, and Tunisia.
This is in itself an indication of the population pressure being experienced by
these countries.

The totals for imports and production of cereals, recorded as cereal
consumption, still reveal considerable variations, some of which are rather
difficult to explain (Table 17.6). At the top of the list are Syria and Turkey,
reflecting the very high production figures of both countries. Israel comes next,
but here 90 per cent of total use is explained by imports. Ten states possess
per capita totals of between 270 and 380 kg/person, including many of the more
populous nations in the Middle East. In some of the poorer countries, like
Jordan, Sudan, and Yemen PDR, total consumption falls below 200 kg/person.

Table 17.6   Cereal production, cereal imports and cereal consumption in the Middle East

	Cereal production (1,000 metric tons) 1980	Cereal imports (1,000 metric tons) 1980	Cereal imports (kg/capita) 1980	Cereal consumption (kg/capita) 1980
Algeria	2,422	2,922	154	282
Bahrain	—	85	272	—
Egypt	8,178	6,082	144	339
Iran	8,107	3,424	90	303
Iraq	2,221	2,476	189	359
Israel	303	1,679	433	479
Jordan	189	451	139	197
Kuwait	—	535	396	—
Lebanon	45	678	255	272
Libya	215	816	274	346
Morocco	4,469	1,821	90	310
Oman	2	85	96	98
Qatar	1	80	337	341
Saudi Arabia	266	3,062	342	372
Sudan	2,933	342	19	179
Syria	3,883	997	111	544
Tunisia	1,196	776	122	310
Turkey	24,414	6	0.13	539
UAE	2	224	215	217
Yemen AR	901	377	65	220
Yemen PDR	117	226	122	185
Total/Average	59,864	27,090	111	356

Source: Cereal production figures and population data obtained from FAO *Population Yearbooks*.

The figure for Oman, at less than 100 kg/person, seems rather anomalous for a relatively rich, oil-producing country and makes one question the reliability of the quoted production figures.

THE FUTURE

In the early 1980s the population of the Middle East was about 270 million. By the year 2000 estimates suggest that the figure will have grown by 57 per cent to a total of approximately 420 million (Table 17.7). Whence these extra 150 million people are to obtain their food presents a serious problem, particularly when it is remembered that the total population of the region was probably less than 150 million around 1950. It is very difficult to quantify the magnitude of the problem with any certainty, but it is interesting to try to provide

Table 17.7   Current population and projected populations for the countries of the Middle East (millions)

	1983	2000
Algeria	20.7	36.5
Bahrain	0.4	0.6
Egypt	45.9	65.5
Iran	42.5	66.0
Iraq	14.5	24.2
Israel	4.1	5.4
Jordan	3.6	6.5
Kuwait	1.6	2.9
Lebanon	2.6	3.7
Libya	3.3	6.1
Morocco	22.9	37.3
Oman	1.0	1.7
Qatar	0.3	0.4
Saudi Arabia	10.4	18.5
Sudan	20.6	33.2
Syria	9.7	18.0
Tunisia	6.8	9.6
Turkey	49.2	70.7
UAE	1.4	2.1
Yemen AR	5.7	9.1
Yemen PDR	2.1	3.4
Total	269.3	421.4

Source: *World Population Reference Bureau, 1983.*

a guide to what needs to be done. As has already been seen, there are basically two possibilities to increase food production. These are either to increase yields or to increase the cultivated area, or a combination of the two. Throughout the region vegetable products supply normally at least 80 per cent of all calories consumed.

Let us assume for the moment that the only feasible way is to increase the cropped area. The next issue is to decide by how much the cultivated area would need to be expanded. An empirical way to achieve this would be to look at the cultivated areas in each of the countries of the region and to discover how much land was being cultivated for each person living in the country. The only problem with this approach is that it is perfectly obvious at the present day that most of the countries are not able to support their own populations, as is witnessed by the huge imports of certain foodstuffs.

Another possibility is to look at the 'cultivated area per capita' figures for a period in the past when most of the countries were largely self-sufficient in terms of foodstuffs. A difficulty here is that the accuracy of statistics declines as one goes back into the past. However, reasonably accurate data are available

Table 17.8   Land availability in hectares/capita in the mid 1950s

Country	Cultivated area (including orchards) ('000 ha)	Population (millions)	Land availability (ha/capita)
Egypt (1951)	2,451	20.872	0.117
Iran (1950)	16,761	18.952	0.886
Iraq (1954)	5,457	4.948	1.104
Israel (1954)	359	1.688	0.213
Jordan (1954)	893	1.384	0.643
Lebanon (1954)	270	1.383	0.194
Saudi Arabia (1952)	210	7.00	0.028
Syria (1954)	4,034	3.670	1.101
Turkey (1954)	21,334	22.949	0.931
Total	51,769	82.846	0.625

Source: Data on cultivated areas and population numbers from Bullard (1961).

for some countries from the early 1950s (Table 17.8). These show wide variations from country to country dependent upon the predominant type of agriculture, whether dry-farming or irrigation. For example, Egypt records a value of 0.117 hectares/person, while for both Syria and Iraq the figure is in excess of 1.1 hectares/person. Saudi Arabia gives an even lower value than Egypt, but the data on which it is based seem questionable. There is also the doubt as to whether Saudi Arabia was self-sufficient in food at this time. For south-west Asia as a whole, the average works out at 0.625 hectares/person.

Using these data it is possible to obtain estimates of how much extra land would be required to support a further 150 million people by the end of the century. Taking the average value leads to a figure of 93.75 million hectares. Interestingly, this is approximately the same value as the cultivated land in the Middle East today. The question must be posed as to where this new land is to be obtained from. Looking round the Middle East it would seem that there is little land, which is not currently being used, that could be cultivated using dry-farming techniques. This means that any new land for cultivation would have to be irrigated. If this is the case, a more realistic figure of the total land needed to support 150 million people would probably be obtained by using the data from Egypt, where all arable land is irrigated. It must be remembered, however, that Egypt obtains more than one crop per year from its lands, which certainly would not be possible in many other parts of the region.

Using the Egyptian value of 0.117 hectares/person, the total land which would be required is 17.55 million hectares, or about seven times the cultivated area of Egypt in the 1950s or about 19 per cent of the region's current cultivated area. Even if this amount of suitable land could be found, there must be doubt as to whether the requisite quantities of water are available. If we assume that each hectare would require an average 10,000 m³/year, the total volume of water required would be 175,500 million m³. This represents approximately twice the annual flow of the Nile or about five-and-a-half times the flow of the Euphrates. There can be no doubt whatsoever that this quantity of water, over and above that already being used in the area, is not available on a long-term basis. It has to be concluded, therefore, that expansion of the cropped area alone is not a feasible solution to the problem of population growth, though small areas of irrigated land almost certainly will be brought into production in the future. Many of these, though, might be irrigated with fossil water which will not be being replaced under present-day environmental conditions.

The other alternative to produce more food is to increase yields. To assess the validity of this proposition it is necessary to discover the yields which were obtained using traditional agricultural methods and then to assess the increases in yields which have occurred as a result of the application of 'modern' farming methods. An immediate problem arises though as to what average yields were. As has already been noted with dry-farming, yields can vary enormously from year to year as a result of fluctuating climatic conditions. Even in irrigated areas yields change as a result of differing water availability in a given year.

Data on yields of wheat, by far the most important cereal crop, from the late 1940s to the mid-1960s allow one to discover the range of production using largely traditional methods. The data show that, with the exception of Egypt and to a lesser extent Israel and the Sudan, almost all the countries had yields between 300 and 1,200 kg/ha. It is interesting to note similarities between countries. For example, Algeria, Tunisia, and Morocco had yields varying from 400 to 800 kg/ha, while Iraq, Syria, and Jordan had a wider range from 300 to 1,000 kg/ha (Table 17.9). The Lebanon, Turkey, and Iran all seemed able to maintain yields above 600 kg/ha. With the Sudan, yields ranged from 760 to 1,700 kg/ha, while in Egypt the minimum values were 1,700 kg/ha. Leaving aside the Sudan and Egypt, it is obvious that in the individual countries there are tremendous variations in yields from year to year which makes a discussion of average values difficult. What is also interesting is that, if one studies the yields over the time period considered, there do not appear to be any significant trends which can be distinguished. The two exceptions to this statement are Egypt and the Sudan.

If wheat yields for the Middle Eastern countries are examined for the early 1980s, it can be seen that many countries have recorded substantial increases compared with the later 1940s to the early 1960s. Perhaps the most significant from the regional viewpoint are Syria and Turkey where yields have been increased to maxima of between 1,600 and 1,900 kg/ha. In terms of the highest yields, Egypt is by far the most impressive with wheat yields over 3,200 kg/ha. Israel provides an interesting example of what modern farming techniques are able to produce, though even here there are still vast changes from one year to another.

Table 17.9   Wheat yields in the Middle East

	Yields 1948–64[a] (kg/ha)	Yields 1980–82[b] (kg/ha)
Algeria	390–800	600–730
Bahrain	—	—
Egypt	1,700–2,760	3,225–3,496
Iran	780–970	1,036–1,083
Iraq	310–830	750–917
Israel	340–2,210	1,282–2,760
Jordan	220–1,170	200–999
Kuwait	— —	
Lebanon	620–1,090	1,278–2,333
Libya	130–280	397–528
Morocco	530–860	542–1,074
Oman	—	500–3,778
Qatar	—	2,182–2,222
Saudi Arabia	830–980	2,000–2,108
Sudan	760–1,710	593–963
Syria	300–970	1,188–1,662
Tunisia	290–710	1,019–1,250
Turkey	630–1,270	1,838–1,908
UAE	—	2,400–3,283
Yemen AR	—	1,021–1,105
Yemen PDR	—	1,500–1,667

[a]FAO, 1966.
[b]FAO *Production Yearbook*, 1982.

The figures for Israel, Syria, and Turkey suggest that better farming techniques applied throughout the region might be able to raise dry farming yields to around 2,000 kg/ha on a long-term basis. Interestingly, this figure is much the same as that currently achieved by Canada and the USA. It should be noted, however, that in Australia, which possesses areas of intense aridity like the Middle East and which already makes use of advanced farming methods, yields vary from 1,000 to 1,800 kg/ha. To maintain yields close to 2,000 kg/ha in the Middle East might, therefore, be rather over-optimistic. Egypt shows that, given intensive irrigation techniques and high fertilizer inputs, wheat yields above 3,000 kg/ha are a possibility. Using these yield values it is possible to predict maximum likely wheat production from lands being cultivated at the present time. It would seem that 2,000 kg/ha is the highest value to be achieved for dry-farming, whereas a figure of 3,000 kg/ha might be achieved in irrigated areas.

The 1982 figures for wheat production in the Middle East show that a total of 33.82 million tonnes was produced from 23.45 million ha. This provides an average yield of 1,418 kg/ha. No data are available as to what proportion of the total wheat crop was grown under dry-farming or irrigated farming techniques. If it is assumed for the basis of making calculations that the wheat crop follows the regional average of 82.6 per cent dry-farming and 17.4 per cent irrigated, dry-farming of wheat is estimated as covering 19.70 million hectares and irrigated wheat 4.22 million hectares. Applying the maximum probable yields of 2000 kg/ha for dry-farming and 3000 kg/ha for irrigated production, a total of 39.40 million tonnes for dry-farming and 12.66 million tonnes for irrigation is arrived at. This gives a grand total of 52.06 million tonnes.

This figure represents the maximum likely production of wheat from the presently cultivated area using modern methods and assuming favourable climatic conditions. It represents an increase of about 54 per cent over the 1982 value, but it must be stressed that it is unlikely to be attained consistently from year to year because climatic conditions in dry-farming areas in particular are not likely to produce high precipitation in a given year. Increases in yields, if they could be maintained, would make a sizeable contribution to the problems of food production in the Middle East. However, the costs, in terms of mechanization and fertilizer use are likely to be high.

In the book *The Agricultural Potential of the Middle East* (Clawson *et al.*, 1971), the authors take what is basically an optimistic view of the future of agriculture in the region, stressing the way forward as being through capital investment and technological changes.

> The technical problems of doubling agricultural output in two or three decades are severe but by no means impossible of solution. The basic resources are there, the expertise can be acquired, the practices adopted to local conditions, and so on down the line.

They go on to quote what they term 'potential' yields which are the maximum which might be expected within a given country. These are much higher figures than those which have been used in the calculations just carried out. For example, in Egypt wheat yields are expected to rise to 6,700 kg/ha and elsewhere in dry-farming areas yields are about 2,700 kg/ha. On the evidence which is available today these estimates appear to have been rather over-optimistic and perhaps the authors have placed too much hope in the technological approach to the solution of agricultural problems.

Given the estimated rise in population by the end of the century to over 420 million, it is interesting to speculate on the likely levels of cereal imports and production which will be needed to feed these people. The simplest calculation to make is to assume that for each of the countries of the region the per capita import and production figures for 1980 remain constant with total population increase being the only changing variable (Tables 17.5 and 17.6). Such calculations suggest that, by the year 2000, imports of cereals will need to be of around 48 million tonnes and indigenous production of cereals of about 100 million tonnes.

How feasible will it be to meet these totals? First it is essential to stress that conditions throughout the world are currently changing very rapidly and, therefore, that it is exceedingly difficult to assess future conditions with any certainty. At the beginning of the 1980s, world export of cereals was around 84 million tonnes. Unless this figure is substantially increased the Middle East woud require about two-thirds of this figure by 2000. Given the fact that many Middle Eastern cereal-importing countries have rich oil revenues it seems possible that they may well be able to purchase these supplies, though it will obviously be at the expense of poorer countries elsewhere in the world. The question of increased production is much more difficult to deal with as so many different crops are involved. However, given the fact that wheat production accounts for about 55 per cent of the total cereal crop, any predictions made for wheat will have a dominant influence on the cereal figures. It has already been suggested that under favourable conditions it might be possible to raise Middle Eastern wheat production to about 52 million tonnes, in increase of 54 per cent above the 1982 figure. If one applies the same percentage increase to total cereal production then the 1982 figure of 59.2 million tonnes would be raised to about 91.2 million tonnes.

Such calculations are, of course, extremely problematical and do assume that optimum conditions will prevail all the time, when they obviously will not. However, it is interesting to note that the estimated need of about 100 million tonnes of cereals each year might be closer to being met from indigenous production if yields could be everywhere raised to the highest values already being achieved in the region. In itself this is a hopeful sign, though the problem of increasing yields consistently especially in dry-farming areas is extremely difficult to solve. It will be very much a case of 'running to stand still'.

CONCLUSION

Given the difficulties with increasing the cultivated areas in the Middle East and with increasing yields rapidly, there seems to be little doubt that for most countries food imports will continue at a high level into the twenty-first century. This will be especially the case in the more populous countries where changes are relatively difficult to implement. In the smaller oil-rich countries the position with regard to food supply may rapidly alter if sufficient capital is invested in agriculture. Saudi Arabia has already shown what can be done if massive funds are available and if food production is given a high priority in terms of national goals.

The question must arise though as to what price Saudi Arabia and countries employing similar policies are having to pay. In most parts of the Middle East the critical production factor is water. In the countries around the Gulf and in parts of Egypt and Libya as well, the water for agricultural development has been provided by groundwater extraction from large aquifer reservoirs. Unfortunately, most of the available evidence suggests that the water in these aquifers is fossil in origin and is not being recharged under present environmental conditions. Therefore, although productive agricultural systems can be established, they will only last for a limited number of years, before depletion of the aquifer takes place.

Long-lasting solutions to Middle Eastern agriculture's problems are not easy to come by. It does seem likely that average yields of many crops will increase over the next decade or so and that the cultivated area will expand somewhat, largely as a result of irrigation. However, these improvements do not seem capable of keeping pace with the rapidly growing population of the region, and so food imports seem destined to continue to play an important role in the foreign trade of the region until the end of the century.

REFERENCES

Bullard, R. (Ed.) (1961) *The Middle East*, Oxford University Press, Oxford.

Clawson, M., Landsberg, H. H., & Alexander, L. T. (1971) *The agricultural Potential of the Middle East*, Elsevier, New York.

FAO (various dates) *Production Yearbook*, FAO, Rome.

FAO (1966) *World Crop Statistics: Area, Production and Yield 1948–64*, FAO, Rome.

MEED (*Middle East Economic Digest*) (1983) 'Water shortage forces drastic measures', in *UAE—Special Report*, December 1983, Middle East Economic digest, London, pp. 70–74.

O'Sullivan, E. (1984) 'The Saudi wheat conundrum', *Middle East Economic Digest*, **28**, 24–25, 27.

World Bank (1981) *World Development Report*, Oxford University Press, Oxford.

World Population Reference Bureau (1983) *World Population Data Sheet 1983*, World Population Reference Bureau, Washington DC.

Index

nomadism, 92
oil, 85
trade, 97
Gulf States, 115, 116
gypsum sub-soils 261

Ha'il (Saudi Arabia), 216, 223
Haj (Saudi Arabia), 90
Hajjar Mountains (United Arab Emirates), 235
Hamadan (Iran), 178
Hamraniyah (United Arab Emirates), 235–236
Hasanabad (Iran), 181
Hawalli (Kuwait), 283
Hawija (Iraq), 195
Hawr al-Hammar (Iraq), 189, 200
Hawr as-Saniyah (Iraq), 189
health, 30, 93, 232, 250
herbicides, 21, 69
hides, 5
high technology farming systems, 3
high yield cereals
    Syria, 270
Hillah (Iraq), 200
Hijaz (Saudi Arabia), 214–215, 217
Hofuf (Saudi Arabia), 19, 216
Homs-Hama canal (Syria), 263
Hong Kong, 197, 202
housing, 30, 151–152, 279, 283
hydro-electric power, 141
    Egypt, 79, 140, 146
    for irrigation, 80
    sale of electricity, 80
    Sudan, 79
    Syria, 261
hydrological observations, 18
hydro-politics of the Nile, 146

Ibn Saud, 218–219
Ibri (Oman), 235
Idfina (Egypt), 80
Ilce (Turkish districts), 290
Imams of Yemen, 244–246, 250
IMF, 296
immigrant labour, 32, 35, 103, 206, 229, 231, 233, 236, 285–286, 297–298
immigration, 57
    Sudan, 146
import dependence
    Saudi Arabia, 222

imported food, 308
income, 29, 30, 37, 44, 46, 86, 98, 151, 161, 166, 172, 178, 180–183, 190, 205, 227, 238, 241, 246–247, 250, 287, 290, 294, 296
India, 16
Indian Ocean, 241, 245
industrial costs, 113
industrial development, 118
industrial investment funds, 113
industrial processing
    agricultural raw materials, 257
industrial sector, 121
    Egypt, 108
    Saudi Arabia, 218
industrial water use
    Egypt, 140
    Saudi Arabia, 222
Industrialization, 113, 296
    Egypt, 127
    Iran, 112, 171, 182
    Turkey, 299
industry
    Bahrain, 231
    'cottage', 229, 243
    Iran, 174, 183–185
    Iraq, 191, 201
    Kuwait, 283
    Oman, 228–229
    textile, 174, 201
    Turkey, 296
    United Arab Emirates, 235
    Yemen, 243, 247, 250
infiltration
    soil property, 13
inflation, 30, 40, 246, 252, 296
infrastructural costs
    new towns, 72
inheritance, 160, 164, 244, 292
integration
    crop and animal husbandry, 66–67
inter-annual variability (precipitation), 4
interest-free loans, 308
interest-rates, 305
International Center for Agricultural Research in Dry Areas (Syria), 277
international migration, 116
investment
    agricultural sector, 68, 111, 212
    capital, 60
investment in irrigation
    Syria, 275

valleys, 189, 244
value of water, 146
  Sudan, 141
Varamin Plain (Iran), 20
vegetable production, 8, 9, 316
  Saudi Arabia, 218
  United Arab Emirates, 309
vegetation, 193, 233
  regeneration, 14
  removal, 67
Vietnam, 252
village co-operatives
  Iran, 119
villages, 8, 9

wadi, 241–244
Wadi Dawasir (Saudi Arabia), 216, 217, 223
Wadi Dhuleil (Jordan), 80
Wadi Halfa (Sudan), 135
Wadi Mujib (Jordan), 70
Wadi Najran, 75, 223
Wadi Sirhan (Saudi Arabia), 220
Wafra (Kuwait), 283
Wahhabi, 218
Wajid aquifer, 216
waqf, 218
war, 231
  effects of, 29, 36, 40, 98, 102–103, 168
    Arab–Israeli, 245
    Gulf, 36, 40, 98, 102, 205, 297
    Lebanese Civil War, 36, 103
    Second World War, 27, 44, 172, 182, 185, 245
    Yemeni, 246
Warba island (Kuwait), 288
Wasit (Iraq), 196, 200
water, 11, 52, 54, 56, 58, 60, 72, 89, 92, 93, 153, 162, 174, 180, 182, 192, 195, 200, 201, 205, 229, 232–233, 236–237, 243–244, 247–249, 251, 279–282, 286–288, 299, 310
  allocation, 77
    Egypt, 125, 140
    Nile, 135
  application rates, 61
    Israel, 82
    Syria, 263
  aridity, 32, 36, 90, 186, 192–193, 200, 238, 241, 243, 285
  availability, 16, 60, 318
    Egypt, 59, 318

competing demands, 78, 79
control and transmission works, 4
cost, 233, 237, 282
demand, 55
  Egypt, 137, 140, 142
  Sudan, 137, 139, 141, 142
demand forecasts
  Egypt, 139
  Sudan, 139
development
  Saudi Arabia, 308
  Syria, 256
development policies
  Syria, 263
distribution systems, 18
domestic demand, 79
drought, 88, 241, 248, 296
duties
  Egypt, 59, 82
  Syria, 261
grid
  Israel, 80
groundwater, 234–245, 247, 280
holding capacity of soils, 69
industrial demand, 79
infrastructures
  United Arab Emirates, 309
intakes (irrigation), 5
integrated development of surface water and groundwater, 80
international competition, 78
irrigation demand, 79
law, 63, 77
  Israel, 80
losses, 69
  Egypt, 86
  reservoirs, 74
management
Egypt, 82
  objectives, 81
  practices, 223
  Syria, 263
marsh, 172–173, 189, 197
nutrient contents, 21
quality changes, 21
permits
  Sudan, 146
pricing policies, 20, 63, 77
resources
  Tunisia, 59
returns, 61